Walter Thirring

A Course in Mathematical Physics

1

Classical Dynamical Systems

Translated by Evans M. Harrell

Springer Science+Business Media, LLC

Dr. Walter Thirring

Institute for Theoretical Physics
University of Vienna
Austria

Dr. Evans Harrell

Massachusetts Institute of Technology
Cambridge, Massachusetts
USA

Library of Congress Cataloging in Publication Data

Thirring, Walter E   1927-
  A course in mathematical physics.

  Translation of Lehrbuch der mathematischen Physik.
  Bibliography: p.
  Includes index.
  CONTENTS: [v.] 1. Classical dynamical systems.
  1. Mathematical physics.   I. Title.
QC20.T4513   530.1′5   78-16172

With 58 Figures

Copyright © 1978 by Springer Science+Business Media New York
Originally published by Springer-Verlag New York Inc. in 1978.
Softcover reprint of the hardcover 1st edition 1978

9 8 7 6 5 4 3 2 1

ISBN 978-3-662-38942-3   ISBN 978-3-662-39892-0 (eBook)
DOI 10.1007/978-3-662-39892-0

Preface

This textbook presents mathematical physics in its chronological order. It originated in a four-semester course I offered to both mathematicians and physicists, who were only required to have taken the conventional introductory courses. In order to be able to cover a suitable amount of advanced material for graduate students, it was necessary to make a careful selection of topics. I decided to cover only those subjects in which one can work from the basic laws to derive physically relevant results with full mathematical rigor. Models which are not based on realistic physical laws can at most serve as illustrations of mathematical theorems, and theories whose predictions are only related to the basic principles through some uncontrollable approximation have been omitted. The complete course comprises the following one-semester lecture series:

   I.   Classical Dynamical Systems
   II. Classical Field Theory
   III. Quantum Mechanics of Atoms and Molecules
   IV. Quantum Mechanics of Large Systems

Unfortunately, some important branches of physics, such as the relativistic quantum theory, have not yet matured from the stage of rules for calculations to mathematically well understood disciplines, and are therefore not taken up. The above selection does not imply any value judgment, but only attempts to be logically and didactically consistent.

General mathematical knowledge is assumed, at the level of a beginning graduate student or advanced undergraduate majoring in physics or mathematics. Some terminology of the relevant mathematical background is

collected in the glossary at the beginning. More specialized tools are introduced as they are needed; I have used examples and counterexamples to
try to give the motivation for each concept and to show just how far each
assertion may be applied. The best and latest mathematical methods to
appear on the market have been used whenever possible. In doing this many
an old and trusted favorite of the older generation has been forsaken, as I
deemed it best not to hand dull and worn-out tools down to the next
generation. It might perhaps seem extravagant to use manifolds in a treatment of Newtonian mechanics, but since the language of manifolds becomes
unavoidable in general relativity, I felt that a course that used them right
from the beginning was more unified.

References are cited in the text in square brackets [ ] and collected at the
end of the book. A selection of the more recent literature is also to be found
there, although it was not possible to compile a complete bibliography.

I am very grateful to M. Breitenecker, J. Dieudonné, H. Grosse, P.
Hertel, J. Moser, H. Narnhofer, and H. Urbantke for valuable suggestions.
F. Wagner and R. Bertlmann have made the production of this book very
much easier by their greatly appreciated aid with the typing, production
and artwork.

Walter Thirring

Note about the Translation

In the English translation we have made several additions and corrections
to try to eliminate obscurities and misleading statements in the German text.
The growing popularity of the mathematical language used here has caused
us to update the bibliography. We are indebted to A. Pflug and G. Siegl
for a list of misprints in the original edition. The translator is grateful to the
Navajo Nation and to the Institute for Theoretical Physics of the University
of Vienna for hospitality while he worked on this book.

Evans M. Harrell

Walter Thirring

Contents

Glossary

Logical Symbols

$\forall$	for every
$\exists$	there exist(s)
$\not\exists$	there does not exist
$\exists!$	there exists a unique
$a \Rightarrow b$	if a then b
iff	if and only if

Sets

$a \in A$	a is an element of A
$a \notin A$	a is not an element of A
$A \cup B$	union of A and B
$A \cap B$	intersection A and B
CA	complement of A (In a larger set B: $\{a: a \in B, a \notin A\}$)
$A \backslash B$	$\{a: a \in A, a \notin B\}$
$A \Delta B$	symmetric difference of A and B: $(A \backslash B) \cup (B \backslash A)$
$\emptyset$	empty set
$C\emptyset$	universal set
$A \times B$	Cartesian product of A and B: the set of all pairs (a, b), $a \in A$, $b \in B$

Important Families of Sets

open sets	contains $\emptyset$ and the universal set and some other specified sets, such that the open sets are closed under union and finite intersection

closed sets	the complements of open sets
measurable sets	contains $\emptyset$ and some other specified sets, and closed under complementation and countable intersection
Borel-measurable sets	the smallest family of measurable sets which contains the open sets
null sets, or sets of measure zero	the sets whose measure is zero. "Almost everywhere" means "except on a set of measure zero."

An equivalence relation is a covering of a set with a non-intersecting family of subsets. $a \sim b$ means that a and b are in the same subset. An equivalence relation has the properties: i) $a \sim a$ for all a. ii) $a \sim b \Rightarrow b \sim a$. iii) $a \sim b, b \sim c \Rightarrow a \sim c$.

Numbers

$\mathbb{N}$	natural numbers
$\mathbb{Z}$	integers
$\mathbb{R}$	real numbers
$\mathbb{R}^+ (\mathbb{R}^-)$	positive (negative) numbers
$\mathbb{C}$	complex numbers
sup	supremum, or lowest upper bound
inf	infimum, or greatest lower bound
I	any open interval
(a, b)	the open interval from a to b
$[a, b]$	the closed interval from a to b
$(a, b]$ and $[a, b)$	half-open intervals from a to b
$\mathbb{R}^n$	$\underbrace{\mathbb{R} \times \cdots \times \mathbb{R}}_{N \text{ times}}$

This is a vector space with the scalar product
$$(y_1, \ldots, y_N | x_1, \ldots, x_N) = \sum_{i=1} y_i x_i$$

Maps ($=$ Mappings, Functions)

$f: A \to B$	for every $a \in A$ an element $f(a) \in B$ is specified
$f(A)$	image of A, i.e., if $f: A \to B$, $\{f(a) \in B : a \in A\}$
$f^{-1}(b)$	inverse image of b, i.e. $\{a \in A : f(a) = b\}$
f^{-1}	inverse mapping to f. Warning: 1) it is not necessarily a function 2) distinguish from $1/f$ when $B = \mathbb{R}$
$f^{-1}(B)$	inverse image of B: $\bigcup_{b \in B} f^{-1}(b)$
f is injective (one-to-one)	$a_1 \neq a_2 \Rightarrow f(a_1) \neq f(a_2)$
f is surjective (onto)	$f(A) = B$
f is bijective	f is injective and surjective. Only in this case is f^{-1} a true function
$f_1 \times f_2$	the function defined from $A_1 \times A_2$ to $B_1 \times B_2$, so that $(a_1, a_2) \to (f_1(a_1), f_2(a_2))$
$f_2 \circ f_1$	f_1 composed with f_2: if $f_1: A \to B$ and $f_2: B \to C$, then $f_2 \circ f_1: A \to C$ so that $a \to f_2(f_1(a))$

1	identity map, when $A = B$; i.e., $a \to a$. Warning: do not confuse with $a \to 1$ when $A = B = \mathbb{R}$.
$f\|_u$	f restricted to a subset $U \subset A$
$f\|_a$	evaluation of the map f at the point a; i.e., $f(a)$.
f is continuous	the inverse image of any open set is open
f is measurable	the inverse image of any measurable set is measurable
supp f	support of f: the smallest closed set on whose complement $f = 0$
C^r	the set of r times continuously differentiable functions
C_0^r	the set of C^r functions of compact (see below) support
χ_A	characteristic function of $A : \chi_A(a) = 1 \ldots$

Topological Concepts

Topology	any family of open sets, as defined above
compact set	a set for which any covering with open sets has a finite subcovering
connected set	a set for which there are no proper subsets which are both open and closed
discrete topology	the topology for which every set is an open set
trivial topology	the topology for which the only open sets are $\varnothing$ and $C\varnothing$
simply connected set	a set in which every path can be continuously deformed to a point
(open) neighborhood of $a \in A$	any open subset of A containing a. Usually denoted by U or V.
(open) neighborhood of $B \subset A$	any open subset of A containing B
p is a point of accumulation ($=$ cluster point)	for any neighborhood U containing p, $U \cap B \neq \{p\}$ or $\varnothing$
$\bar{B}$	closure of B: the smallest closed set containing B
B is dense in A	$\bar{B} = A$
B is nowhere dense in A	$A \setminus \bar{B}$ is dense in A
metric (distance function) for A	a map $d : A \times A \to \mathbb{R}$ such that $d(a, a) = 0$; $d(a, b) = d(b, a) > 0$ for $b \neq a$; and $d(a, c) \leq d(a, b) + d(b, c)$ for all a, b, c in A. A metric induces a topology on A, in which all sets of the form $\{b : d(b, a) < \eta\}$ are open.
separable space	a space with a countable dense subset
homeomorphism	a continuous bijection with a continuous inverse
product topology on $A_1 \times A_2$	the family of open sets of the form $U_1 \times U_2$, where U_1 is open in A_1 and U_2 is open in A_2, and unions of such sets

Mathematical Conventions

$f_{,i}$	$\partial f / \partial q_i$
$\dot{q}(t)$	$\dfrac{dq(t)}{dt}$

$\text{Det } |M_{ij}|$	determinant of the matrix M_{ij}
$\text{Tr } M$	$\sum_i M_{ii}$
δ^i_j, δ_{ij}	1 if $i = j$, otherwise 0
$\varepsilon_{i_1,\dots,i_m}$	the totally antisymmetric tensor of degree m, with values ± 1.
M^t	transposed matrix: $(M^t)_{ij} = M_{ji}$
M^*	Hermitian conjugate matrix: $(M^*)_{ij} = (M_{ji})^*$
$\mathbf{v} \cdot \mathbf{w}, (\mathbf{v}|\mathbf{w}),$ or $(\mathbf{v} \cdot \mathbf{w})$	scalar (inner, dot) product
$\mathbf{v} \times \mathbf{w}$	cross product
∇f	gradient of f
$\nabla \times \mathbf{f}$	curl of $\mathbf{f}$
$\nabla \cdot \mathbf{f}$	divergence of $\mathbf{f}$
$\|\mathbf{v}\|$ (in 3 dimensions, $|\mathbf{v}|$)	length of the vector $\mathbf{v}$: $\|\mathbf{v}\| = (\sum_{i=1} v_i^2)^{1/2} = d(\mathbf{0}, \mathbf{v})$
$d\mathbf{s}$	differential line element
$d\mathbf{S}$	differential surface element
$d^m q$	m-dimensional volume element
$\perp$	is perpendicular (orthogonal) to
$\parallel$	is parallel to
$\angle$	angle
$d\Omega$	element of solid angle

Groups

GL_n	group of $n \times n$ matrices with nonzero determinant
O_n	group of $n \times n$ matrices M with $MM^t = \mathbf{1}$ (unit matrix)
SO_n	subgroup of O_n with determinant 1
E_n	Euclidean group
S_n	group of permutations of n elements
U_n	group of complex $n \times n$ matrices M with $MM^* = \mathbf{1}$ (unit matrix)

Physical Symbols

m_i	mass of the i-th particle
$\mathbf{x}_i$	Cartesian coordinates of the i-th particle
$t = x^\circ/c$	time
s	proper time
q_i	generalized coordinates
p_i	generalized momenta
e_i	charge of the i-th particle
κ	gravitational constant
c	speed of light
$\hbar = h/2\pi$	Planck's constant divided by 2π
F^α_β	electromagnetic field tensor
$g_{\alpha\beta}$	gravitational metric tensor (relativistic gravitational potential)
$\mathbf{E}$	electric field strength
$\mathbf{B}$	magnetic field strength in a vacuum
$\sim$	is on the order of
$\gg$	is much greater than

Symbols Defined in the Text

$T^*(\Phi)$	transposed derivative	(2.4.17)
$(\Phi^{-1})^*$	pull-back, or inverse image of the covariant tensors	(2.4.24)
$E_p(M)$	set of p-forms	(2.4.27)
$\wedge$	wedge (outer, exterior) product	(2.4.28)
$*$	$*$-mapping	
V_k^ℓ	contraction	(2.4.31)
i_X	interior product	(2.4.33)
$[\ ]$	Lie bracket	(2.5.9; 6)
Θ, ω	canonical forms	(3.1.1)
Ω	Liouville measure	(3.1.2; 3)
X_H	Hamiltonian vector field	(3.1.9)
b	bijection associated with ω	(3.1.9)
$\{\ \}$	Poisson brackets	(3.1.11)
M_e	generalized configuration space	(3.2.12)
$\mathscr{H}$	Hamiltonian on M_e	(3.2.12)
(I, φ)	action-angle variables	(3.3.14)
$\Omega_\pm$	Møller transformations	(3.4.4)
S	scattering matrix	(3.4.9)
$d\sigma$	differential scattering cross-section	(3.4.12)
$\mathbf{L}$	angular momentum	(4.1.3)
$\mathbf{K}$	boost	(4.1.9)
$\eta_{\alpha\beta}$	Minkowski space metric	(5.1.2)
γ	$1/\sqrt{1 - v^2/c^2}$ (relativistic dilatation)	(5.1.3; 2)
F	electromagnetic 2-form	(5.1.10; 1)
A	1-form of the potential	(5.1.10; 1)
Λ	Lorentz transformation	(5.1.12)
r_0	Schwarzschild radius	(5.7.1)

Introduction 1

1.1 Equations of Motion

The foundations of the part of mechanics that deals with the motion of point-particles were laid by Newton in 1687 in his *Philosophiae Naturalis Principia Mathematica*. This classic work does not consist of a carefully thought-out system of axioms in the modern sense, but rather of a number of statements of various significance and generality, depending on the state of knowledge at the time. We shall take his second law as our starting point: "Force equals mass times acceleration." Letting $x_i(t)$ be the Cartesian coordinates of the i-th particle as a function of time, this means

$$m_i \frac{d^2 x_i(t)}{dt^2} = F_i(x_i), \qquad i = 1, 2, \ldots, N, \tag{1.1.1}$$

where F_i denotes the force on the i-th particle. In nature, so far as we know, there are just four fundamental forces: the strong, weak, electromagnetic, and gravitational forces. In physics books there are in addition numerous other forces, such as friction, exchange forces, forces of constraint, fictitious forces (centrifugal, etc.), and harmonic forces, with which we shall only be peripherally concerned. The first two fundamental forces operate at the subatomic level, outside the realm of classical mechanics, so in fact we shall only discuss gravitation and electromagnetism.

The exact expressions for these forces are rather complicated in their full generality, but, surprisingly, they both simplify greatly in the limit where the velocities of the particles are much less than the speed of light. They are the gradients of the Newtonian and Coulombic potentials, i.e.,

$$F_i(x_i) = \sum_{j \neq i} \frac{x_j - x_i}{|x_j - x_i|^3} (\kappa m_i m_j - e_i e_j), \tag{1.1.2}$$

where κ is the gravitational constant and e_i is the charge of the i-th particle.

For the elementary constituents of matter, e^2 and κm^2 are of very different orders of magnitude: for protons, $e^2 \sim 10^{36} \, \kappa m_p^2$. The reason that gravitation is nonetheless significant is that all masses are positive and add constructively in a large object, whereas the overall charge can be neutral. In astronomical bodies ($N \sim 10^{57}$ for the sun), only gravitation contributes significantly to (1.1.2). One might hesitate to apply (1.1.1) to such bodies, because a star is hardly a point particle, and it is unclear what meaning should be attached to $\mathbf{x}_i$. But it is noteworthy that (1.1.1) also applies to the center of mass of the whole body, which moves according to Newton's law in response to the net force. In practice there is no difficulty with the meaning of $\mathbf{x}_i$, since heavenly bodies are usually rather small, compared with typical distances between them.

To get a feeling for the meaning of the constants of nature just introduced, let us look at their orders of magnitude in the framework of (1.1.1) and (1.1.2). Suppose a particle orbits a star with $N \sim 10^{57}$ protons, with period τ at radius R. Then from (1.1.1) and (1.1.2), essentially

$$\frac{R^3}{\tau^2} = N\kappa m_p, \qquad (1.1.3)$$

in which the mass of the orbiting particle has dropped out, with a purely gravitational force. In cgs units, $\kappa m_p \sim 10^{-32}$, so for a given R we expect period $\tau \sim 10^{16} \, R^{3/2} \, N^{-1/2}$ and velocity $v \sim 10^{-16} \, R^{-1/2} \, N^{1/2}$. For typical cosmic distances and $N \sim 10^{57}$:

	R(cm)	τ(sec)	v(cm/sec)
Earth's orbit	10^{13}	10^7	10^6
double star	10^{11}	10^4	10^7
black hole	10^5	10^{-5}	10^{10}

We see that in a planetary system typical speeds are 10–100 km/sec, which may seem rather fast, but is modest compared with the speed of light. It is only when the dimensions are roughly those of a black hole, in which the mass of a star is compressed to within a few kilometers, that gravitation can lead to speeds approaching the speed of light. At that point the equations of motion (1.1.1) lose their validity and must be replaced with their relativistic version, discussed below.

As already noted, the electrical force between protons is 10^{36} times stronger than their gravitational force. For a proton-electron system this number is raised by three orders of magnitude, the ratio of the proton's mass to the electron's mass, giving 10^{39}. Correspondingly, the relationships between R, τ, and v become $\tau \sim 10^{-7/2} \, R^{3/2} \, N^{-1/2}$ and $v \sim 10^{7/2} \, R^{-1/2} \, N^{1/2}$. On the atomic scale ($R \sim 10^{-8}$ cm), and for $N \sim 1$, we now find impressive speeds, $v \sim 10^{7.5}$ cm/sec and $\tau \sim 10^{-15.5}$ sec. It is thus relatively easy to accelerate charged elementary particles to nearly the speed of light, which necessitates a generalization of Newton's equation of motion.

The law that replaces (1.1.1) and (1.1.2) in these cases is best formulated if one regards ct and $\mathbf{x}$ as dependent variables x^α, $\alpha = 0, 1, 2, 3$, and introduces a parameter s, the proper time, as the independent variable, defined so that $ds^2 = c^2 dt^2 - |d\mathbf{x}|^2$. The electromagnetic field is no longer a vector, but a tensor field of the second degree. The equation of motion generalizing (1.1.1) for a charged particle in an electromagnetic field then reads

$$m \frac{d^2 x^\alpha}{ds^2} = e F^\alpha_\beta(x) \frac{dx^\beta}{ds}, \tag{1.1.4}$$

where by convention the repeated index β is summed over.

The force in (1.1.2) can be written as the gradient of a potential. In the relativistic case the electromagnetic field may be expressed with derivatives of a vector potential as

$$F_{\alpha\beta} = \frac{\partial}{\partial x_\alpha} A_\beta - \frac{\partial}{\partial x_\beta} A_\alpha. \tag{1.1.5}$$

Since A_α depends on the positions, or more precisely the trajectories, of the charged particles, the relativistic formula (1.1.5) is rather more complicated than (1.1.2) and requires the use of field theory. At present we must content ourselves with the restricted problem of a particle in a specified external field $F_{\alpha\beta}$.

The utility of (1.1.4) is further reduced, because macroscopic objects rarely approach c, while the motion of elementary particles actually belongs to the quantum theory. Nonetheless the classical equation (1.1.4) gives the essential behavior in many cases.

The equations of motion which generalize (1.1.1) for fast-moving bodies in a gravitational field are even more complicated than (1.1.4). As in the non-relativistic theory the force is proportional to the mass, but one now needs an equation with three indices:

$$\frac{d^2 x^\alpha}{ds^2} = -\Gamma^\alpha_{\beta\gamma}(x) \frac{dx^\beta}{ds} \frac{dx^\gamma}{ds}. \tag{1.1.6}$$

Gravitation is generalized through $\Gamma^\alpha_{\beta\gamma}$, which again can be written with derivatives of a potential, though now a symmetric tensor of the second degree:

$$\Gamma^\alpha_{\beta\gamma} = \tfrac{1}{2}(g^{-1})^{\alpha\sigma}\left(\frac{\partial g_{\sigma\beta}}{\partial x_\gamma} + \frac{\partial g_{\sigma\gamma}}{\partial x_\beta} - \frac{\partial g_{\beta\gamma}}{\partial x_\sigma}\right). \tag{1.1.7}$$

Once more we must resort to field theory at this point if we wish to determine $g_{\alpha\beta}(x)$ for a given distribution of mass. We shall only study these equations of motion for certain g's; it turns out that despite a mathematical structure similar to (1.1.1), the physics enters a completely different world.

1.2 The Mathematical Language

Formula (1.1.1) is an ordinary differential equation of second order for a vector in $\mathbb{R}^{3N}$. However, since the forces (1.1.2) have a singularity when $\mathbf{x}_i = \mathbf{x}_j$, $i \neq j$, it is advisable to remove those points and work in an open subset of $\mathbb{R}^{3N}$. In doing this one gives up all information about what happens after a collision, but that is just as well, for otherwise the equations would undoubtedly be pushed beyond their physical validity. The equations could in fact be regularized by the introduction of another variable in place of t, so that the solutions would extend beyond the collision (see [6], [7]). There is indeed some physical interest in these regularizations, but only in the possibility of more accurate numerical analyses of near misses; they cannot describe true catastrophes.

We shall, however, broaden the mathematical domain of definition of the equations of motion on open sets of $\mathbb{R}^{3N}$ somewhat further. The process of differentiation depends only on local properties of a Euclidean space, and thus carries over to anything that looks just like a Euclidean space to a near-sighted observer. In this way we are led to introduce differentiable manifolds, for the following reasons:

1. When one deals with a three-dimensional space with the origin removed, polar coordinates are preferable to Cartesian coordinates for many purposes. The space does not then appear as an open subset of $\mathbb{R}^3$, but as (positive numbers) × (surface of a sphere). Hence it is desirable to formulate a differential calculus for spherical surfaces, which are not open subsets of $\mathbb{R}^n$.
2. If we know a constant of the motion K, we may restrict the equations of motion to the surface $K = $ constant, which is a manifold. This might typically be motion on a torus, which has quite different properties from free motion in $\mathbb{R}^n$.
3. Equation (1.1.6) and problems with constraints are generally set up on manifolds in the first place.
4. It is essential to distinguish local and global quantities in order to understand the mathematical structure of classical mechanics. A Hamiltonian system with n degrees of freedom will always locally have $2n - 1$ time-independent constants. The crucial question is how many of these may be defined globally. The concept of manifold serves to clarify this distinction.

In the second chapter we shall develop the necessary mathematical methods. The almost infinitesimal ratio of the number of propositions to the number of definitions is plain evidence that it is less a question of obtaining deep results than of generalizing and sharpening our knowledge of elementary mathematics, or simply common sense. Elementary mechanics gets extended to a more flexible scheme. The various infinitely small quantities like "infinitesimal variations" and "virtual displacements" disappear and are replaced more precisely with mappings of the tangent spaces. The tangent spaces and their associated bundles are the real stage for dynamics, where, roughly speaking, the tangent bundle is the space of q and $\dot{q}$, and the cotangent bundle is the space of q and p, that is, phase space. After a little necessary preparation we thus arrive at Cartan's symbolism, in which all the rules of elementary differential and integral calculus are written down with a very few symbols. At first it may seem only an exercise in the abstract style of writing. But the reward is that this abstract notation succeeds in reducing the general assertions of classical mechanics to trivialities.

1.3   The Physical Interpretation

In order to interpret the formalism it must first be agreed what the observable quantities are. The observables generally correspond to the coordinates and momenta of the particles. There is of course no reason that the coordinate system should necessarily be Cartesian; for example, in astronomy it is usually angles that are directly measured. We should therefore allow arbitrary functions of coordinates and momenta as observables, subject only to boundedness and, for mathematical convenience, differentiability. Such functions form an (Abelian) algebra, and the time-evolution defined by the equations of motion gives an automorphism of the algebra, since sums are transformed to sums and products to products. It is well to distinguish this algebra of observables conceptually from the state in which a particular specimen of the system is to be found; the state has nothing to do with the laws of nature, but only reflects our knowledge of the initial conditions that happened to be realized.

Whereas the observables are functions on phase space, the states are construed as probability measures on it. For each state there is a probability distribution $\rho(q, p)$ such that the average of many measurements of an observable $f(q, p)$ is predicted to be

$$\bar{f} = \int d^{3N}q \, d^{3N}p \, \rho(q, p) f(q, p) \tag{1.3.1}$$

Note that $\overline{f + g} = \bar{f} + \bar{g}$, but $\overline{f \cdot g} \neq \bar{f} \cdot \bar{g}$. This means that fluctuations arise so that $\overline{(f - \bar{f})^2} \neq 0$, unless the measure $d^{3N}q \, d^{3N}p\rho$ is concentrated at a point. Such "extremal" states amount to complete knowledge of all coordinates and momenta. With the solution $q(0), p(0) \rightarrow q(t), p(t)$ of the equations of motion, the automorphism mentioned above is $f(q(0), p(0)) \rightarrow f(q(t), p(t))$.

Although this conceptual distinction between observables and states is avoidable until one encounters quantum mechanics, it draws attention to the essential nature of the problem even in classical mechanics. It is not sufficient to solve the equations of motion for a few initial conditions which happen to arise; instead, they must be solved for arbitrary initial conditions.

In particular the stability of the solutions under small perturbations of the initial conditions, which are never exactly known in reality, becomes an essential question. Above all this point of view is well suited to statistical physics, where only a small amount of information is given for a system of many degrees of freedom, and the critical facts are the absence of stability and mixing properties of the time-evolution.

To be sure, the execution of this program for realistic forces (1.1.2) creates some difficulties. As mentioned, when there is a collision the trajectory leaves the domain of definition of the problem, at which point we can look no further into the time-evolution. Since initial conditions can always be found so that a collision takes place within an arbitrarily short time, we do not really have an automorphism of the algebra. In the two-body problem it happens that the situation may be remedied by removing the region with angular momentum zero from phase space, since in the rest of phase space no collisions can occur. However, in a three-body system this only avoids triple collisions, and it is necessary to regularize the equations of motion with a new time variable if one wants to get an automorphism. In the relativistic case (1.1.4, 6) the situation is even more hazardous, and even in the two-body problem particles that have nonzero angular momentum can be pulled into the singularity. Popularly speaking, there is a black hole and not just a black point. Hence we must moderate our demands and be contented to examine smaller pieces of phase space. The central questions become: Which con-figurations are stable? Will collisions ever occur? Will particles ever escape to infinity? Will the trajectory always remain in a bounded region of phase space? The words "always" and "ever" make it hard to give exact answers. Computer calculations and, often, mathematical existence theorems provide answers only for the not-too-distant future, and predictions for longer times are notoriously inaccurate. In any case, an assertion that something will happen loses its interest for physics when the time in question is longer than the age of the universe.

For (1.1.1; 2) with two particles it is known that all finite orbits are periodic. But this is a degenerate case, which does not hold relativistically (1.1.4; 6) or when there are three particles. Instead, almost-periodic orbits are more typical, where the system returns arbitrarily close to the starting point, but the orbits are not closed. Rather, they intertwine densely in some higher-dimensional shape (a Lissajou figure). Between these almost-periodic orbits are no doubt embedded an infinite number of others that are strictly periodic. For (1.1.2), $e_j = 0$, and more than two particles, there is a strong suspicion that the trajectories for which particles are sent off to infinity fill up most of phase space for all energies. This is certainly energetically possible, since the remaining particles can use potential energy to compensate for the loss. In fact computer studies [8] show that fairly soon two particles will come so near that they can release enough energy to accelerate one of them off to infinity. It is apparent that this process is of great significance for planetary and stellar systems.

The book closes with an investigation of how the physical space-time manifold is determined by the laws of mechanics. At first the structure of space and time appears to be given a priori. Yet it is determined by real rulers and clocks, which are themselves subject to the equations of motion.† Thus it will be necessary to study whether the relationship between rulers and clocks that comes out of the equations of motion is consonant with our original assumptions about space and time. We shall see, for example, that space-time loses its pseudo-Euclidean nature through equations (1.1.6) and gets in its place a Riemannian structure. In other words, gravitation affects rulers so that the space they measure appears curved.

The attraction of the mechanics of point particles is that despite the simplicity of the basic laws, the trajectories that are possible produce such a large and complex picture that it is difficult to survey it all. It is already evident that the consequences of these laws of nature, which can be expressed so briefly, are hugely complicated.

---

† Of course real matter is governed by the quantum theory, so we must anticipate some later material

2 Analysis on Manifolds

2.1 Manifolds

The intuitive picture of a smooth surface becomes analytic with the concept of a manifold. On the small scale a manifold looks like a Euclidean space, so that infinitesimal operations like differentiation may be defined on it.

A function f from an open subset U of $\mathbb{R}^n$ into $\mathbb{R}^m$ is differentiable at a point $x \in \mathbb{R}^n$ if it may be approximated there with a linear mapping $Df: \mathbb{R}^n \to \mathbb{R}^m$. We can make this notion more precise by requiring that for all $\varepsilon > 0$ there exists a neighborhood U of x such that

$$\| f(x') - f(x) - Df(x)(x - x') \| < \varepsilon \|x - x'\| \ \forall x' \in U.$$

Here x and f are respectively vectors in $\mathbb{R}^n$ and $\mathbb{R}^m$, and $\|v\|$ is the length of the vector v. (We shall always make use of vector and matrix notation with the indices dropped, unless there is some reason to write them out.) Written out in components, Df is the matrix of the partial derivatives,

$$(Df)_{ij} = \frac{\partial f_i}{\partial x_j}, \qquad i = 1, \ldots, m, \quad j = 1, \ldots, n \qquad (2.1.1)$$

Remarks (2.1.2)

1. The function f must be given in a neighborhood of x. If we speak simply of differentiability (at all points), we have to deal with a mapping of some open set.
2. At every point the derivative Df is a linear mapping $\mathbb{R}^n \to \mathbb{R}^m$, which has the following significance: if the curve $u: I \to \mathbb{R}^n$ passes through x, then

Df transforms the direction of the curve into the direction of the image of the curve under f. $(df_i(x(t))/dt = f_{i,j}\, dx_j/dt)$.

3. Df can also be regarded as a function, specifically as a mapping into the linear transformations. As such it can itself be differentiable, which simply means that the $f_{i,j}$ are further differentiable. We denote the set of p-times continuously differentiable functions by C^p, the set of infinitely-often differentiable functions by C^∞, and the set of C^∞-functions of compact support by C_0^∞.

In this section we extend the idea of differentiability to sets M which resemble open sets in $\mathbb{R}^n$ only locally. In §2.2 we can then look for the spaces' which are mapped linearly by the derivative. First we introduce some concepts which should be perfectly clear due to their geographical flavor.

Definition (2.1.3)

Let M be a topological space. A **Chart** (V, Φ) is a homeomorphism Φ of an open set V (the **domain** of the chart) of M to an open set in $\mathbb{R}^m$. Two charts are **compatible** in case $V_1 \cap V_2 = \varnothing$, or if the mappings $\Phi_1 \circ \Phi_2^{-1}$ and $\Phi_2 \circ \Phi_1^{-1}$, restricted in the obvious way, are C^∞-mappings of open sets in $\mathbb{R}^m$ (Figure 1).

Definition (2.1.4)

An **atlas** is a set of compatible charts that cover M. Two atlases are called **compatible** if all their charts are compatible.

Remarks (2.1.5)

1. Compatibility of atlases is obviously an equivalence relationship: every atlas is compatible with itself, and the definition is symmetric. Suppose $\bigcup_i (V_{1i}, \Phi_{1i})$ is compatible with $\bigcup_i (V_{2i}, \Phi_{2i})$, which is compatible with $\bigcup_i (V_{3i}, \Phi_{3i})$. Cover $V_{1i} \cap V_{3j}$ with the V_{2k}, and recall that $f \circ g$ is differentiable when f and g are.
2. Assuming that all the charts map M into an $\mathbb{R}^m$ with the same m, m is called the dimension of M. Occasionally this definition is also used when $m = 0$, although $\mathbb{R}^0$ is a point, for which there can be nothing to differentiate.
3. If the V's are chosen small enough, we can suppose that they are all connected sets.

Definition (2.1.6)

A differentiable **manifold** is a separable, metrizable space M with an equivalence class of atlases.

Examples (2.1.7)

1. $M = \mathbb{R}^n = V$. $\Phi = 1$. Only one chart is necessary in this case. This is also true for the somewhat more general case of an open subset of $\mathbb{R}^n$.

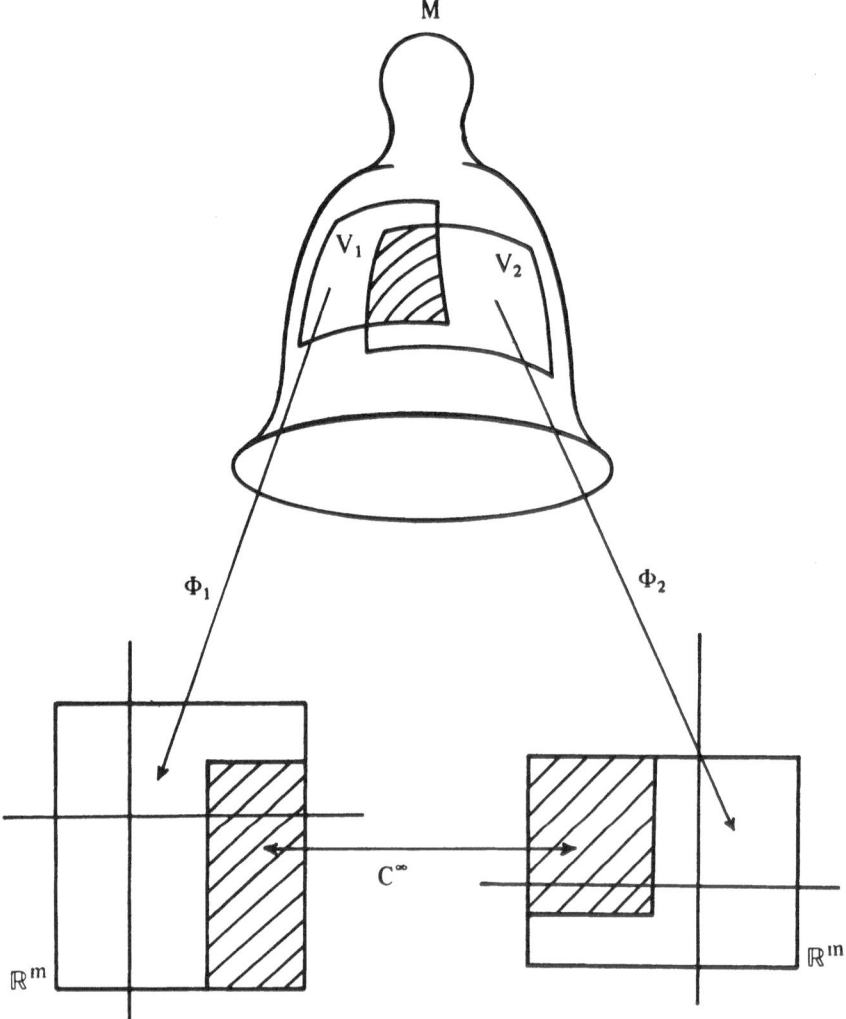

Figure 1   Compatibility of two charts.

2. $M = \{(x_1, x_2) \in \mathbb{R}^2 : x_1^2 + x_2^2 = 1\}$ is called the one-dimensional sphere S^1 or the one-dimensional torus T^1. M is compact and therefore not homeomorphic to an open subset of $\mathbb{R}$. At least two charts are needed:

$$V_1 = T^1 \backslash \{(-1, 0)\}, \qquad \Phi_1^{-1} : \varphi \to (\cos \varphi, \sin \varphi), \quad -\pi < \varphi < \pi$$
$$V_2 = T^1 \backslash \{(1, 0)\}, \qquad \Phi_2^{-1} : \varphi \to (\cos \varphi, \sin \varphi), 0 < \varphi < 2\pi$$

The compatibility of Φ_1 and Φ_2 is trivial (Problem 1); but they can not be replaced with a single mapping (see Figure 2).

$$\underbrace{T^1 \times T^1 \times \cdots \times T^1 = T^n}_{n \text{ times}}$$

is the n-dimensional torus, and the n-dimensional sphere is defined as

$$S^n = \{(x_i) \in \mathbb{R}^{n+1} : x_1^2 + x_2^2 + \cdots + x_{n+1}^2 = 1\}.$$

3. $M = \mathbb{R}^2 \setminus \{(0,0)\} = U$, $\Phi = 1$. As an open subset of some $\mathbb{R}^n$, M needs only one chart. However, M is also homeomorphic to $\mathbb{R}^+ \times S^1$, which suggests an atlas with two charts (polar coordinates).

4. Let $f \in C^1$ and $M = \{x \in \mathbb{R}^n : f(x) = 0$, and $\forall x \, \exists j : f_{,j}(x) \neq 0\}$. This generalizes Example 2, and the implicit function theorem guarantees the existence of suitable charts so that M becomes an $n - 1$-dimensional manifold. The condition on the derivative is obviously necessary, for suppose f is a constant function; then the inverse image of 0 is either the empty set or all of $\mathbb{R}^n$.

5. The n^2 elements of an $n \times n$ matrix define a point in $\mathbb{R}^{n^2}$. Hence the $n \times n$ matrices may be identified with $\mathbb{R}^{n^2}$ and inherit its structure as a manifold (and also as a vector space). The invertible matrices M, Det $M \neq 0$, are an open subset, and form the group $GL(n)$. The unimodular matrices M, Det $M = 1$, are characterized by a condition as in Example 4, and are thus a manifold.

6. $M = \{x \in \mathbb{R}^2 : |x_1| = |x_2|\}$. This can not be a manifold, since every neighborhood of $\{(0,0)\}$ decomposes M without that point into four

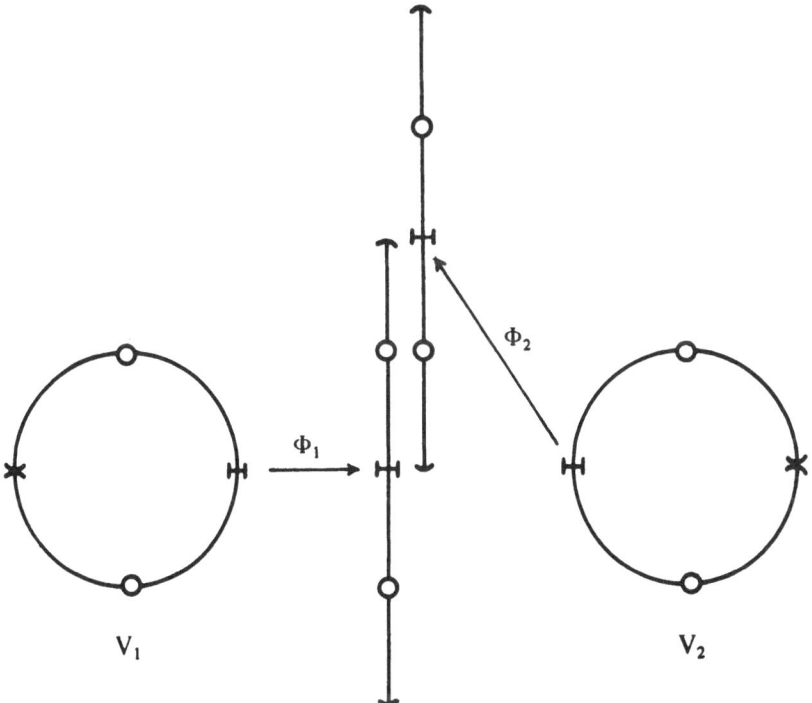

Figure 2   An atlas for T^1. ⊢ and ○ label corresponding points

rather than two components, and consequently can not be mapped homeomorphically onto an open interval.

M: M \ {(0, 0)} :

7. $M = \bigcup_{n=1,2,\dots} \{(1/n, \mathbb{R})\} \cup \{(0, \mathbb{R})\} \subset \mathbb{R}^2$ is certainly no manifold, since it is not locally connected at $\{(0, 0)\}$.
8. Given two manifolds one can define the product manifold $M_1 \times M_2$ (cf. Examples 2 and 3). This set comes equipped with the product topology, and the product chart $(V_1, \Phi_1) \times (V_2, \Phi_2) = (V_1 \times V_2, \Phi_1 \times \Phi_2)$ uses the mapping $(q_1, q_2) \to (\Phi_1(q_1), \Phi_2(q_2))$ into $\mathbb{R}^{m_1 + m_2}$. It is clear that the product of two atlases is another atlas, since the conditions of covering and compatibility are fulfilled.

Remarks (2.1.8)

1. In Examples 1 to 4, M is given directly as a subset of $\mathbb{R}^n$ with the induced topology. It is not always done this way. More obviously, manifolds can be constructed by piecing overlapping regions together. This determines the global structure, while locally everything is determined by the dimension. However, it can be shown [(1), chapters 16, 25] that every m-dimensional manifold is homeomorphic to a subset of $\mathbb{R}^{2m+1}$.
2. It must be assumed that M is separable in order to exclude a number of pathologies; it is not implied by M's being locally Euclidean. This is why we require a topology on M, rather than simply defining one with the charts.

    Example: $M = \mathbb{R} \times (\mathbb{R}$ with the discrete topology), $V_y = \mathbb{R} \times \{y\}$, $\Phi_y : (x, y) \to x$. By this devious construction a plane becomes a one-dimensional manifold.

3. We shall usually suppose that the manifolds are C^∞, which is not an excessively burdensome restriction. Of course, many results could be obtained with fewer assumptions, but it is not our goal to figure out what the optimal assumptions are. Moreover, in the future we will not always check whether all the assumptions of differentiability are satisfied. This is left to the conscientious reader, who will find that there are no real difficulties, since for these local questions everything works as in $\mathbb{R}^n$. For this same reason we shall simply say "manifold" rather than "C^∞-manifold."
4. Since in the cases that will concern us, most of the functions that crop up are analytically continuable, it is sometimes convenient to work with complex manifolds. In this $\mathbb{C}^n$ is substituted for $\mathbb{R}^n$, and analyticity for all degrees of differentiability. For an example of a complex manifold, think of the Riemann surface for $\sqrt{z}$ or for $\ln z$.

5. Physicists are used to the terms "local coordinate system" or "parametrization" instead of charts. That M is not defined with any particular atlas, but with an equivalence class of atlases, is a mathematical formulation of "general covariance." Every suitable coordinate system is equally good. A Euclidean chart may well suffice for an open subset of $\mathbb{R}^n$, but this coordinate system is not to be preferred to the others, which may require many charts (as with polar coordinates), but are more convenient in other respects.

As we have seen (Examples 6 and 7), not all subsets of $\mathbb{R}^n$ may be used as manifolds. They need not necessarily be open subsets of $\mathbb{R}^n$, but one should at least be able to define differentiation on them. The question now arises of when a subset can inherit the structure of a manifold.

Definition (2.1.9)

$N \subset M$ is an n-dimensional **submanifold** iff $\forall q \in N$ there exists a chart (U, Φ), where $q \in U$ and $\Phi: U \to (x_1, \ldots, x_m)$, such that $\Phi|_{U \cap N}: U \cap N \to (x_1, \ldots, x_n, 0, \ldots, 0)$.

Examples (2.1.10)

1. N is an open subset of M. This is the trivial case with $m = n$.
2. $N = S^1$, $M = \mathbb{R}^2$. The charts in (2.1.7; 2) are not of the form (2.1.9), but charts of that form are easy to find (Problem 2).
3. Let f_i, $i = 1, \ldots, k \leq m$, be differentiable functions $\mathbb{R}^m \to \mathbb{R}$ such that the vectors Df_i at each point where $f_i = 0$, $i = 1, \ldots, k$, are linearly independent, or, equivalently, the rank of the matrix $f_{i,j}$, $i = 1, \ldots, k$ and $j = 1, \ldots, m$, is maximal. Then according to the rank theorem [(1), section X.3], $N = \{x \in \mathbb{R}^m : f_i(x) = 0 \ \forall i\}$ is a closed submanifold of dimension $m - k$ of $\mathbb{R}^m$. In particular the orthogonal matrices M, $MM^t = 1$, are a submanifold of the invertible matrices (cf. (2.1.7; 5)).
4. $M = \mathbb{R}^2$, $N = \{x \in M : x_2 = |x_1|\}$ can be equipped with a manifold structure but is not a submanifold of M.† There is a kink in M, which cannot be put into the differentiable form required in the definition, even with a new set of charts. Yet the atlas $(U = N, \Phi: (x_1, x_2) \to x_1)$ makes N a manifold.

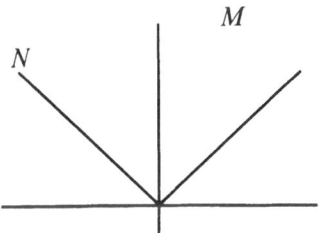

† However, N is the union of three submanifolds.

Remarks (2.1.11)

1. It is easy to see that the atlas in (2.1.9) gives N the structure of a differentiable manifold: the differentiability required for compatibility is unaffected, since only the first n coordinates vary.
2. The last example shows that even when a manifold M is a set-theoretical and topological subspace of $\mathbb{R}^n$, it does not necessarily have to be a submanifold of $\mathbb{R}^n$. However, with the imbedding of M in $\mathbb{R}^{2m+1}$ mentioned in (2.1.8; 1), M is in fact a submanifold of $\mathbb{R}^{2m+1}$.
3. We produced submanifolds of $\mathbb{R}^n$ by requiring that $f_i(x) = 0$, $f_i \in C^\infty$, and that the Df_i were linearly independent. The existence of such functions f is implicit in the definition, at least locally.
4. The following proposition can be proved: Let Y be a submanifold of X, and Z a subset of Y. Then Z is a submanifold of Y iff it is a submanifold of X [(1), 16.8.7].

Now we are ready to generalize the concept of a differentiable mapping of open sets in $\mathbb{R}^n$ to manifolds by following the usual custom in physics: something is called differentiable when it is differentiable in local coordinates.

Definition (2.1.12)

A mapping $f: M_1 \to M_2$ is **p-times differentiable** iff for all charts of an atlas for M_1 and of an atlas for M_2, the obvious restriction of $\Phi_2 \circ f \circ \Phi_1^{-1}$ is a p-times differentiable mapping from $\Phi_1(U_1) \subset \mathbb{R}^{m_1}$ into $\mathbb{R}^{m_2}$ (see Figure 3).

Examples (2.1.13)

1. If M_1 is a submanifold of M_2, then the natural injection is infinitely-often differentiable, because a projection in $\mathbb{R}^n$ is (cf. 2.1.9).
2. Addition and multiplication of two $n \times n$ matrices are C^∞-mappings of the manifold of $n \times n$ matrices.
3. If f_1 and $f_2 \in C^p$, then their composition $f_1 \circ f_2: M_3 \overset{f_2}{\to} M_2 \overset{f_1}{\to} M_1$, is also a C^p-mapping (Problem 6).
4. If M is the product manifold $M_1 \times M_2$ and $f = f_1 \times f_2$, then when the f_i are C^p-mappings, so is f.
5. $M_1 = I \subset \mathbb{R}$. If $f \in C^p$, f is called a C^p-curve.
6. $M_2 = \mathbb{R}$. The p-times differentiable functions in this case are denoted by $C^p(M_1)$. They form an algebra with the usual product in $\mathbb{R}$ because of the elementary rules for differentiation of sums and products. For any f which never vanishes, $1/f$ also belongs to the algebra.

Remarks (2.1.14)

1. Definition (2.1.12) dealt with only one atlas. The condition of compatibility guarantees that differentiability is defined equivalently for all atlases of an equivalence class. This means that a differentiable mapping

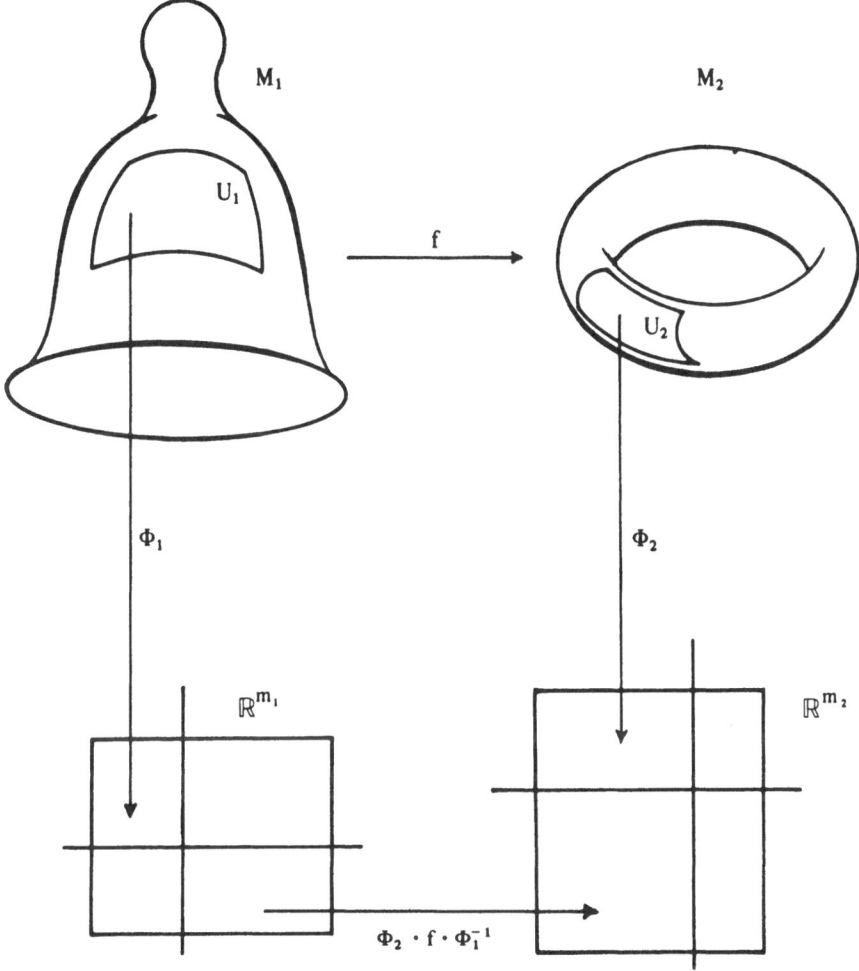

Figure 3   Differentiability of a mapping of manifolds.

remains differentiable under a change of charts, and a nondifferentiable mapping can never become differentiable (cf. (2.1.10; 4)).

2. If N_1 is a submanifold of M_1, then $f_{|N_1}$ is as differentiable as f. On the chart (2.1.9) the restriction only amounts to holding the last $m - n$ coordinates fixed, which does not adversely affect differentiability.

The topological concept of homeomorphism can now be somewhat sharpened for manifolds.

Definition (2.1.15)

A **diffeomorphism** f of two manifolds is a bijection for which both f and $f^{-1} \in C^\infty$. Two manifolds are **diffeomorphic** iff there exists a diffeomorphism between them.

Examples (2.1.16)

1. A chart (U, Φ) provides a diffeomorphism Φ between the submanifold U and $\Phi(U) \subset \mathbb{R}^m$, since $\mathbf{1} \in C^\infty$.
2. $M_1 = \mathbb{R}$ topologically, but has the atlas $U = \mathbb{R}$, $\Phi: x \to x^3$ (only one chart). According to Example 1, Φ is a diffeomorphism $M_1 \overset{\Phi}{\to} \mathbb{R}$.
3. $\Phi: x \to x^3$ is not a diffeomorphism $\mathbb{R} \to \mathbb{R}$, since $\Phi^{-1} \notin C^\infty$.
4. Two manifold structures on the same set M are identical (i.e., are defined with equivalent atlases) iff $\mathbf{1}$ is a diffeomorphism.

Remarks (2.1.17)

1. These examples show that over $\mathbb{R}$ there exist diffeomorphic manifold structures that are not identical, since $M_1 \overset{1}{\to} \mathbb{R}$ is not a diffeomorphism. It should be borne in mind, if one wants to identify manifolds related by a diffeomorphism, that they are not necessarily the same manifold in the sense of the definition. When we refer to $\mathbb{R}^n$ as a manifold without qualification, we mean $\mathbb{R}^n$ with the standard chart $(\mathbb{R}^n, \mathbf{1})$.
2. On complicated topological spaces there are manifold structures which are not even diffeomorphic. But of the connected one-dimensional manifolds the compact ones are all diffeomorphic to S^1 and the non-compact ones to $\mathbb{R}$.

Although an open interval (a, b) and $\mathbb{R}$ can be made into diffeomorphic manifolds, care must be taken in integrating by parts, since (a, b) has boundary points $\{a, b\}$, but $\mathbb{R}$ has none. To emphasize this distinction and highlight the boundary points, we introduce the somewhat more general concept of a manifold with a boundary. It is modeled on a half-space with its boundary.

Definition (2.1.18)

$\mathbb{R}^n_+ \equiv \{x \in \mathbb{R}^n : x_1 \geq 0\}$, $\partial\mathbb{R}^n_+ = \{x \in \mathbb{R}^n : x_1 = 0\}$. A mapping f of an open subset $U \subset \mathbb{R}^n_+$ into $\mathbb{R}^m$ is **differentiable** iff there exist an open subset of $\mathbb{R}^n$ $\tilde{U}$ which contains U, and a differentiable mapping $\tilde{f}: \tilde{U} \to \mathbb{R}^m$, such that $\tilde{f}_{|U} = f$.

Remarks (2.1.19)

1. U need not be open in $\mathbb{R}^n$ and can contain parts of $\partial\mathbb{R}^n_+$.
2. $\mathbb{R}^n_+$ is not a submanifold of $\mathbb{R}^n$, though $\partial\mathbb{R}^n_+$ is.

Just as a manifold is composed of the inverse images of open sets of $\mathbb{R}^n$, a manifold with a boundary is likewise composed of inverse images from $\mathbb{R}^n_+$.

Definition (2.1.20)

Let M be a separable metrizable space. M has the structure of a **manifold with a boundary** when there exist an open covering $\{U_i\}$ and homeomorphisms

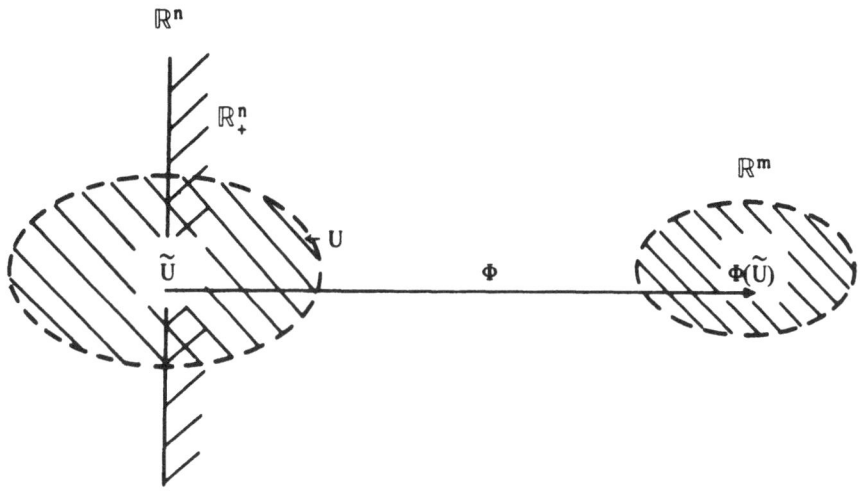

Figure 4   A differentiable mapping of $\mathbb{R}^n_+$.

$\Phi_i \colon U_i \to$ open subsets of $\mathbb{R}^n_+$, for which $\forall i, j,\ \Phi_i \circ \Phi_{j|\Phi_j(U_i \cap U_j)}^{-1} \in C^\infty$. The **boundary** of M is $\partial M = \bigcup_i \Phi_i^{-1}(\Phi_i(U_i) \cap \partial\mathbb{R}^n_+)$ (see Figure 5).

Examples (2.1.21)

1. $M = [a, b] \colon U_1 = [a, b),\ \ \Phi_1 \colon x \to x - a,\ \ U_2 = (a, b],\ \ \Phi_2 \colon x \to b - x$; then $\partial M = \{a\} \cup \{b\}$.
2. $M = \{x \in \mathbb{R}^2 \colon x_1^2 + x_2^2 \leq 1\} \colon\ \ U_1 = \{x_1^2 + x_2^2 < 1\},\ \ \Phi_1 \colon (x_1, x_2) \to (x_1 + 1, x_2)$, and let charts be introduced on $U_2 = \{\frac{1}{2} < x_1^2 + x_2^2 \leq 1\} = T^1 \times (\frac{1}{2}, 1]$ as in Example 1 and (2.1.7; 2). $\partial M = T^1$.

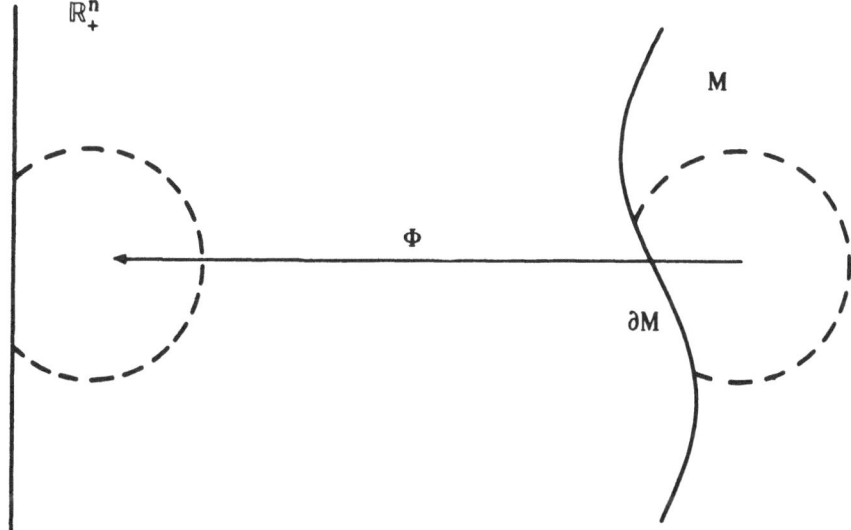

Figure 5   A manifold with a boundary.

3. $M = \{x \in \mathbb{R}^2 : |x_1| \leq 1, |x_2| \leq 1\}$. This is not a manifold with a boundary, because it has corners. However, its interior is a manifold, while its boundary alone is not.†

Remarks (2.1.22)

1. The boundary ∂M is to be distinguished from a topological boundary, which depends on an imbedding. Topologically $\partial \mathbb{R}^n_+$ as a subset of $\mathbb{R}^n$ is its own boundary, but as $\mathbb{R}^{n-1}$ it has no boundary. In $\mathbb{R}^n$ the topological boundary of $\mathbb{R}^n_+$ is $\partial \mathbb{R}^n_+$.
2. Once again, a system of charts such that $M = \bigcup_i U_i$ suffices to determine a manifold, but this system is not to be preferred to other compatible ones. Structures which possess compatible atlases will henceforth be identified.
3. If $\partial M = \varnothing$, this definition reduces to (2.1.6), and in this case we speak simply of a manifold, and only otherwise of a manifold with a boundary.
4. Since the $\Phi_i \circ \Phi_j^{-1}|_{\Phi_j(U_i \cap U_j)}$ are homeomorphisms, boundary points of $\mathbb{R}^n_+$ are mapped to boundary points, and thus compatible atlases define the same boundary.
5. A manifold with a boundary need not be compact (e.g., $(0, 1]$) and a compact manifold may not have a boundary (e.g., T^1).

At the same time the U_i in (2.1.20) provide charts both for the interior, $M \setminus \partial M$, and for the boundary ∂M. These are mapped respectively into open subsets of $\mathbb{R}^n$ and of $\mathbb{R}^{n-1}$, and it is easy to prove that the condition of compatibility continues to hold under these restrictions. Thus we conclude

Proposition (2.1.23)

Both $M \setminus \partial M$ and ∂M have the structure of manifolds (without boundary).

Remarks (2.1.24)

1. That a boundary has no boundary will also follow from a generalization of Stokes's integral theorem (cf. (2.6.7)).
2. ∂M may not be connected, even when M is, as for example $M = [a, b]$. Conversely ∂M may be connected while M is not: $M = [a, b) \cup (b, c)$, $\partial M = \{a\}$.

Problems (2.1.25)

1. Prove the compatibility of the charts of (2.1.7), Example 2.

2. Show that S^1 is a submanifold of $\mathbb{R}^2$ in the sense of (2.1.9).

3. Is $\bigcup_{n=1,2,\dots} \{(1/n, \mathbb{R})\}$ a submanifold of $\mathbb{R}^2$? How many charts of the kind in (2.1.9) are necessary?

---

† It is for this reason that we did not model manifolds with boundaries on $\{x \in \mathbb{R}^n : x_i \geq 0 \ \forall i\}$.

4. What are the minimal necessary and sufficient conditions for a mapping of two open sets in $\mathbb{R}^n$ to be a diffeomorphism?

5. Are addition and multiplication of two matrices C'-mappings from $GL(n) \times GL(n) \to GL(n)$?

6. Show that if $f_1, f_2 \in C^p$, then so is $f_1 \circ f_2$.

7. Show that the "diagonal" $\{(q, q') \in M \times M : q = q'\}$ is a submanifold of $M \times M$.

8. Let $g : \mathbb{R}^m \to \mathbb{R}$ be C^∞ with $Dg \neq 0$ for all x at which $g(x) = 0$. Show that $\{x \in \mathbb{R}^m : g(x) \leq 0\}$ is a manifold with a boundary.

9. Show: $q \in \partial M \Leftrightarrow \exists$ a chart (U, Φ) such that $\Phi(q) = (0, 0, \ldots, 0)$.

Solutions (2.1.26)

1. $U_1 \cap U_2 = T^1 \setminus \{(-1, 0)\} \setminus \{(0, 1)\} \xrightarrow{\Phi_2} (0, \pi) \cup (\pi, 2\pi) \xrightarrow{\Phi_1 \circ \Phi_2^{-1}} (0, \pi) \cup (-\pi, 0)$, by $\varphi \to \varphi$, or as the case may be $\varphi \to \varphi - 2\pi$ (see Figure 2). Likewise for $\Phi_2 \circ \Phi_1^{-1}$.

2. Use four charts:
$$U_{\frac{1}{2}} = \{(x_1, x_2) \in S^1 : x_2 \gtrless 0\}, \Phi_{\frac{1}{2}} : (x_1, x_2) \to x_1$$
$$U_{\frac{3}{4}} = \{(x_1, x_2) \in S^1 : x_1 \gtrless 0\}, \Phi_{\frac{3}{4}} : (x_1, x_2) \to x_2.$$
Compatible, since for $x \neq 1$, $x \to \sqrt{1 - x^2}$ is C^∞.

3. Yes. Infinitely many: $U_n = I_n \times \mathbb{R}$, where I_n is an open interval containing $1/n$, and $I_m \cap I_n = \emptyset$. $\Phi_n : (x, y) \to (y, x - 1/n)$.

4. Let the mapping be $x \to y(x)$. Minimal conditions are that $\forall x$, $y_{i,j} \in C^\infty$ and Det $y_{i,j} \neq 0$ (see [(1), 10.2.5]).

5. Yes, since they are restrictions (2.1.14; 2) of differentiable mappings (2.1.13; 2).

6. $\Phi_1 \circ f_1 \circ \Phi_2^{-1} \circ \Phi_2 \circ f_2 \circ \Phi_3^{-1} = \Phi_1 \circ f_1 \circ f_2 \circ \Phi_3^{-1}$.

7. Introduce the variables $q_k^\pm = q_k \pm q_k'$ on the product chart $m_1 \times m_2 \to (q_1, \ldots, q_m, q_1', \ldots, q_m')$; this gives a chart of the type of (2.1.9).

8. On the charts for which the submanifold $g = 0$ of $\mathbb{R}^m$ has the form (2.1.9), $g \leq 0$ is of the form (2.1.20).

9. If $\Phi(q) = (0, 0, \ldots, 0)$, then q is a boundary point by (2.1.20). Conversely if $q \in \partial M$, then on every chart, $\Phi(q) = (0, x_2, \ldots, x_m)$. By changing the chart so that $\Phi \to \Phi - (0, x_2, \ldots, x_m)$, q is mapped to the origin.

2.2 Tangent Spaces

A smooth surface may be approximated at a point by the tangent plane. The generalization of this concept is the tangent space.

For a mapping $f : \mathbb{R}^n \to \mathbb{R}^m$ the derivative Df (2.1.2) is a linear transformation $\mathbb{R}^n \to \mathbb{R}^m$. Since in general manifolds do not possess a linear structure, we must construct a linear space in order to define the derivative of a mapping of manifolds $M_1 \xrightarrow{f} M_2$. To do this we first generalize the concept of a tangent plane. Then later we shall also distinguish a tangential direction by means of the derivative in this direction of the numerical functions.

Differentiability was defined through the existence of $D(\Phi_2 \circ f \circ \Phi_1^{-1})$, a chart-dependent quantity; with a change of charts on M_2, $\Phi_2 \rightarrow \bar{\Phi}_2$, it is multiplied by $D(\bar{\Phi}_2 \circ \Phi_2^{-1})$. This is a linear transformation, for which only certain kinds of statements have an invariant meaning. For example, for curves (i.e., the special case $\mathbb{R} \supset I = M_1, f : t \rightarrow u(t) \in M_2 = M$), which pass through the point $q = u(0)$, the statement that they are tangent, or osculate, at that point does not depend on the chart used: $D(\Phi_2 \circ u)_{|0}$ is a vector tangent to the image of the curve in $\mathbb{R}^m$, and the statement that the curves are tangent means that the corresponding vectors are parallel—a statement which is invariant under linear transformations. To be more precise, let us call two curves $\mathbb{R} \supset I \xrightarrow{u_{1,2}} M$ tangent at q when $D(\Phi \circ u_1)_{|0} = D(\Phi \circ u_2)_{|0}$; then we may collect all the curves passing through q into equivalence classes of mutually tangent curves, independently of the charts. Each such class can be associated with the tangent vector of the images of the curves at $\Phi(q)$, that is, with $D(\Phi \circ u) \in \mathbb{R}^m$. Thus on any chart $C = (U \ni q, \Phi)$ a bijection $\Theta_C(q)$ is created between the equivalence classes and vectors in $\mathbb{R}^m$.

Definition (2.2.1)

The **mapping** $\Theta_C(q)$ sends a curve u that passes through q to the following vector:

$$u \xrightarrow{\Theta_C(q)} D(\Phi \circ u)(0) \in \mathbb{R}^m.$$

Conversely, for any $v \in \mathbb{R}^m$ the inverse mapping defines a representative curve u of the appropriate class:

$$v \xrightarrow{\Theta_C^{-1}(q)} u \equiv \{t \in \mathbb{R} \rightarrow \Phi^{-1}(\Phi(q) + tv) \in M\}. \tag{2.2.2}$$

Remarks (2.2.3)

1. Different curves of one class correspond to the same vector by $\Theta_C(q)$, but on different charts the same class is associated to different vectors:
2. It might be supposed that a tangent vector directed along the curve u could be defined simply as $\lim_{n \to \infty}(n(u(1/n) - u(0))$, thereby avoiding abstract mental acrobatics. The unfortunate drawback is that this difference is undefined for finite n.
3. The mapping $\Theta_C(q)$ provides the equivalence classes with the structure of a vector space. This is independent of the choice of charts, since under a change of charts $\Theta_C(q)$ is multiplied by $D(\bar{\Phi} \circ \Phi^{-1})(q)$. The chain rule for D then implies that, as the derivative of a bijection, this is an invertible linear transformation (Problem 1), and thus preserves the vector-space

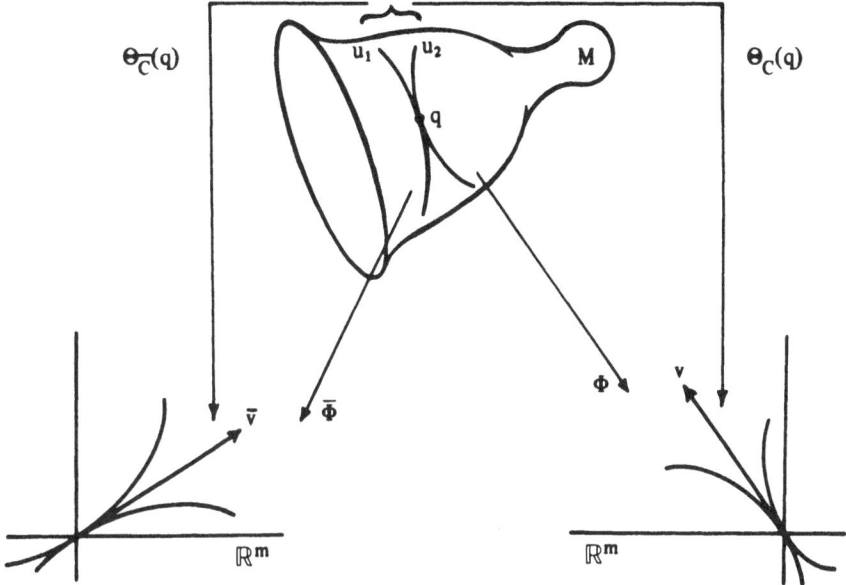

Figure 6   Action of the bijection $\Theta_C(q)$.

structure. Hence the bijection Θ_C allows the desired characteristics of a tangent plane to be preserved, although in fact in the absence of a canonical imbedding of M into $\mathbb{R}^m$ no tangent plane is defined.

Definition (2.2.4)

The space of equivalence classes of curves tangent at q is called the **tangent space** of M at q and denoted $T_q(M)$. It has the structure of a vector space when by definition for $v, w \in T_q(M)$ and $\alpha, \beta \in \mathbb{R}$ we set

$$\alpha v + \beta w = \Theta_C^{-1}(q)(\alpha\Theta_C(q)(v) + \beta\Theta_C(q)(w)),$$

in which Θ_C is given by (2.2.1).

Examples (2.2.5)

1. $M = \mathbb{R}^n, \Phi = 1$. In this case to any vector v, $\Theta_C^{-1}(q)$ assigns the line which passes through q and is parallel to v. $T_q(M)$ may be identified naturally with M, crudely writing $\Theta = 1$.†

2. Suppose a surface F in $\mathbb{R}^3$ is given by a parametrization $g: \mathbb{R}^2 \to \mathbb{R}^3$, $(u, v) \xrightarrow{g} (x(u, v), y(u, v), z(u, v))$. If $g^{-1}|_F$ is used as a chart, then the coordinate lines $u = $ constant and $v = $ constant are just sent to the two axes in $\mathbb{R}^2$ by $\Theta_C(q)$.

† This notation is deprecated by the pedants, but is perfectly all right among friends.

Remarks (2.2.6)

1. Definition (2.2.4) may seem abstract, but it really only formalizes the intuitive notion of vectors in a tangent plane as arrows pointing in the directions of the curves passing through the point. This makes them elements of $\mathbb{R}^n$ on the charts used to make M a manifold.
2. It is clear that with a change of charts Θ is multiplied by $D(\bar{\Phi} \circ \Phi^{-1})$, which is the matrix that specifies how the images of the curves u are twisted around when put in different charts. This is closely related to the usual transformation relations for vectors under a coordinate transformation; if $\bar{x} \overset{\bar{\Phi}}{\leftarrow} q \overset{\Phi}{\to} x$, then $D(\bar{\Phi} \circ \Phi^{-1})$ is simply expressed as the matrix $\partial\bar{x}_i/\partial x_j$, and a vector of the tangent space is transformed as $v_i \to \bar{v}_i = v_j\,\partial\bar{x}_i/\partial x_j$. This property is frequently made the definition of a vector, in the sense of "a vector is a vector that transforms like a vector."
3. If there is no distinguished coordinate system for M, then there is also no distinguished basis for $T_q(M)$ (and hence no scalar product, either). It is only due to their structure as vector spaces that we can identify $\mathbb{R}^n$ and $T_q(\mathbb{R}^n)$, as we do from now on.

Now it is possible to interpret the derivative at a point q of a mapping of manifolds $M_1 \overset{f}{\to} M_2$ as a linear transformation of the tangent space $T_q(M_1) \to T_{f(q)}(M_2)$. To accomplish this, Θ is used to write everything in local coordinates.

Definition (2.2.7)

The **derivative** of f at the point q, written $T_q(f)$, is defined as $T_q(f) = \Theta_{C_2}^{-1}(f(q)) \circ D(\bar{\Phi}_2 \circ f \circ \Phi_1^{-1}) \circ \Theta_{C_1}(q)$.

Remarks (2.2.8)

1. Figure 7 is an attempt to clarify this difficult notation for the action of $T_q(f)$.
2. Although this definition makes reference to a chart, it is in fact chart-independent, since the tangent spaces and $D(\Phi_2 \circ f \circ \Phi_1^{-1})$ transform so as to compensate. If, for example, the chart on M_1 is changed to $\bar{\Phi}_1$, $D(\Phi_2 \circ f \circ \Phi_1^{-1})$ is multiplied on the right by $D(\Phi_1 \circ \bar{\Phi}_1^{-1})$ by the chain rule, while, as we already know, $\Theta_{C_1}(q)$ gains a factor $D(\bar{\Phi}_1 \circ \Phi_1^{-1})$ on the left.

Examples (2.2.9)

1. $M_i = \mathbb{R}^{n_i}$, $i = 1, 2$. The canonical identification of the tangent spaces implies $\Theta = 1$, and so $T_q(f) = D(f)(q)$. This test case brings us back to the old definition.
2. $M_1 = \mathbb{R} = T_0(M_1)$, $\Phi_1 = 1$. With this identification we denote the basis

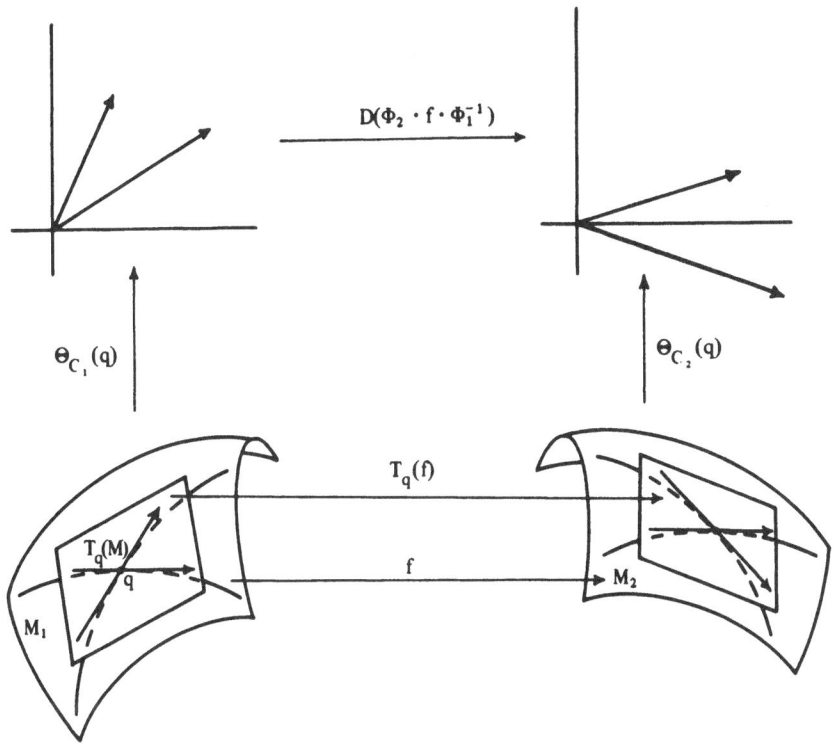

Figure 7   The action of $T_q(f)$.

vector of $T_0(\mathbb{R})$ by 1. Then $T_0(f) \cdot 1 = \Theta_C^{-1} \circ D(\Phi_2 \circ f) = \Theta_C^{-1} \circ \Theta_C(f) \in$ $T_{f(0)}(M_2)$ is just the representative vector of the equivalence class of f, which in this case consists of a single curve. Schematically,

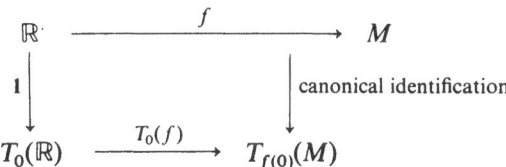

3. If $v \in T_q(M_1)$ is determined by the curve u, then $T_q(f) \cdot v$ is determined by $f \circ u$, because

$$T_q(f) \cdot v = \Theta_{C_2}^{-1}(f(q))D(\Phi_2 \circ f \circ \Phi_1^{-1})D(\Phi_1 \circ u) = \Theta_{C_2}^{-1}(f(q))D(\Phi_2 \circ f \circ u).$$

In words: f transforms the curve u into $f \circ u$, and $T_q(f)$ maps the tangent vectors to the curve u at the point q to the tangent vectors to $f \circ u$ at the point $f(q)$. This statement reads the same in all charts, showing that

$T_q(f)$ is chart-independent. Schematically,

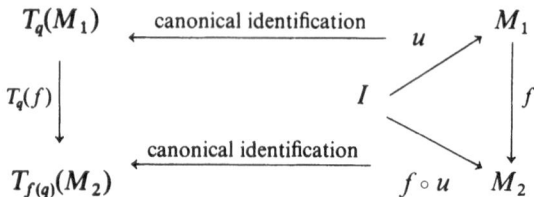

4. $M_2 = \mathbb{R}$. The f's form an algebra, and the derivative behaves as usual for the algebraic operations: from $(f_1 + f_2) \circ \Phi^{-1} = f_1 \circ \Phi^{-1} + f_2 \circ \Phi^{-1}$ and $(f_1 \cdot f_2) \circ \Phi^{-1} = (f_1 \circ \Phi^{-1}) \cdot (f_2 \circ \Phi^{-1})$ it follows that

   (a) $T_q(\text{constant}) = 0$
   (b) $T_q(f_1 + f_2) = T_q(f_1) + T_q(f_2)$
   (c) $T_q(f_1 \cdot f_2) = f_1(q) \cdot T_q(f_2) + f_2(q) \cdot T_q(f_1)$.

5. M_1 is a submanifold of M_2 and f is the natural injection. By using a single chart it can be shown (Problem 6) that $T_q(f)$ is injective, and we can identify $T_q(M_1)$ as a subspace of $T_q(M_2)$.

6. We concluded in (2.1.13; 3) that the composition of two differentiable maps is differentiable. This can also be verified for T (Problem 7):

$$T_q(f_1 \circ f_2) = T_{f_2(q)}(f_1) \circ T_q(f_2),$$

using the chain rule for D.

These examples illustrate how the essential rules of differentiation for manifolds can be expressed independently of the charts.

In order to progress from evaluating the derivative at a point to treating it as a function of q, we have to connect the tangent spaces at different points. At this stage they still have nothing to do with each other; there is no way to say that vectors at different points q are "parallel at a distance." Yet within the domain of a single chart one could identify $T(U) = \bigcup_{q \in U} T_q(M)$ with $U \times \mathbb{R}^m$, and then extend the mapping $\Theta_C(q)$ to

$$\Theta_C : T(U) \to \mathbb{R}^m \times \mathbb{R}^m, (q, v) \to (\Phi(q), \Theta_C(q) \cdot v). \qquad (2.2.10)$$

It is possible to compare tangent vectors within this "tangent bundle" over U. The mapping Θ_C is plainly a bijection, and $T(U)$ can be topologized so that it becomes a homeomorphism. One can even make it into a diffeomorphism and thereby confer a manifold structure on $T(U)$. The atlas then has only one chart, so there are no conditions of compatibility to verify. To extend the tangent bundle over all of M, one constructs it for all U_i of an atlas $\bigcup_i (U_i, \Phi_i)$. It suffices to show the compatibility of these charts, which also verifies the compatibility of the product topologies on the individual $T(U_i) = U_i \times \mathbb{R}^m$. Now,

$$\Theta_{\bar{C}}(q) \circ \Theta_C^{-1}(q) : v \to \frac{d}{dt} \bar{\Phi} \circ \Phi^{-1}(\Phi(q) + vt)_{|t=0} = D(\bar{\Phi} \circ \Phi^{-1}) \cdot v, \qquad v \in \mathbb{R}^m,$$

and thus

$$\Theta_{\bar{C}} \circ \Theta_C^{-1} : (x, v) \to (\bar{\Phi}(\Phi^{-1}(x)), D(\bar{\Phi} \circ \Phi^{-1})(x) \cdot v), \qquad (2.2.11)$$

where $\bar{\Phi} \circ \Phi^{-1}$ is assumed to be C^∞, coming from an atlas. As regards the second factor, it is linear (and therefore C^∞) in v, and also C^∞ with respect to x by the assumption on $\bar{\Phi} \circ \Phi^{-1}$. This proves compatibility of the Θ's, allowing us to make

Definition (2.2.12)

$T(M) = \bigcup_{q \in M} T_q(M)$ is called the **tangent bundle** of M. It is a manifold with the atlas $\bigcup_i (U_i \times \mathbb{R}^m, \Theta_{C_i})$.

Examples (2.2.13)

1. $M = U \subset \mathbb{R}^m$, $C = (U, 1)$, $\Theta_C : (x_i, x_i + tv_i) \to (x_i, v_i)$, where the second argument stands for the curve $t \to x + tv$. $T(M) = U \times \mathbb{R}^m$. As we see, for open subsets of a Euclidean space the tangent bundle is a Cartesian product.
2. $M = S^1$, charts as in (2.1.7; 2). $\Theta_{C_{1,2}} : (\cos \varphi, \sin \varphi; \cos(\varphi + \omega t), \sin(\varphi + \omega t)) \to (\varphi, \omega)$. The two charts C_i are simply the products of charts C_i of S^1 and the identity chart of $\mathbb{R}$. Combined, they give a diffeomorphism between $T(S^1)$ and $S^1 \times \mathbb{R}$. The tangent bundle is again a product.
3. $M = S^2 = \{x \in \mathbb{R}^3 : x_1^2 + x_2^2 + x_3^2 = 1\}$, $C_\pm = (S^2 \backslash (0, 0, \pm 1); (x_1, x_2, x_3) \to (x_1, x_2)/(1 \mp x_3))$ (stereographic projection),

$$\Theta_{C_\pm} : (x_1, x_2, x_3; x_1 + v_1 t, x_2 + v_2 t, x_3(t)) \to$$

$$\left(\frac{(x_1, x_2)}{1 \mp x_3} ; \frac{(v_1, v_2)}{1 \mp x_3} \pm \frac{(x_1, x_2)}{(1 \mp x_3)^2} \frac{x_1 v_1 + x_2 v_2}{x_3} \right).$$

The mappings in the region of overlap are different and can not be extended continuously. For example, as $x_3 \to 1$, in the second position Θ_{C_-} becomes the mapping: (left side) $\to (\ ; (v_1, v_2)/2)$; thus it acts essentially like the identity. On the other hand Θ_{C_+} becomes the mapping (left side) $\to (\ ; (2/(x_1^2 + x_2^2))((v_1, v_2) - 2(x_1, x_2)(x_1 v_1 + x_2 v_2)/(x_1^2 + x_2^2)))$, acting like a dilatation by $1/|x|^2$ followed by a reflection about x. This is singular at $x = 0$, and can not be continuously joined to Θ_{C_-}. A deeper analysis shows that all charts show similar behavior, because of which $T(S^2) \neq S^2 \times \mathbb{R}^2$ topologically.
4. Obviously, $T(M_1 \times M_2) = T(M_1) \times T(M_2)$, with the product charts.

Remarks (2.2.14)

1. At present $T(M)$ is defined abstractly and not given concretely as a submanifold of some $\mathbb{R}^n$. The meaning of the tangent bundle becomes more intuitive, however, if we think of it as the space of the positions and velocities of particles.

2. If $T_q(M)$ is thought of as the pair $\{q\} \times \mathbb{R}^m$, then in a purely set-theoretical sense $T(M) = \bigcup_{q \in M} T_q(M) = \bigcup_{q \in M} (\{q\} \times \mathbb{R}^m) = (\bigcup_{q \in M} \{q\}) \times \mathbb{R}^m = M \times \mathbb{R}^m$ is always a product. However, with the Θ_{C_i} it could be topologized as, say, a Möbius strip (cf. (2.2.16; 3)), so that $T(M) \neq M \times \mathbb{R}^m$ topologically. If it happens moreover that $T(M) = M \times \mathbb{R}^m$ in the sense of the manifold structure (and therefore topologically as well), we say that M is **parallelizable**, because it is possible to define parallelism of tangent vectors at different points in the product topology. The only n-spheres that are parallelizable are S^1, S^3, and S^7. Locally, $T(M)$ is always a product manifold.

3. M may be identified with the submanifold of $T(M)$ corresponding to the point $\{0\}$ in $\mathbb{R}^m$, because of which there exists, even globally, a projection $\Pi: T(M) \to M$, $(q, v) \to q$, onto a distinguished submanifold. Note that for a Cartesian product $(q, v) \to v$ would also be given canonically, but that it is chart-dependent.

4. As a manifold $T(M)$ admits various other charts, though it is for the so-called bundle chart used up to now that Π has the simple form $(q, v) \to q$. A change of charts for M induces a transition from one bundle chart to another on $T(M)$.

This additional structure is the motivation for

Definition (2.2.15)

A **vector bundle** consists of a manifold X, a submanifold M, a surjection $\Pi: X \to M$, and a vector space F, such that every point $q \in M$ has a neighborhood $U \subset M$ for which $\Pi^{-1}(U)$ is diffeomorphic to $U \times F$. M is called the **basis** and $\Pi^{-1}(q)$ a **fiber**. X is said to be **trivializable** iff on some chart $X = M \times F$, and **trivial** iff it is given as a Cartesian product.

Examples (2.2.16)

1. $X = \mathbb{R} \times \mathbb{R}$, $M = F = \mathbb{R}$, $\Pi: (x, y) \to x$. $X = M \times F$ is trivial. There are many coordinate systems for $\mathbb{R} \times \mathbb{R}$ as a manifold, though the product structure distinguishes the Cartesian one.

2. $X = T(M)$, $F = \mathbb{R}^m$, $\Pi: (q, v) \to q$. The fibers are the tangent spaces $T_q(M)$. Trivializable iff parallelizable.

3. $X = [0, 2\pi) \times \mathbb{R}$ (as sets), with two charts $C_i = (U_i, \Phi_i)$, which also define the topology on X:

$$C_1: ((0, 2\pi) \times \mathbb{R}, \mathbf{1}),$$

$$C_2: \left([0, \pi) \cup (\pi, 2\pi) \times \mathbb{R}, (\varphi, x) \to \begin{array}{ll} (\varphi, x) & \text{if } 0 \leq \varphi < \pi \\ (\varphi - 2\pi, -x) & \text{if } \pi < \varphi < 2\pi \end{array} \right),$$

$$M = S^1, F = \mathbb{R}, \Pi: (\varphi, x) \to \varphi.$$

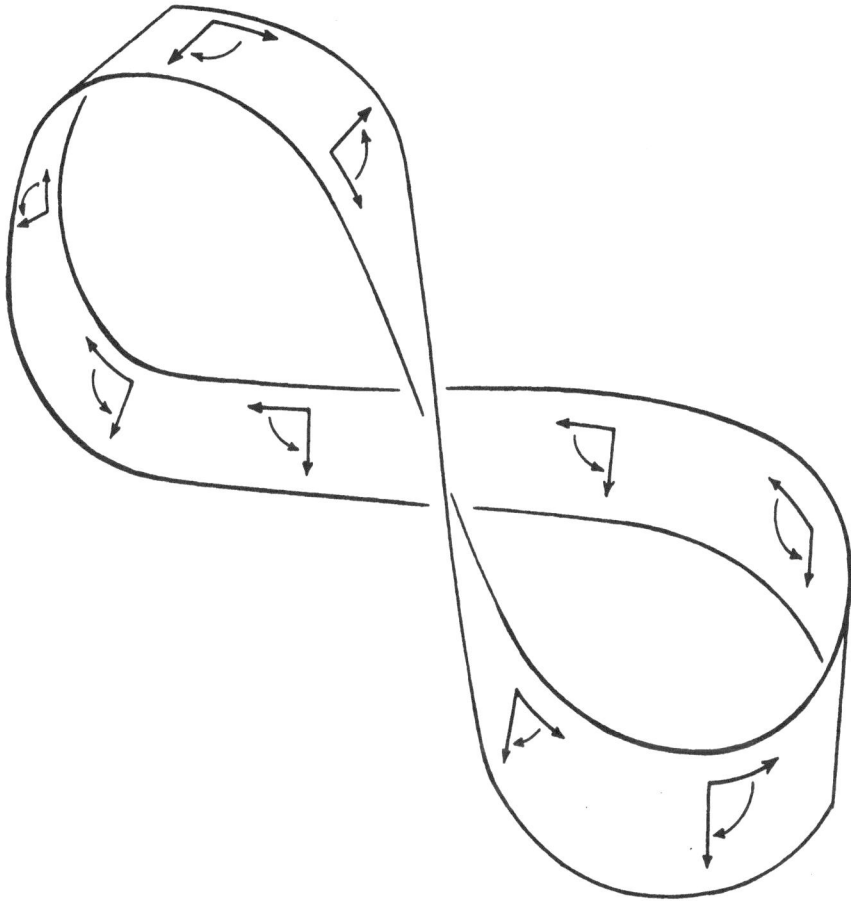

Figure 8   A non-trivializable vector bundle.

With C_2, $\phi \to 2\pi$ is joined to $\varphi = 0$, but in such a way that the sign of x is reversed. X is then an infinitely wide Möbius strip. We shall later see that it can not be trivialized with any charts, since it is not orientable.

The derivative defined at a point (2.2.17) can now be generalized as a mapping of the tangent bundle.

Definition (2.2.17)

The mapping $T(M_1) \to T(M_2):(q, v) \to (f(q), T_q(f) \cdot v)$ is called the **derivative** $T(f)$ of the function $f: M_1 \to M_2$.

Remarks (2.2.18)

1. If $f \in C^r$, then $T(f) \in C^{r-1}$.
2. If f is a diffeomorphism, then so is $T(f)$ (cf. Problem 1).

3. The diagram

$$
\begin{array}{ccc}
M_1 & \xrightarrow{\;f\;} & M_2 \\
\Big\uparrow{\Pi_1} & & \Big\uparrow{\Pi_2} \\
T(M_1) & \xrightarrow{\;T(f)\;} & T(M_2)
\end{array}
\quad\text{, which means}\quad
\begin{array}{ccc}
q & \xrightarrow{\;f\;} & f(q) \\
\Big\uparrow{\Pi_1} & & \Big\uparrow{\Pi_2} \\
(q,v) & \xrightarrow{\;T(f)\;} & (f(q),\,T_q(f)\cdot v)
\end{array}
\quad,
$$

commutes.

4. The chain rule (2.2.9; 6) is now written more conveniently $T(f \circ g) = T(f) \circ T(g)$ (Problem 7).

5. With Cartesian products everything factorizes, including the derivative: $T(f \times g) = T(f) \times T(g)$.

The manifold structure of $T(M)$ defines when the transition between vectors at neighboring points is continuous. This allows

Definition (2.2.19)

A C^r-**vector field** X is a C^r-mapping $M \to T(M)$ such that $\Pi \circ X = 1$. The set of all vector fields for M is denoted $\mathcal{T}_0^1(M)$.

Remarks (2.2.20)

1. On the bundle chart of $T(M)$, $\Pi \circ X = 1$ means that $X: q \to (q, v(q))$; usually only the vector part $v(q)$ is written.
2. $T(M)$ is trivializable iff there exist m linearly independent C^∞-vector fields.[†] E.g., on S^2 this is not the case; there is not even one vector field that never vanishes. (See (2.6.17; 6).)

A change of charts induces a transformation of a vector field which is locally a diffeomorphism according to (2.2.18; 2). More generally formulated:

Definition (2.2.21)

A diffeomorphism $\Phi: M_1 \to M_2$ induces a **mapping** $\Phi^*: \mathcal{T}_0^1(M_1) \to \mathcal{T}_0^1(M_2)$ defined by the commutativity of the diagram:

$$
\begin{array}{ccc}
M_1 & \xrightarrow{\;\Phi\;} & M_2 \\
\Big\downarrow{X} & & \Big\downarrow{\Phi^*X} \\
T(M_1) & \xrightarrow{\;T(\Phi)\;} & T(M_2)
\end{array}
$$

I.e., $\Phi^*X = T(\Phi) \circ X \circ \Phi^{-1}$.

[†] If these fields are X_i, $i = 1, \ldots, m$, this means that the $X_i(q)$ are linearly independent $\forall q \in M$.

Examples (2.2.22)

1. $M = \mathbb{R}^n$, $\Phi: \mathbf{x} \to \mathbf{x} + \mathbf{a}$, $T(\Phi): (\mathbf{x}, \mathbf{u}) \to (\mathbf{x} + \mathbf{a}, \mathbf{u})$, $X: \mathbf{x} \to (\mathbf{x}, \mathbf{v}(\mathbf{x}))$, $\Phi^*X:$ $\mathbf{x} \to (\mathbf{x}, \mathbf{v}(\mathbf{x} - \mathbf{a}))$. A vector remains unchanged under a displacement, but one must take care to talk only about a vector at one particular point, which has different coordinates in the new system.

2. $M = \mathbb{R}^n$, $\Phi: x_i \to L_{ik} x_k$, $T(\Phi): (x_i, u_j) \to (L_{ik} x_k, L_{jm} u_m)$,† $\Phi^*X: x_i \to$ $(x_i, L_{ik} v_k(L^{-1}x))$. Under linear transformations v transforms like x.

3. The following diagram of a transformation holds in general:

$$
\begin{array}{ccc}
q_i & \xrightarrow{\;\;\Phi\;\;} & \bar{q}_i \\[2pt]
\Big\downarrow{\scriptstyle X} & & \Big\downarrow{\scriptstyle \Phi^*X} \\[6pt]
(q_i, v_i(q)) & \xrightarrow{\;\;T(\Phi)\;\;} & \left(\bar{q}_i, \dfrac{\partial \bar{q}_i}{\partial q_j} v_j(q)\right)
\end{array}
$$

Remark (2.2.23)

The derivative of $g \in C^r(M)$ at the point $q \in M$ is a mapping $T_q(M) \to T_{g(q)}(\mathbb{R})$, which on a bundle chart is

$$
(q, u) \to \left(g(q), u_i \frac{\partial g}{\partial q_i}(q)\right).
$$

The second component is the derivative of g in the direction of u, and consequently an ordinary number. Given a particular vector field $X: q \to (q, u(q))$, this component is in $C^{r-1}(M)$. In this way a vector field provides a mapping of the space $C^\infty(M)$ into itself. Its direct significance is as the rate of change of the function along the curve which defines the vector. It is denoted L_X, and according to (2.2.9; 4) it has the properties;

(a) $L_X(f + g) = L_X(f) + L_X(g)$
(b) $L_X(f \cdot g) = f \cdot L_X(g) + g \cdot L_X(f)$.

Putting $f = 1$ in b), we find $L_X(1) = 0$, and so $L_X(\text{constant}) = 0$.

These properties in fact completely characterize the vector fields. Thus one can also define directions on a manifold so as to specify the rate of change of the C^∞-functions. We will not go into the easy proof of this fact here (cf. Problem 8), but only cite

Theorem (2.2.24)

A mapping $L: C^\infty(M) \to C^\infty(M)$ with the properties

(a) $L(f + g) = L(f) + L(g)$
(b) $L(f \cdot g) = f \cdot L(g) + g \cdot L(f)$

determines a unique vector field such that $L = L_X$.

† Here x_i stands for $x_i e_i \in \mathbb{R}^n$, where $\{e_i\}$ is a basis for $\mathbb{R}^n$, and similarly for v.

Remarks (2.2.25)

1. L_X is called the Lie derivative associated with X. In mechanics it is known as the Liouville operator, so the notation L is doubly justified.
2. Since L_X is defined with a local operation, it suffices to know the action of L_X on the C^∞-functions of compact support in order to determine X.
3. Since the diagram

$$
\begin{array}{ccccc}
M_1 & \xrightarrow{\;\;\Phi\;\;} & M_2 & \xrightarrow{\;\;f\;\;} & \mathbb{R} \\[2pt]
\Big\downarrow{\scriptstyle X} & & \Big\downarrow{\scriptstyle \Phi^*X} & & \\[6pt]
T(M_1) & \xrightarrow{\;T(\Phi)\;} & T(M_2) & \xrightarrow{\;T(f)\;} & T(\mathbb{R})
\end{array}
$$

commutes, that is, $T(f) \circ T(\Phi) \circ X = T(f) \circ \Phi^*X \circ \Phi$, we conclude that

$$L_X(f \circ \Phi) = (L_{\Phi \cdot X}(f)) \circ \Phi.$$

Thus the image of X acts on a function as X acts on its inverse image. This fact becomes obvious when one thinks of L_X (and $L_{\Phi*X}$) as the rate of change along a curve in the direction of X (respectively along the image of the curve).

The Natural Basis (2.2.26)

The connection between vector fields and differential operators also allows the basis for $T_q(M)$ induced by $\Theta_c(q)$ to be written symbolically as $\partial/\partial q_i$, or ∂_i: Let e_i be the basis vectors of $\mathbb{R}^m$ and $\Phi: q \to \sum_i e_i q_i \in \mathbb{R}^m$. To any function $g \in C^\infty(M)$ on this chart is assigned the mapping $g \circ \Phi^{-1}: \mathbb{R}^m \to \mathbb{R}$, for brevity written $g(q_i)$. By (2.2.23) the derivation $\partial/\partial q_i$ is assigned to the Lie derivative with respect to the inverse image of e_i under $\Theta_c(q)$, $L_{\Theta_c^{-1}(q)e_i} g = \partial g/\partial q_i$, because the images under Φ of the curves corresponding to $\Theta_c^{-1}(q)e_i$ are just the e_i-axes in q. To an arbitrary vector $\sum_i u^i e_i \in \mathbb{R}^m$ there corresponds the differential operator $u^i \partial/\partial q_i$, the connection being simply the replacement of e_i with $\partial/\partial q_i$.

Problems (2.2.27)

1. Show that for a diffeomorphism Ψ, $T(\Psi^{-1}) = (T(\Psi))^{-1}$.

2. Show that on a chart Φ^*X produces the usual transformation law for vectors (2.2.6; 2).

3. Write $L_X g$ out explicitly on a chart.

4. Show the chart-independence of (2.2.4).

5. Show that if $M_1 \xrightarrow{f} M_2 \supset N_2$, $T(f)$ is surjective, and N_2 is a submanifold of M_2, then $f^{-1}(N_2)$ is a submanifold of M_1. (If N_2 is a one-point space, this reduces to (2.1.10; 3).)

6. Show that for the natural injection to a submanifold, $T(f)$ is injective.

7. Verify the chain rule.

8. Show that a mapping $L: C^\infty \to C^\infty$ with the properties (i) $L(f_1 + f_2) = L(f_1) + L(f_2)$ and (ii) $L(f_1 \cdot f_2) = L(f_1) \cdot f_2 + f_1 \cdot L(f_2)$ must be of the form $L(f)(p) = (X \,|\, df)_{|p}$. For the definition of df, see (2.4.3).

Solutions (2.2.28)

1. This follows from the chain rule applied to $\Psi \circ \Psi^{-1} = 1$ and from $T(1) = 1$.

2. Let $X: q \to (q, v^i(q)\partial_i)$ and $\Phi: q \to q(\bar{q})$. Then $\Phi^* X: \bar{q} \to (\bar{q}, v^i(q)(\partial \bar{q}_i/\partial q_j)\bar{\partial}_{i|\bar{q}})$. Observe that the components v^i transform the same way as the differentials dq_i and the other way around from the basis ∂_i.

3. Let $g: q \to g(q)$. Then $L_X g: q \to v^i(q)(\partial g(q)/\partial q_i)$ when $X: q \to (q, v(q))$.

4. $\Theta_C^{-1}(q)(\alpha\Theta_C(q)(v) + \beta\Theta_C(q)(w)) = \Theta_{\bar{c}}^{-1}(D(\Phi \circ \bar{\Phi}^{-1}))^{-1}(\alpha D(\Phi \circ \bar{\Phi}^{-1})\Theta_{\bar{c}}(v)$
$$+ \beta D(\Phi \circ \bar{\Phi}^{-1})\Theta_{\bar{c}}(w)) = \Theta_{\bar{c}}^{-1}(\alpha\Theta_{\bar{c}}(v) + \beta\Theta_{\bar{c}}(w)).$$

5. Use a chart homeomorphism Φ on M_2 of the type (2.1.9). If $\Phi \circ f = \sum e_i f_i$, then $f^{-1}(N_2) = \{x \in M_1 : f_i(x) = 0, i = n_2 + 1, \ldots, m_2\}$. That $T(f)$ is surjective means that $f_{i,k}, i = 1, \ldots, m_2, k = 1, \ldots, m_1 \geq m_2$, has maximal rank, that is, the vectors $f_{i,k}, i = n_2 + 1, \ldots, m_2$ must be linearly independent. Then we have the situation of (2.1.10; 3).

6. On the chart of (2.1.9), $f: (x_1, \ldots, x_m) \to (x_1, \ldots, x_n, 0, \ldots, 0)$

$$T(f): m \left\{ \begin{bmatrix} 1 & & & & & & \\ & 1 & & & & & \\ & & 1 & & & & \\ & & & 1 & & & \\ & & & & 1 & & \\ & & & & & 1 & \\ & & & & & & 1 \end{bmatrix} \right\} n \qquad T(f)v = 0 \Leftrightarrow v = 0.$$

over bracket: n

7.
$$T_q(f_2 \circ f_1) = \Theta_{C_3}^{-1}(f_2 \circ f_1(q)) \circ D(\Phi_3 \circ f_2 \circ f_1 \circ \Phi_1^{-1})\Theta_{C_1}(q)$$
$$= \Theta_{C_3}^{-1}(f_2 \circ f_1(q))D(\Phi_3 \circ f_2 \circ \Phi_2^{-1})\Theta_{C_2}(f_1(q))$$
$$\times \Theta_{C_2}^{-1}(f_1(q))D(\Phi_2 \circ f_1 \circ \Phi_1^{-1}) \circ \Theta_{C_1}(q).$$

8. On a chart which maps f to the origin, f is of the form
$$f(x) = f(0) + x_i f_{,i}(0) + \tfrac{1}{2}x_i x_j f_{,ij}(0) + \cdots.$$

From (i) and (ii) it follows that $L(f(0)) = L(\text{constant}) = 0$, and then from (ii), $L(x_i f_{,i}(0)) = L(x_i)f_{,i}(0)$ and $L(x_i x_j f_{,ij})/2 = x_i L(x_j)f_{,ij}(0)$. Therefore $L(f)(0) = f_{,i}L(x_i)$. The $L(x_i)$ are the components of the vector field X. It is obviously necessary that $L(f) \in C^\infty$ for the components to be C^∞.

2.3 Flows

A vector field X defines a motion in the direction of X at all points of a manifold. Under the right circumstances this defines a flow, that is, a one-parameter group of diffeomorphisms of M.

A vector field X is to be regarded as a field of direction indicators: to every point of M it assigns a vector in the tangent space at that point. A curve $u: I \to M$, $t \to u(t)$ will be called an **integral curve**† of X if it has the same direction as X at every point; or more precisely if the tangent vector determined by it equals X at every point. Pictorially:

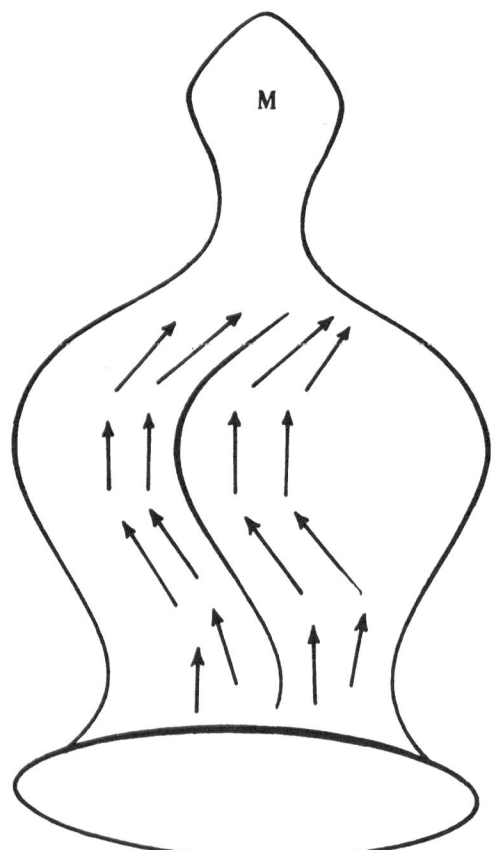

Figure 9   An integral curve of a vector field.

If $\dot{u} = T(u) \cdot (\mathbf{1} \times$ the unit tangent vector), then this is equivalent to the formula:

$$\dot{u} = X \circ u \tag{2.3.1}$$

† When an integral curve is the path a physical system follows, i.e., the solution of the equations of motion, we shall generally refer to it as a **trajectory**.

And expressed schematically it is the following commutative diagram:

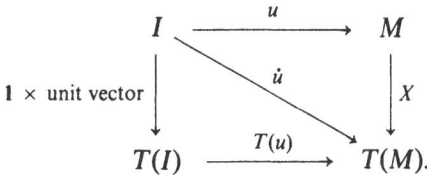

On a chart for which $\Phi \circ u: t \rightarrow u_i(t)$ and $T(\Phi) \circ X \circ \Phi^{-1}: q_i \rightarrow (q_i, X_i(q))$, this is commonly written in the form

$$\dot{u}_i(t) = X_i(u(t)), \tag{2.3.2}$$

an ordinary differential equation of first order in m dimensions.

Remarks (2.3.3)

1. Because of our physical orientation we refer to the parameter t as time, so that (2.3.2) describes motion on M.
2. It is no real restriction that (2.3.2) is of first order, since a higher-order equation can always be reduced to a first-order equation (on a space of higher dimension) by introducing new variables.
3. It is also no loss of generality that the independent variable t only occurs implicitly on the right side of (2.3.2) through the dependent variables. One could easily take t as a dependent variable and introduce a new independent variable s; we shall look into this possibility later. In the meantime let us discuss the general properties of (2.3.1).

Recall that if $X_i(q)$ is continuous, then the existence of a local solution is guaranteed, and a Lipschitz condition would imply that it is unique. Therefore, exactly one integral curve passes through every point, and different integral curves can never cross. Since we are dealing with C^∞ vector fields, for which these conditions are fulfilled, we can make the following statement about the solution with arbitrary intitial conditions:

Theorem (2.3.4)

Let X be a C^∞ vector field on a manifold M. Then for all $q \in M$, there exist $\eta > 0$, a neighborhood V of q, and a function $u: (-\eta, \eta) \times V \rightarrow M, (t, q(0)) \rightarrow u(t, q(0))$ such that

$$\dot{u} = X \circ u, \qquad u(0, q(0)) = q(0) \quad \forall q(0) \in V.$$

For all $|t| < \eta$, the mapping $q(0) \rightarrow u(t, q(0))$ is a diffeomorphism Φ_t^X between V and some open set of M.

Proof: See [(1), 10.8].

Examples (2.3.5)

1. $M = \mathbb{R}^n$, $X:(x_1, \ldots, x_n) \to (x_1, \ldots, x_n; v, 0, \ldots, 0)$. $V = \mathbb{R}^n$, $\eta = \infty$, $u(t, x(0)):(t, x_i(0)) \to (x_1(0) + vt, x_2(0), \ldots, x_n(0))$. A constant vector field induces a linear field of motion.
2. $M = \mathbb{R}^n \backslash \{(0, 0, \ldots, 0)\}$, $X:(x_1, \ldots, x_n) \to (x_1, \ldots, x_n; v, 0, \ldots, 0)$, V arbitrary, but η is the smallest value of t for which $V + (vt, 0, \ldots, 0)$ contains the origin. $u(t, x(0))$ is again as in Example 1. The constant field of motion may leave M, in a length of time that depends on V.

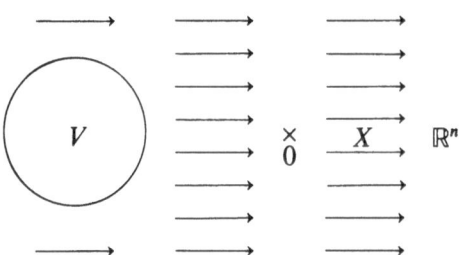

3. $M = \mathbb{R}^n \backslash \mathbb{R} \times \{(0, 0, \ldots, 0)\}$, that is, the x_1-axis is removed; and X is as above. Once again $V = M$, $\eta = \infty$.
4. $M = \mathbb{R}$, $X:x \to (x, x^3/2)$, V arbitrary, $\eta = \inf_{u \in V} 1/u^2$, $u(t, x(0)):(t, x(0)) \to x(0)(1 - tx^2(0))^{-1/2}$. For large x the vector field becomes so strong that every point except the origin is sent to infinity in a finite time.

Remarks (2.3.6)

1. Theorem (2.3.4) states that trajectories that are near neighbors can not suddenly be separated. There is a well-known estimate [(1), 10.5] according to which points can not diverge faster than exponentially in time if the derivative of X is uniformly bounded.
2. In Example 1, X provides a one-parameter group of diffeomorphisms Φ_t^X on M. Because $u(t_1 + t_2, q(0)) = u(t_2, u(t_1, q(0)))$, its existence is equivalent to the possibility of letting $V = M$ and $\eta = \infty$. It can be shown that this is possible, for instance, when X is of compact support. This is intuitively clear, since the worst eventuality is for some trajectories to leave M in a finite time. But if X equals zero outside some compact subset of M, the trajectories can not leave $M[(1), 18.2.11]$.
3. In Example 2 there is no diffeomorphism of all of M, and in Example 3 we saved the group of diffeomorphisms by getting rid of the trajectories that go through the origin. This is not always possible; in Example 4 only one point of the manifold would be left after a similar operation.

These possibilities are delineated by

Definition (2.3.7)

If the diffeomorphisms Φ_t^X of theorem (2.3.4) form a one-parameter group of bijections $M \to M$, X is said to be **complete** and the group is called a **flow**. If the relationship

$$\Phi_{t_1}^X \circ \Phi_{t_2}^X = \Phi_{t_1 + t_2}^X$$

holds only for sufficiently small neighborhoods of any point and sufficiently short times, Φ_t^X is called a **local flow**.

As mentioned in §1.3 we would like to construe time-evolution as a group of automorphisms of the algebra of observables. Choosing the algebra as C_0^∞, the C^∞-functions of compact support, the local flow of a vector field provides an automorphism for short times by

$$\tau_t^X(f) \equiv f \circ \Phi_t^X, \qquad f \in C_0^\infty. \tag{2.3.8}$$

If X is complete, the τ_t^X are a one-parameter group:

$$\tau_{t_1}^X \circ \tau_{t_2}^X = \tau_{t_1 + t_2}^X, \quad \forall t_1, t_2 \in \mathbb{R}. \tag{2.3.9}$$

In any case the mapping $t \to \tau_t^X(f)(q)$ is differentiable for t in some neighborhood of 0, the size of which depends on f. As can be seen by using a chart (Problem 4), the time-derivative is the same as the Lie derivative associated with X (cf. (2.5.7)).

$$\frac{d}{dt} \tau_t^X(f)|_{t=0} = L_X f, \quad \forall f \in C_0^\infty. \tag{2.3.10}$$

Remarks (2.3.11)

1. Thus a vector field determines a local flow, which then determines the automorphisms of C_0^∞ given by (2.3.8). By (2.3.10) and (2.2.24) the automorphisms determine in turn a vector field, so we can combine the three concepts into one.
2. If X is an analytic vector field and f is analytic, then $t \to \tau_t^X(f)|_q$ is analytic in t in a complex neighborhood of 0. The power series in t may be written as

$$\tau_t^X(f) = e^{tL_X} f \equiv \sum_{n=0}^\infty \frac{t^n}{n!} (L_X)^n f.$$

3. It may happen that the flows of two vector fields approach each other asymptotically, so that the limit

$$\lim_{t \to \infty} \Phi_t^X \circ \Phi_{-t}^{\tilde{X}} = \Omega$$

exists. A (pointwise) limit of diffeomorphisms might not be a diffeomorphism; e.g., the limit of the mappings $x \to x/t$ on $\mathbb{R}$ is $\mathbb{R} \to \{0\}$. However, if Ω is a diffeomorphism, then from the above equation it follows that

$$\Phi_t^X = \Omega \circ \Phi_t^{\tilde{X}} \circ \Omega^{-1}, \quad \forall t.\dagger$$

Therefore the flows induced by X and $\tilde{X}$ must also be diffeomorphic. According to (2.2.25; 3) and (2.2.24), taking the time-derivative of $f \circ \Phi_t^X \circ \Omega = f \circ \Omega \circ \Phi_t^{\tilde{X}}$, $f \in C^\infty$, yields $L_{\Omega \cdot \tilde{x}} = L_X$, or $\Omega^* \tilde{X} = X$: The diffeomorphism Ω transforms the vector fields into each other.

The trivial case (2.3.5; 1) is typical in that in the neighborhood of any point q where $X(q) \neq 0$ (i.e., other than at a point of equilibrium) the general case may be reduced to it by a suitable change of coordinates:

Theorem (2.3.12)

At every point $q \in M$ where $X(q) \neq 0$ there exists a chart (U, Φ) such that $\Phi(U) = I \times V$, $V \subset \mathbb{R}^{m-1}$; for all $x \in V$, $t \to \Phi^{-1}(t \times \{x\}) \, \forall t \in I$ is an integral curve for X; and $\Phi^* X : (x_1, \ldots, x_m) \to (x_1, \ldots, x_m; 1, 0, \ldots, 0)$.

Proof:

Since $X(q) \neq 0$, a chart (U, ψ) with $\psi(q) = 0 \in \mathbb{R}^m$ can be found such that $\psi^* X(0) = (1, 0, \ldots, 0)$. Since $\psi^* X \in \mathcal{T}_0^1(\mathbb{R}^m)$ is continuous, there is an open, relatively compact neighborhood U of 0 on which the first component of the image of X is greater than $\frac{1}{2}$: $(\psi^* X)^1(x) > \frac{1}{2} \, \forall x \in U$. If $X_0 \in \mathcal{T}_0^1(\mathbb{R}^m) : x \to (x; 1, 0, \ldots, 0)$, then we define the vector field

$$\tilde{X} = f \cdot \psi^* X + (1 - f) X_0 \in \mathcal{T}_0^1(\mathbb{R}^m)$$

with

$$1 \geq f \in C^\infty(\mathbb{R}^m) : f = \begin{cases} 0 \text{ on } CU \\ 1 \text{ on } \tilde{U} \subset U \end{cases}, \quad q \in \tilde{U} = \text{some open set.}$$

Clearly $(\tilde{X})^1(x) > \frac{1}{2} \, \forall x \in \mathbb{R}^m$, and $\tilde{X}$ induces a flow, because it agrees with X_0 outside some compact set (see Figure 10). Hence

$$\Omega = \lim_{t \to \infty} \Phi_{-t}^{X_0} \circ \Phi_t^{\tilde{X}}$$

also exists, for $(\Phi_t^{\tilde{X}}(x))^1 \geq x^1 + t/2$, and $\Phi_t^{X_0}$ and $\Phi_t^{\tilde{X}}$ are identical on

$$\left\{ x \in \mathbb{R}^m : x^1 > \sup_{\bar{x} \in U} \bar{x}^1 \right\}.$$

Therefore the limit is attained on compact sets after a finite time, and Ω is a diffeomorphism. According to (2.3.11; 3) Ω transforms $\tilde{X}$ into X_0, and $\tilde{X}$ and $\psi^* X$ are equal on U. The mapping Φ of the theorem is $\Omega \circ \psi$.   □

$\dagger$ Because $\Omega \circ \Phi_t^{\tilde{X}} = \lim_{t \to \infty} \Phi_t^X \circ \Phi_t^X \circ \Phi_{-t}^{\tilde{X}} = \Phi_t^X \circ \Omega$.

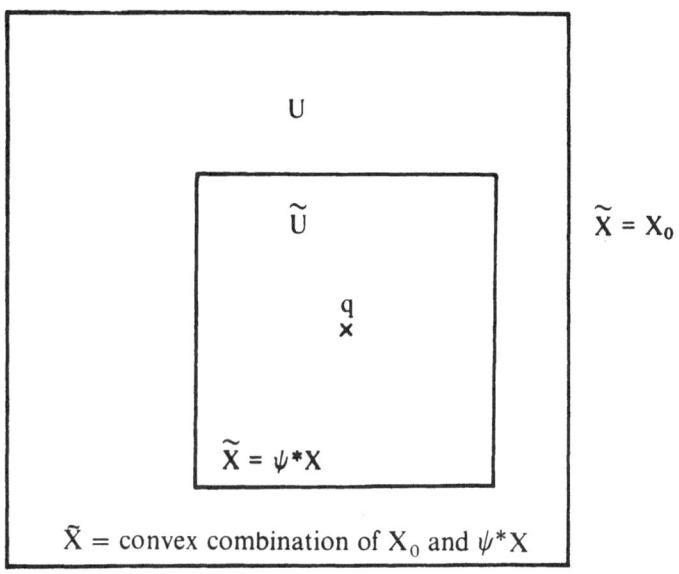

Figure 10   The interpolating vector field $\tilde{X}$.

Remarks (2.3.13)

1. The idea of a comparison diffeomorphism used in the proof plays an important role in physics. For a direct proof see Problem 5.
2. Points q at which $X(q) = 0$ are fittingly called critical points; they are fixed points of the flow. In section §3.4 we investigate what happens in their vicinity.
3. The theorem displays the m local integrals of motion, $m - 1$ of which are independent of time: $x_1 - t, x_2, \ldots, x_m$. However, it ought to be borne in mind that the x_i are only functions $U \to \mathbb{R}$. It is not said whether they extend to C^r-functions $M \to \mathbb{R}$.

Example (2.3.14)

$M = T^2 = S^1 \times S^1, X:(\varphi_1, \varphi_2) \to (\varphi_1, \varphi_2; \omega_1, \omega_2), \omega_i \in \mathbb{R}. u:(t, \varphi_1(0), \varphi_2(0))$
$\to (\varphi_1(0) + \omega_1 t, \varphi_2(0) + \omega_2 t)$. The two constants $\varphi_1 - \omega_1 t$ and $\varphi_1/\omega_1 - \varphi_2/\omega_2$ can not be extended to all of M. Only if the ratio of the frequencies is rational, $\omega_i = g_i \omega, g_i$ an integer, can a global, time-independent constant like $\sin(g_2 \varphi_1 - g_1 \varphi_2)$ be constructed.

Remarks (2.3.15)

1. Later we shall prove that for an irrational ratio of the ω_i every trajectory is dense. Because of this there can be no C^∞-function K constant in time and with $dK \neq 0$. (Cf. (2.4.3).)

2. The phrase "integrals of motion" will be reserved for C^r-functions $(r \geq 1)M \to \mathbb{R}$, leaving their existence and number open questions.
3. Example (2.3.14) is typical in that it can be shown [(3), 25.17] that every vector field can be approximated arbitrarily well by one that does not have any constants (= integrals) of motion. This fact is of physical interest only when the uncertainty of our knowledge allows appreciable alterations of the solutions in physically relevant times.

Since $T(M)$ is a manifold and $\dot{u}$ is a mapping: $I \to T(M)$, a second-order equation $\ddot{u} = X \circ \dot{u}$ can be formulated on $T(M)$ without specifying a co-ordinate system, by means of a vector field $X: T(M) \to T(T(M))$. For this purpose X is not at all arbitrary, as can be seen by the following argument: On the bundle chart of $T(M)$, $\dot{u}$ is the mapping $t \to (u(t), du/dt)$, so

$$\Pi_M \circ \dot{u} = u, \tag{2.3.16}$$

where Π_M is the projection $T(M) \to M$ of (2.2.14). Therefore the first-order equation requires that $\Pi_M \circ X = 1$; X is thus a vector field rather than simply an arbitrary mapping $M \to T(M)$. An analogous and likewise trivial condition arises from the fact that on the corresponding chart $\ddot{u}: t \to (u, du/dt; du/dt, d^2u/dt^2)$. As a result $T(\Pi_M) \circ \ddot{u} = \dot{u}$, as also follows from differentiating euation (2.3.16). Consequently for the second-order equation X must satisfy $T(\Pi_M) \circ X = 1$; of course, as a vector field, X also satisfies $\Pi_{T(M)} \circ X = 1$.

Remark (2.3.17)

The two mappings $\Pi_{T(M)}$ and $T(\Pi_M)$ are to be distinguished. Both map $T(T(M)) \to T(M)$, but if on the usual chart we write an element of $T(M)$ as $(q, \dot{q})$ and one of $T(T(M))$ as $(q, \dot{q}, \partial q, \partial \dot{q})$, then the mappings are as shown below:

$$\Pi_{T(M)}: (q, \dot{q}; \partial_q, \partial_{\dot{q}}) \to (q, \dot{q}; 0, 0)$$
$$T(\Pi_M): (q, \dot{q}; \partial_q, \partial_{\dot{q}}) \to (q, 0; \partial_q, 0)$$

$$
\begin{array}{ccc}
T(M) & \xrightarrow{\;\Pi_M\;} & M \\
\Big\uparrow {\scriptstyle \Pi_{T(M)}} & & \Big\uparrow {\scriptstyle \Pi_M} \\
T(T(M)) & \xrightarrow{\;T(\Pi_M)\;} & T(M)
\end{array}
$$

According to (2.2.18; 3) this diagram is permutable. In this way two different bundle structures can be defined for $T(T(M))$, both with the basis $T(M)$. In the future we shall always mean the one with the projection $\Pi_{T(M)}$.

At present a trajectory is given by a specification of the initial point in $T(M)$, or, physically speaking, by the initial position and velocity, $\dot{u}(t, (q_0, \dot{q}_0))$.

It can be shown that the diffeomorphism mentioned in Theorem (2.3.4) creates another diffeomorphism between the tangent space of a point and a neighborhood of that point in M [(1), 18.3.4].

Theorem (2.3.18)

Let $\dot{u}(t, (q_0, \dot{q}_0))$ be the solution of $\ddot{u} = X \circ \dot{u}$ with initial condition

$$\dot{u}(0, (q_0, \dot{q}_0)) = (q_0, \dot{q}_0).$$

Then for sufficiently small $t \neq 0$ the mapping

$$\dot{q}_0 \rightarrow \Pi_M \circ \dot{u}(t, (q_0, \dot{q}_0))$$

is a diffeomorphism between a neighborhood of the origin of $T_{q_0}(M)$ and a neighborhood of q_0 in M.

Remark (2.3.19)

Our feeling that by the proper choice of the initial velocity one could get anywhere and that the final point depends continuously on the initial velocity is made more precise in this theorem. For free motion in $\mathbb{R}^3$, $\ddot{x} = 0$, this is obvious: for $x(t) = x(0) + t\dot{x}(0)$, the mapping $\dot{x}(0) \rightarrow x(t)$ is even a diffeomorphism $\mathbb{R}^3 \rightarrow \mathbb{R}^3$. But if the manifold were, for example, $\mathbb{R}^3$ with certain pieces removed, the holes would cast shadows, and the theorem would only hold locally.

The differential equations of mechanics, (1.1.1) through (1.1.6), are somewhat special, as they are the Euler–Lagrange equations of a variational problem, to wit, the requirement that the (Fréchet) derivative DW of a functional

$$W = \int dt \, L(x(t), \dot{x}(t)) \tag{2.3.20}$$

of $x(t)$ vanishes. This has the advantage of a coordinate-free formulation, since the requirement that $DW = 0$ does not single out any particular coordinate system. We will not delve further into this matter, because later we shall prove the more general invariance of the equations of motion under canonical transformations. For that end we need only the elementary fact that with the Lagrangian

$$L = \sum_{i=1}^{N} m_i \frac{|\mathbf{x}_i|^2}{2} - \sum_{i>j}(e_i e_j - \kappa m_i m_j)|\mathbf{x}_i - \mathbf{x}_j|^{-1}, \tag{2.3.21}$$

the Euler–Lagrange equations

$$\frac{d}{dt}\frac{\partial L}{\partial \dot{\mathbf{x}}_i} = \frac{\partial L}{\partial \mathbf{x}_i}, \qquad i = 1, \ldots, N \tag{2.3.22}$$

produce the equations of motion (1.1.1) and (1.1.2). With generalized coordinates $q_i(x)$, $i = 1, \ldots, 3N$, L may be written

$$L = \sum_{i,k=1}^{3N} m_{ik}(q) \frac{\dot{q}_i \dot{q}_k}{2} - V(q) \qquad (2.3.23)$$

Because m_{ik} is a nonsingular matrix for all q, the q_i can be expressed with the conjugate momenta $p_i = \partial L / \partial \dot{q}_i = m_{ik}(q)\dot{q}_k$, and the Euler–Lagrange equations

$$\frac{d}{dt} \frac{\partial L}{\partial \dot{q}_i} = \frac{\partial L}{\partial q_i}, \qquad i = 1, \ldots, 3N, \qquad (2.3.24)$$

can equally well be written in the Hamiltonian form

$$\frac{dq_i}{dt} = \frac{\partial H}{\partial p_i}, \qquad \frac{dp_i}{dt} = -\frac{\partial H}{\partial q_i}, \qquad (2.3.25)$$

where

$$H(q, p) = \sum_i p_i \dot{q}_i - L = \sum_{i,k} \frac{p_i p_k}{2} (m^{-1}(q))_{ik} + V(q). \qquad (2.3.26)$$

The Legendre transformation leading from L to the Hamiltonian H is invertible:

$$L = \sum_i p_i \frac{\partial H}{\partial p_i} - H. \qquad (2.3.27)$$

To see these equations in the framework of the structure we have constructed up to now we need the concept of a cotangent bundle, which we develop in the next chapter. Briefly, L furnishes a vector field on the tangent bundle (coordinates $(q, \dot{q})$) and H furnishes one on the cotangent bundle (coordinates (q, p)), which will be called phase space, whereas the underlying manifold will be called configuration space.

Problems (2.3.28)

1. In what sense is equation (2.3.1) formulated "invariantly" (or "covariantly")?

2. Discuss the integrals of motion for the one-dimensional and two-dimensional harmonic oscillators:

$$H_1 = p^2 + \omega^2 q^2, \qquad H_2 = p_1^2 + p_2^2 + \omega_1^2 q_1^2 + \omega_2^2 q_2^2.$$

3. Same problem for $M = T^2$, $X:(\varphi_1, \varphi_2) \to (\varphi_1, \varphi_2; \omega, \alpha \sin \varphi_1)$.

4. Derive (2.3.10).

5. Prove (2.3.12) by using the streamlines of X as coordinate lines.

6. Give an example of a vector field for $M = \mathbb{R}$, which is continuous but not C^∞, such that (2.3.1) does not have a unique solution.

Solutions (2.3.29)

1. With a diffeomorphism $\Phi: M_1 \to M_2$, the commutativity of the diagram (cf. (2.2.21))

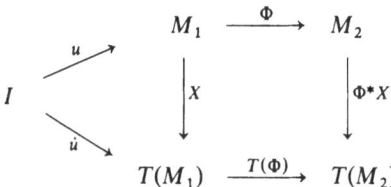

implies for $\bar{u} \equiv \Phi \circ u$ that $\dot{\bar{u}} = \Phi^* X \circ \bar{u}$.

2. The general solution of the equation of motion is

$$(q_i(t), p_i(t)) = (A_i \sin(\omega_i t + \varphi_i), A_i \omega_i \cos(\omega_i t + \varphi_i)).$$

For H_1 the constant $A^2 = p^2/\omega^2 + q^2$ is defined globally, but $\varphi = \arctan(q/p) - \omega t$ only locally. Similarly for H_2 there are two integrals, A_1^2 and A_2^2, and again the φ_i exist only locally, and likewise for the third time-independent constant, $\varphi_1/\omega_1 - \varphi_2/\omega_2 = (1/\omega_1)\arctan(q_1/p_1) - (1/\omega_2)\arctan(q_2/p_2)$. If $\omega_i = g_i\omega$, with g_i integral, then once more there is a global constant $\sin(\varphi_1 g_2 - \varphi_2 g_1)$.

3. Locally, $\varphi_1 - \omega t$ and $\varphi_2 + (\alpha/\omega)\cos \varphi_1$ are constant. In this case there is a global time-independent constant, $\sin(\varphi_2 + (\alpha/\omega)\cos \varphi_1)$.

4. Let $q(t) = u(t, q)$ be the solution of (2.3.1). Then $\tau_t^X f_{|q} = f \circ \Phi_{t|q}^X = f(q(t))$. Consequently,

$$\frac{d}{dt} \tau_t^X f \big|_{t=0} = \frac{\partial f}{\partial q_i} \frac{\partial q_i}{\partial t} \big|_{t=0} = \frac{\partial f}{\partial q_i} X_i(q) = L_X f.$$

5. In the notation of the proof of Theorem (2.3.12), let U_1 be the domain of ψ and let $\psi(U_1) = I_1 \times V_1, I_1 \subset \mathbb{R}, V_1 \subset \mathbb{R}^{m-1}$. Theorem (2.3.4) guarantees the existence of a local solution $u(t; x_1, \ldots, x_m)$ of the equation $\psi^* X \circ u = \dot{u}$, using this chart. At the origin the function $f(t, x_2, \ldots, x_m) \equiv u(t, 0, x_2, \ldots, x_m): I_2 \times V_2 \to \mathbb{R}^m, I_2 \subset I_1, V_2 \subset V_1$, has the derivative $Df(0) = 1: \mathbb{R}^m \to \mathbb{R}^m$, because

$$\frac{\partial f_i}{\partial t} \big|_0 = X_i(0) = (1, 0, 0, \ldots, 0), \frac{\partial f_i}{\partial x_2} \big|_0 = \delta_{i2}, \text{ etc.}$$

Therefore Df is invertible on a neighborhood $I_3 \times V_3$, where $I_3 \subset I_2$ and $V_3 \subset V_2$, and consequently f is a diffeomorphism there [(1), 10.2.5]. Because $f(0, x_2, \ldots, x_m) = (0, x_2, \ldots, x_m)$, $\psi(U) = I_3 \times V_3 \cap f(I_3 \times V_3) \neq \emptyset$, it is possible to introduce $(U, f^{-1} \circ \psi_{|U})$ as a new chart, Pictorially,

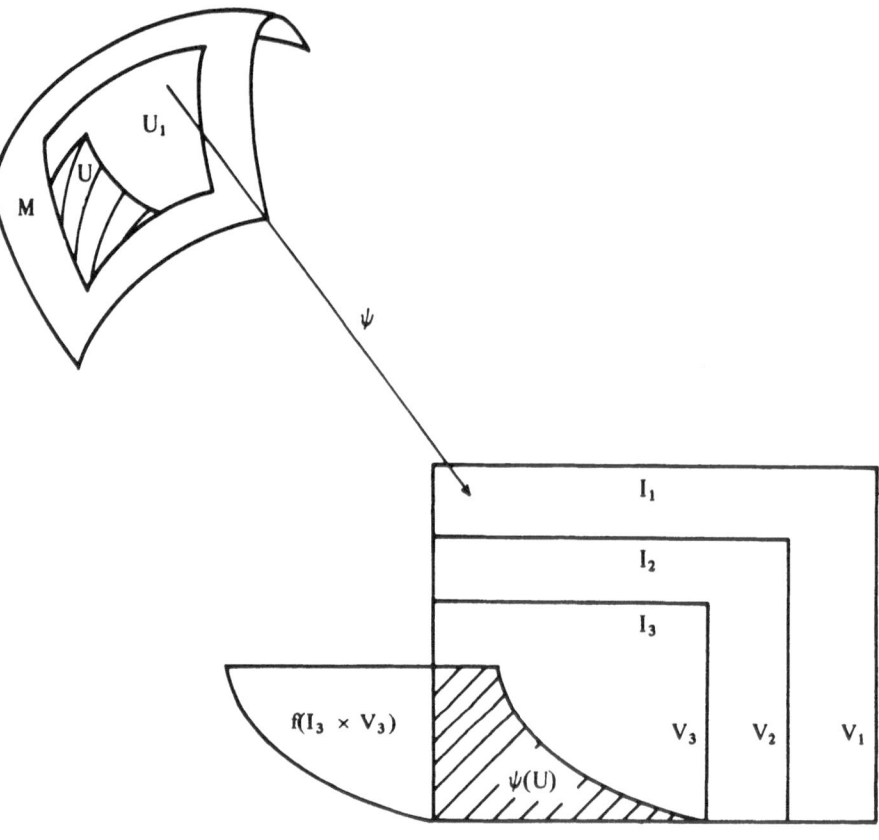

Figure 11   The relationship of the domains.

On this chart the vector field has the form $\Phi^* X = T(\Phi) \circ X \circ \Phi^{-1} = T(f^{-1}) \circ \psi^* X \circ f = T(f^{-1}) \circ \dot{f} = T(f^{-1}) \circ T(f) \circ (1, 0, \ldots, 0) = (1, 0, 0, \ldots, 0)$. Therefore the $I \times \{x\}$ are integral curves.

6. $X : x \to (x, \sqrt{x}$ for $x > 0$, and otherwise $0)$. For $u(0) = 0$ there are two solutions, $u(t) = 0$ and $u(t) = t^2/4$.

2.4   Tensors

The tangent space $T_q(M)$ admits multilinear mappings, the tensors at the point q. This linear structure is extended in the global definition of the bundle of tensors over M, which is then used to define tensor fields.

If E is a (finite-dimensional) vector space, then the space of linear mappings $E \to \mathbb{R}$ (or $\mathbb{C}$) is called its dual space E^*. We may write these mappings as scalar products: For any $V^* \in E^*$ there is a mapping $V \to (V^*|V) \in \mathbb{R} \: \forall V \in E$. A linear structure is naturally defined on E^*, and to any linear mapping L of E to another vector space F we can bijectively (see Problem 1) associate a linear mapping from F^* to E^*, called the transposed mapping L^t:

$$\left.\begin{array}{l} L: E \to F \\ L^t: F^* \to E^* \end{array}\right\} \text{such that } (L^t W^*|V) \equiv (W^*|LV) \: \forall V \in E, \: W^* \in F^*.$$

Definition (2.4.1)

The dual space $T_q^*(M)$ of $T_q(M)$ is known as the **cotangent space** of M at the point q.

Remarks (2.4.2)

1. With an orthogonal basis $\{e_i\}$, $(e_i|e_j) = \delta_{ij}$, $\mathbb{R}^n$ can be identified with its dual space. But if the basis is transformed with a non-orthogonal bijection L, $e_i \to Le_i$, the dual basis $\{e_i^*\}$ must be transformed with $(L^{-1})^t$ so as to satisfy $(e_i^*|e_j) = \delta_{ij}$. Since the transformation $T_q(\Phi)$ induced on $T_q(M)$ by a diffeomorphism Φ is not generally orthogonal, and since no coordinate system is distinguished, the statement that a vector in $T_q(M)$ is the same as one in $T_q^*(M)$ has no chart-independent meaning. With a change of charts they transform differently and become unequal. Hence it is necessary to distinguish $T_q(M)$ from $T_q^*(M)$ unless, as discussed later, a metric is specified on $T(M)$.
2. The identification of the dual space $T_q^{**}(M)$ of $T_q^*(M)$ with $T_q(M)$ is unaffected by a change of charts, since $(((L^{-1})^t)^{-1})^t = L$.

Examples (2.4.3)

1. In elementary vector calculus one encounters the gradient of a function f as an example of a vector. Here we shall recognize it as an element of $T_q^*(M)$. A function $f \in C^\infty(M)$ defines a mapping $T_q(f): T_q(M) \to T_{f(q)}(\mathbb{R}) = \mathbb{R}$, which is therefore an element of $T_q^*(M)$. We may denote this mapping by $df_{|q}$ and call it the differential of f at the point q. On a chart we write the usual formula

$$df_{|q}(v) = v^i \left.\frac{\partial f}{\partial q^i}\right|_q, \qquad \forall v \in T_q(M).$$

If a vector of $T_q(M)$ is specified by the vector field X, this mapping becomes $df_{|q}(X(q)) = (L_X f)(q)$.
2. Given a chart $C: (U, \Phi)$, $\Phi(q) = \sum e_i q^i \in \mathbb{R}^m$, the inverse mapping $\Theta_C^{-1}(q)$ transplants the basis $\{e_i\}$ from $\mathbb{R}^m$ to $T_q(U)$, where we write it symbolically as $\{\partial/\partial q^i\}$. Similarly, $\Theta_C^t(q)$ transforms the e_i into the basis of $T_q^*(U)$ dual to the $\partial/\partial q^i$. In the notation of Example 1, this is written as dq^i, if we consider q^i as a C^∞-function on M:

$$(dq^i|\Theta_C^{-1}(q)e_j) = dq^i{}_{|q}(\Theta_C^{-1}(q)e_j) = L_{\Theta_C^{-1}(q)e_j}(q^i) = \frac{\partial q^i}{\partial q^j} = \delta_{ij},$$

The dq^i, the differentials of the coordinates,† are referred to as the natural basis of $T_q^*(U)$ (cf. (2.2.26)).

Taking these algebraic considerations a step further produces the idea of the space of tensors at a point, that is, the tensor product of the tangent and cotangent spaces, defined as follows: If $T_q(M)$ is identified with $T_q^{**}(M)$, then $T_q(M)$ can be considered as a linear mapping $T_q^*(M) \to \mathbb{R}$. A mapping

$$\underbrace{T_q^*(M) \times T_q^*(M) \times \cdots \times T_q^*(M)}_{r \text{ times}} \to \mathbb{R}$$

which is linear in every factor is called a contravariant tensor of degree r. Like any set of multilinear mappings, the tensors form a vector space. Similarly, a covariant tensor is a multilinear mapping $T_q(M) \to \mathbb{R}$, and more generally we make

Definition (2.4.4)

A mapping

$$\underbrace{T_q^*(M) \times T_q^*(M) \times \cdots \times T_q^*(M)}_{r \text{ times}} \times \underbrace{T_q(M) \times T_q(M) \times \cdots \times T_q(M)}_{s \text{ times}} \to \mathbb{R}$$

at the point q, which is linear in every factor is a **tensor contravariant of degree r and covariant of degree s**. The m^{r+s}-dimensional vector space of these tensors is denoted by $T_q{}^r_s(M)$. (Thus $T_q{}^1_0(M) = T_q(M)$ and $T_q{}^0_1(M) = T_q^*(M)$.)

Remark (2.4.5)

On $\mathbb{R}^n$ with the basis $\{e_1, e_2, \ldots, e_n\}$, each $e_{i_1} \otimes e_{i_2} \otimes \cdots \otimes e_{i_r}$ is a tensor of degree r, where $\otimes$ is defined so that

$$(v_1^*, v_2^*, \ldots, v_r^* | e_{i_1} \otimes e_{i_2} \otimes \cdots \otimes e_{i_r}) \equiv (e_{i_1} \otimes e_{i_2} \otimes \cdots \otimes e_{i_r})(v_1^*, v_2^*, \ldots, v_r^*)$$
$$\equiv (v_1^* | e_{i_1})(v_2^* | e_{i_2}) \cdots (v_r^* | e_{i_r}).$$

If we let $(i_1, i_2, \ldots, i_r)$ run through all r-tuples of indices, we obtain a basis for the vector space of tensors of degree r. Using this basis we can identify the space of tensors of degree r with $\mathbb{R}^{nr}$. Every such tensor can be written as

$$t = \sum_{(i)} c^{i_1, \ldots, i_r} e_{i_1} \otimes e_{i_2} \otimes \cdots \otimes e_{i_r}, \quad c^{i_1, \ldots, i_r} \in \mathbb{R},$$

so that

$$(v_1^*, v_2^*, \ldots, v_r^* | t) = \sum_{(i)} c^{i_1, \ldots, i_r} \prod_{k=1}^{r} (v_k^* | e_{i_k}).$$

The $c^{i_1, \ldots, i_r}$ are the components of the tensors, and are the quantities usually called tensors in physics. The tensor product of r arbitrary vectors is defined

---

† Following the usual convention we use subscripts for the bases and superscripts for the components in a tangent space, and do it the other way around in a cotangent space. This does not fix what to do about coordinates, which are not vectors.

similarly to the tensor product of basis elements. Not every tensor can be written in the form $v_1 \otimes v_2 \otimes \cdots \otimes v_r$, but only as a linear combination of such expressions. Note that with the Cartesian product the dimensions add, but with the tensor product they multiply.

The next step is to collect all the tensors at different points into a bundle over M. We see again that on the domain of a chart C of M, $(\Theta_C^{-1})^t$ provides a chart for the cotangent bundle,

$$T^*(U) = \bigcup_{q \in U} T_q^*(M) = U \times \mathbb{R}^m \qquad (2.4.6)$$

through

$$T^*(U) \to \mathbb{R}^m \times \mathbb{R}^m : (q, v^*) \to (\Phi(q), (\Theta_C^{-1}(q))^t v^*). \qquad (2.4.7)$$

As noted above (2.2.10), for different U's these charts are compatible; $D(\bar{\Phi} \circ \Phi^{-1})$ is merely replaced with $D(\Phi \circ \bar{\Phi}^{-1})^t$, which does not destroy the required differentiability. The bundle structure carries over directly to the tensors. The mapping

$$\underbrace{\Theta_C(q) \otimes \Theta_C(q) \otimes \cdots \otimes \Theta_C(q)}_{r \text{ times}} \otimes \underbrace{(\Theta_C^{-1}(q))^t \otimes (\Theta_C^{-1}(q))^t \otimes \cdots \otimes (\Theta_C^{-1}(q))^t}_{s \text{ times}}$$

sends T_{qs}^r at every point $q \in U$ into $\mathbb{R}^{m(s+r)}$. As a bijection this mapping can be used for the charts of the tensor bundle, leading us to make a general

Definition (2.4.8)

Let M be a manifold with the atlas $\bigcup_i C_i = \bigcup_i (U_i, \Phi_i)$. The vector bundle over M defined by the atlas $\bigcup_i (U_i \times \mathbb{R}^{m(s+r)}, (q; u_1, \ldots, u_r, v_1, \ldots, v_s)) \to (\Phi_i(q); \Theta_{C_i}(q)u_1, \ldots, \Theta_{C_i}(q)u_r, (\Theta_{C_i}^{-1}(q))^t v_1, \ldots, (\Theta_{C_i}^{-1}(q))^t v_s)$ on $T_s^r(M) = \bigcup_q T_{qs}^r(M)$ is called the **bundle of r-fold contravariant and s-fold covariant tensors**.

Remarks (2.4.9)

1. With this definition $T(M) \equiv T_0^1(M)$ and $T^*(M) \equiv T_1^0(M)$.
2. The linear structure required by definition (2.2.15) is that of the tensors, and the projection is $\Pi : (q; u_1, \ldots, u_r, v_1, \ldots, v_s) \to (q; 0, \ldots, 0)$. As with $T(M)$, the topology used on $T_s^r(M)$ is the product topology of $U \times \mathbb{R}^{m(s+r)}$.
3. In order to specify the mapping which defines a chart, it suffices to specify it on a basis (see (2.4.4)), which is what was done in definition (2.4.8) by writing out the way r contravariant and s covariant vectors are transformed.

Examples (2.4.10)

1. Suppose M is an m-dimensional, linear topological space: $T_s^r(M) = M \times \mathbb{R}^{m(r+s)}$. Then $T^*(M)$ and $T(M)$ are both of the form $M \times \mathbb{R}^m$, but

can not be identified, because no basis has been provided that is distinguished as orthogonal. If $M = \mathbb{R} \times \cdots \times \mathbb{R}$, an orthogonal basis would exist, because of the additional Riemannian structure we discuss later.

2. $M = S^1$, $T_s^r(M) = M \times \mathbb{R}^{r+s}$. Again no canonical identification of $T^*(M)$ is given; a canonical identification of $T^*(M)$ and $T(M)$ would amount to a specification of arc length, but as a manifold S^1 could be a circle of any radius.

3. $M = S^2$: $T_s^r(M)$ is not a Cartesian product, any more than $T(M)$ is.

There is an immediate generalization of the concept of a vector field.

Definition (2.4.11)

A C^∞-mapping $t: M \to T_s^r(M)$ such that $\Pi \circ t = 1$ is an **r-fold contravariant and s-fold covariant tensor field**. The set of all such tensor fields is denoted by $\mathcal{T}_s^r(M)$.

Remarks (2.4.12)

1. The terms "vector field" and "1-fold contravariant tensor field" are synonymous, as are "covariant vector field" and "1-fold covariant tensor field."

2. A tensor field can be written locally in the natural basis of a chart as

$$\sum_{(i)(j)} c_{j_1,\ldots,j_s}^{i_1,\ldots,i_r}\, \partial_{i_1} \otimes \cdots \otimes \partial_{i_r} \otimes dq^{j_1} \otimes \cdots \otimes dq^{j_s},$$

with $c_{(j)}^{(i)} \in C(M)$ (cf. (2.2.26) and (2.4.3; 2)). In the physical literature the components $c_{(j)}^{(i)}$ are referred to as the tensor fields.

Examples (2.4.13)

1. In (2.4.3) there was an example of a covariant vector field, the differential $df \in \mathcal{T}_1^0(M)$. Thus df symbolizes the rate of change of f in some direction (to be specified later), and is not some infinitely small quantity.

2. At every point $q \in M$ a covariant vector field $g \in \mathcal{T}_2^0(M)$ maps $T_q(M) \times T_q(M)$ into $\mathbb{R}$. If $\{e_i\}$ is a basis for $T_q(M)$, $v = v^i e_i$ and $w = w^i e_i$, and $\{e^{*i}\}$ is the dual basis of $T^*(M)$, $g = e^{*i} \otimes e^{*j} g_{ij}(q)$, then the mapping g becomes $(v, w) \to \langle v|w \rangle \equiv v^i w^k g_{ik}$. It is linear in both factors and can be used as a scalar product so long as

$$\langle v|w \rangle = \langle w|v \rangle, \quad \text{and} \quad \langle v|v \rangle = 0 \Leftrightarrow v = 0 \quad \forall v, w \in T_q(M).$$

This is the case if the matrix g_{ik} is positive (i.e., $g_{ik} = g_{ki}$, and all eigenvalues are positive). Then $\|v\| \equiv \langle v|v \rangle^{1/2}$ can be interpreted as the length of the vector v.

If all the eigenvalues of $g_{ik} = g_{ki}$ are different from zero but not necessarily positive, then we can still make the weaker statement

$$\langle v|w \rangle = 0 \quad \forall v \in T_q(M) \Leftrightarrow w = 0.$$

If this holds for all $q \in M$, g is said to be nondegenerate. By the equation $\langle v | w \rangle = (v^* | w) \ \forall w \in T_q(M)$, to every v is associated a $v^* = e^{*i} g_{ik} v^k$.

The additional structure introduced in the preceding example has far-reaching consequences, and justifies

Definition (2.4.14)

If a manifold M is given a nondegenerate, symmetric tensor field $g \in \mathcal{T}_2^0(M)$, it is called a **pseudo-Riemannian** space. If g is in fact positive, M is a **Riemannian** space, and g is called its **metric**.

Examples (2.4.15)

1. $\mathbb{R}^n$ becomes a pseudo-Riemannian space with $g = \sum_{i,k} dx^i \otimes dx^k g_{ik}$, where g_{ik} is a constant symmetric matrix with all nonzero eigenvalues. The matrix g can be diagonalized with some orthogonal transformation $x^i \to m^{ij} x^j$, and then the eigenvalues can all be normalized to $g_{ii} = \pm 1$ with a dilatation $x^i \to x^i/(|g_{ii}|)^{1/2}$. These charts have a special status, because they are determined up to pseudo-Euclidean transformations. (For $n = 4$ and $g_{ii} = (-1, 1, 1, 1)$, the transformations would form the Poincaré group.) When all $g_{ii} = 1$, $\mathbb{R}^n$ becomes a Riemannian space. On other charts the g_{ij} of this space do not have to be either diagonal or constant. For example, in the Riemannian case on $\mathbb{R}^2$, and using polar coordinates, $g = dr \otimes dr + r^2 \, d\varphi \otimes d\varphi$.
2. If N is a submanifold of M, and therefore $T(N)$ is a submanifold of $T(M)$, a nondegenerate $g \in \mathcal{T}_2^0(M), g > 0$, induces a Riemannian structure on N, because g also provides a nondegenerate mapping $T_q(N) \times T_q(N) \to \mathbb{R}$. The metric $g_{ik} = \delta_{ik}$ on $\mathbb{R}^m$ induces the usual metric on S^n or $T^n \subset \mathbb{R}^{n+1}$. Since every m-dimensional manifold can be imbedded as a submanifold of $\mathbb{R}^{2m+1}$, it is always possible to find a Riemannian structure for any manifold.
3. The Riemannian structure of $\mathbb{R}^n$ shows up in mechanics because of the kinetic energy, which we wrote as $m_{ik}(q) \dot{q}_i \dot{q}_k/2$ in (2.3.23). Up to a factor, this mapping $T(M) \times T(M) \to \mathbb{R}$ is exactly the metric. In the last chapter of the book we shall discuss why a concept of length that was introduced purely mathematically should be the physically measured interval. The bijection $T(M) \to T^*(M)$ mentioned above (2.4.10; 1), induced by the metric, sends $\dot{q}_i$ to $m_{ik}(q) \dot{q}_k = \partial L / \partial \dot{q}_i$, that is, to the canonically conjugate momentum p_i: (q, p) represents a point of $T^*(M)$.

Remark (2.4.16)

On a pseudo-Riemannian space, g creates a bijection $T_q(M) \to T_q^*(M)$ $\forall q \in M: v = v^i e_i \to v_i e^{*i}$, $v_i = g_{ik} v^k$, which allows one to speak of the contravariant components v^i and covariant components v_i of the vector v. But even in the Riemannian case neither type of component is a component of v in the direction of e. Instead, for example, $\|v^1 e_1\| = |v^1|(g_{11})^{1/2}$ is the length of v in the direction e_1. This is in fact the geometric mean of both types of components, if the e_i diagonalize the metric. Henceforth, the $|v^i||g_{ii}|^{1/2}$ will be called the components† of v, because of their intuitive significance. When written out in these components, many formulas lose their simplicity.

A diffeomorphism $\Phi: M_1 \to M_2$ induces a diffeomorphism $T(\Phi): T(M_1) \to T(M_2)$. Now we introduce another diffeomorphism $T^*(M_1) \to T^*(M_2)$, such that the scalar product, $(\,|\,): T^*(M) \underset{\pi}{\times} T(M) \to M \times \mathbb{R}$, remains invariant,‡ and consequently dual bases are mapped to dual bases.

Definition (2.4.17)

For any diffeomorphism $\Phi: M_1 \to M_2$, we define another diffeomorphism $T^*(\Phi): T^*(M_1) \to T^*(M_2)$, so as to make the diagrams

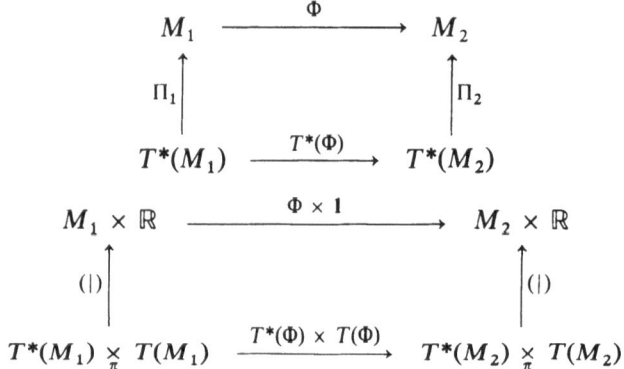

commute.

Clarification (2.4.18)

Although the mapping $(\,|\,)$ is surjective without being injective, and thus does not have an inverse, it determines $T^*(\Phi)$ uniquely, because $v^* = 0$ follows from $(v^*|w) = 0 \ \forall w \in T_q(M)$. On the domain of a chart, $T^*(\Phi)$ is evidently $(q, u) \to (\Phi(q), (T(\Phi^{-1}))^t(q) \cdot u)$, the differentiability of which is obvious. We shall often denote this simply by $T^*(\Phi) = T(\Phi^{-1})^t$, with the understanding that t refers to the transposition of a linear mapping at constant q.

Constructing the tensor product

$$\underbrace{T(\Phi) \otimes T(\Phi) \otimes \cdots \otimes T(\Phi)}_{r \text{ times}} \otimes \underbrace{T^*(\Phi) \otimes T^*(\Phi) \otimes \cdots \otimes T^*(\Phi)}_{s \text{ times}}$$

shows how $T_s^r(M)$ is altered under a diffeomorphism. The vectors of the basis $\partial_{i_1} \otimes \cdots \otimes \partial_{i_r} \otimes dq^{j_1} \otimes \cdots \otimes dq^{j_s}$ must be transformed by $T(\Phi)$ or

---

† Often also called "orthogonal components."

‡ The notation $\underset{\pi}{\times}$ means to take the pair with the same basis point.

respectively $T^*(\Phi)$, and linearity extends this to all of $T^r_s(M)$. In this way we obtain the transformation law for tensor fields under diffeomorphisms of a manifold, and in particular under a change of charts.

Definition (2.4.19)

A diffeomorphism $\Phi: M_1 \to M_2$ induces a mapping $\Phi^*: \mathcal{T}^r_s(M_1) \to \mathcal{T}^r_s(M_2)$ defined by the permutability of the diagram:

$$
\begin{array}{ccc}
 & \Phi & \\
M_1 \ni q & \xrightarrow{\hspace{3cm}} & \Phi(q) \\
\Big\downarrow{\scriptstyle t(q)} & \quad\overbrace{}^{r}\quad\overbrace{}^{s} & \Big\downarrow{\scriptstyle \Phi^* t(\Phi(q)),} \\
T^r_{q\,s} & \xrightarrow[T_q(\Phi)\otimes\cdots\otimes T_q(\Phi)\otimes T^*_q(\Phi)\otimes\cdots\otimes T^*_q(\Phi)]{} & T^r_{\Phi(q)\,s}
\end{array}
$$

where $t \in \mathcal{T}^r_s(M)$:†

$$\Phi^* t = \underbrace{T(\Phi)\otimes\cdots\otimes T(\Phi)}_{r\text{ times}}\otimes\underbrace{T^*(\Phi)\otimes\cdots\otimes T^*(\Phi)}_{s\text{ times}}\circ t\circ \Phi^{-1}.$$

Examples (2.4.20)

1. $M = \mathbb{R}^n$, $\Phi: \mathbf{x} \to \mathbf{x} + \mathbf{a}$, $\mathbf{a} \in \mathbb{R}^n$. $T^*(\Phi): (\mathbf{x}, \mathbf{v}) \to (\mathbf{x} + \mathbf{a}, \mathbf{v})$, and $T(\Phi)$ has already been given in (2.2.22). Under displacements a tensor t remains component invariant.

$$t: \mathbf{x} \to (\mathbf{x}, t^{i_1, \ldots, i_r}_{j_1, \ldots, j_s}(\mathbf{x})\partial_{i_1}\otimes\cdots\otimes\partial_{i_r}\otimes dx^{j_1}\otimes\cdots\otimes dx^{j_s})$$

$$\Phi^* t: \mathbf{x} \to (\mathbf{x}; t^{i_1, \ldots, i_r}_{j_1, \ldots, j_s}(\mathbf{x} - \mathbf{a})\partial_{i_1}\otimes\cdots\otimes\partial_{i_r}\otimes dx^{j_1}\otimes\cdots\otimes dx^{j_s}).$$

2. $M = \mathbb{R}^n$, $\Phi: x_i \to L_{ik}x_k$, $T^*(\Phi): (x_i; dx^j v_j) \to (L_{ik}x_k, dx^j L^{-1}_{kj}v_k)$. For a transformation of the covariant indices L must be replaced with $(L^{-1})^t$. With t as above,

$$\Phi^* t: \mathbf{x} \to (\mathbf{x}; \partial_{i_1}\otimes\cdots\otimes\partial_{i_r}\otimes dx^{j_1}\otimes\cdots$$
$$\otimes dx^{j_s}L_{i_1 m_1}\cdots L_{i_r m_r}L^{-1}_{n_1 j_1}\cdots L^{-1}_{n_s j_s}t^{m_1, \ldots, m_r}_{n_1, \ldots, n_s}(L^{-1}x)).$$

3. $g \in C^\infty(M_1)$: $\Phi^* \, dg = d(g \circ \Phi^{-1})$. It is intuitively clear that the image of the differential of a function must be the differential of the image of the function. When dg is applied to a vector v determined by a curve $u: I \to M_1$, it yields the rate of change of g along u. But that is the same as the rate of change of $g \circ \Phi^{-1}$ along $\Phi \circ u$,

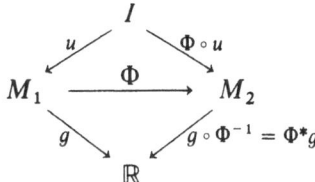

† This $*$ has nothing to do with the one at the beginning of §2.4 or the one of (2.4.29; 4).

and the latter curve determines the image of the vector v under $T(\Phi)$. Formally, and written out in pedantic detail, $T(g) = T(g) \circ T(\Phi^{-1}) \circ T(\Phi)$ $= T(g \circ \Phi^{-1}) \circ T(\Phi) \equiv T(\Phi^*g) \circ T(\Phi)$. Now, $T(g):(q, v) \to (g(q), T_q(g) \cdot v)$ $\equiv (g(q), (dg(q)|v))$,† but on the other hand, $T(g \circ \Phi^{-1}) \circ T(\Phi):(q, v) \to$ $T(g \circ \Phi^{-1})(\Phi(q), T_q(\Phi) \circ v) = (g(q), (d(g \circ \Phi^{-1})(\Phi(q))| T_q(\Phi) \cdot v))$. Therefore $(dg(q)|v) = (d(\Phi^*g)(\Phi(q))| T_q(\Phi) \cdot v) \; \forall v \in T_q(M_1)$. However, $\Phi^* \, dg$ is defined so that

$$(\Phi^* \, dg(\Phi(q))|w) = (dg(q)| T_{\Phi(q)}(\Phi^{-1}) \cdot w) = (dg(q)|v) \quad \forall w \in T_{\Phi(q)}(M_2),$$

if we set $w = T_q(\Phi) \cdot v$. It follows that

$$d(\Phi^*g)(\Phi(q)) = (\Phi^* \, dg)(\Phi(q)) \; \forall q \in M_1,$$

and hence $d(\Phi^*g) = \Phi^*(dg)$.

To sum up concerning the mapping Φ^*: The image of a vector is determined by the images of the curves that define it. The image of a covector is such that its product with the image of any vector equals the original product of the vector and the covector. These conditions fix the relationships among the bases and, because of the permutability of Φ^* with algebraic operations:

$$\Phi^*(t_1 + t_2) = \Phi^*(t_1) + \Phi^*(t_2), \quad \Phi^*(t_1 \otimes t_2) = \Phi^*(t_1) \otimes \Phi^*(t_2),$$

among all tensors. As for the tensor fields, at every point they transform in the same way as the tensors at that point.

For compositions, $(\Phi_1 \circ \Phi_2)^* = \Phi_1^* \circ \Phi_2^*$.

Remark (2.4.21)

Until now we have only investigated Φ^* for diffeomorphisms. In case Φ is not bijective, it is only possible to define the inverse images of covariant tensor fields. Even if Φ is injective, for example if it is the injection j of a submanifold, $j: N \to M \supset N$, $T(j): T(N) \to T(M) \supset T(N)$ (cf. (2.2.27; 6)), neither the image nor the inverse image of a vector field is defined; the image fails to be defined everywhere, and the inverse image lacks a distinguished subspace of $T_q(M)$ complementary to $T_q(N)$ unless M is given a metric.

Example (2.4.22)

$M = \mathbb{R}^2$ without a scalar product, $N = \mathbb{R}^1$, $j: x \to (x, 0)$, $T(j):(x, v) \to$ $(x, 0; v, 0)$, and $X \in \mathcal{T}_0^1(M): (x, y) \to (x, y; 1, 1)$. While $Y \in \mathcal{T}_0^1(N): x \to (x, 1)$ reproduces the components of $X_{|N}$ in $T(N)$, if we look at the basis: $e_1 = (1, 0)$, $e_2 = (1, 1)$, and we let (x, y) mean the vector $xe_1 + ye_2$, then we see that j is still $x \to (x, 0)$; but $X: (x, y) \to (x, y; 0, 1)$ has the component 0 in $T(N)$. A glance at Figure 12 shows that $j^*X = X_{|N}$ is not uniquely determined.

---

† $(dg(q)|v)$ means dg at the point q applied to v, and so on.

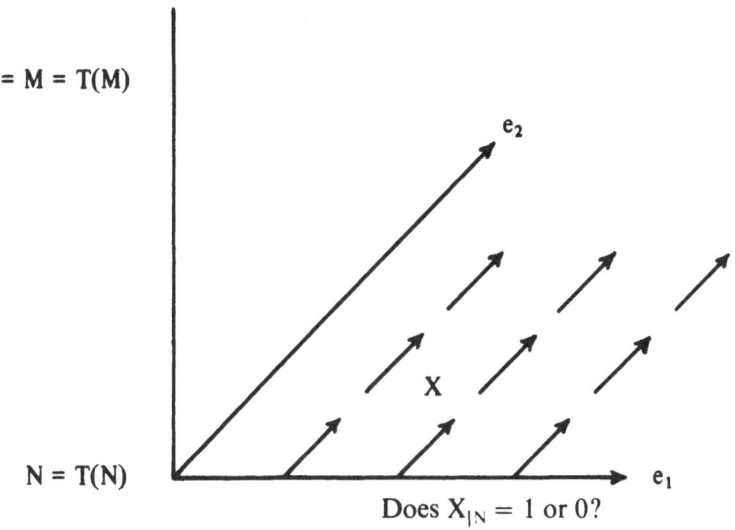

Figure 12   Restriction of a vector field to a subspace.

The inverse image of X under j would be a restriction $X_{|N}$, which is undefined, unless on N the vector field X takes its values in $T(N)$. These difficulties do not occur for covariant tensors, because they are defined as multilinear mappings of $T(M)$. Their restrictions are merely the mappings of the vectors of $T(N) \subset T(M)$. By (2.4.19) the inverse image, or **pull-back**, of a covariant vector field can be diagrammed as follows:

$$(\Phi^{-1})^* X = (T(\Phi))' \circ X \circ \Phi, \quad (\Phi^{-1})^* X \Bigg| \qquad \qquad \Bigg| X \qquad (2.4.23)$$

$$
\begin{array}{ccc}
N & \xrightarrow{\ \Phi\ } & M \\
\Big\downarrow & & \Big\downarrow \\
T^*(N) & \xleftarrow{\ T(\Phi)'\ } & T^*(M)
\end{array}
$$

The pull-back of a general covariant tensor field t is defined so that it always acts on vectors exactly as t acts on their images: †

$$(\Phi^{-1})^* t(v_1, \ldots, v_s) = t(T(\Phi)v_1, \ldots, T(\Phi)v_s). \qquad (2.4.24)$$

Examples (2.4.25)

1. Let us return to Example (2.4.22) and reinterpret the vector fields introduced at the beginning as $X \in \mathcal{T}^0_1(M)$ and $Y \in \mathcal{T}^0_1(N)$. This time $(Y(q)|v) = (X(q)|T_q(j)v) \ \forall v \in T_q(N)$. Suppose we transform the coordinates linearly with the matrix

$$L = \begin{bmatrix} 1 & -1 \\ 0 & 1 \end{bmatrix}.$$

† By abuse of notation we write $(\Phi^{-1})^*$, although no mapping Φ^{-1} in fact exists.

Then X transforms with

$$(L^{-1})^\ell = \begin{bmatrix} 1 & 0 \\ 1 & 1 \end{bmatrix},$$

becoming $(x, y) \rightarrow (x, y; 1, 2)$. The components in the direction of $T(N)$ remain unchanged.

2. If $t = dg \in \mathscr{T}_1^0(M)$, then according to (2.4.20; 3), $(\Phi^{-1})^* \, dg = d(g \circ \Phi)$, where Φ does not have to be either injective or surjective.

After discussing the transformation relations of tensor fields, we must define some more algebraic operations on them.

The sum of two tensors is only defined when they belong to the same $T_q^r(M)$. If we let $T_s^r(M)$ inherit a linear structure from $T_q^r(M)$,† then we must add tensors pointwise: $(t_1 + t_2)(q) \equiv t_1(q) + t_2(q)$. In making linear combinations, we can even multiply by functions in $C^\infty(M)$, because this does not send the tensors out of $T_s^r(M)$.‡

The tensor product provides a mapping $\mathscr{T}_{s_1}^{r_1} \times \mathscr{T}_{s_2}^{r_2} \overset{\otimes}{\rightarrow} \mathscr{T}_{s_1+s_2}^{r_1+r_2}$, in which we multiply at every point as in (2.4.5). The mapping is associative and distributive,

$$(t_1 \otimes t_2) \otimes t_3 = t_1 \otimes (t_2 \otimes t_3)$$

$$t_1 \otimes (\alpha t_2 + \beta t_3) = \alpha t_1 \otimes t_2 + \beta t_1 \otimes t_3, \qquad \alpha, \beta \in C^\infty(M). \quad (2.4.26)$$

There are no additional symmetry properties: $t_1 \otimes t_2$ is different from $t_2 \otimes t_1$.

Definition (2.4.27)

The space of p-fold covariant, totally antisymmetric tensor fields is denoted by $E_p(M)$, and its elements are called **p-forms**. For $p = 0$ and 1, $E_0(M) = C^\infty(M)$, and $E_1(M) = \mathscr{T}_1^0(M)$.

An antisymmetric tensor product $\wedge$ ("wedge")§ can be defined on $E_p(M)$. The action of an n-fold wedge product of 1-forms ω_i on n vectors $v_i \in T_q(M)$ is given by

$$((\omega_1 \wedge \omega_2 \wedge \cdots \wedge \omega_n)(q)|v_1, v_2, \ldots, v_n) = \sum_P (-1)^P \prod_{i=1}^n (\omega_i(q)|v_{p_i}), \quad (2.4.28)$$

in which $(p_1, p_2, \ldots, p_n)$ is a permutation P of the numbers $(1, 2, \ldots, n)$, and the sum is taken over all $n!$ permutations. The $\wedge$-product is associative

---

† The dimension of $T_s^r(M)$ is taken as that of $T_q^r(M)$.

‡ This makes $T_s^r(M)$ a **module** over $C^\infty(M)$.

§ Also known as the exterior, outer, or alternating product.

and distributive and thus can be extended to a mapping $(E_{p_1} \times E_{p_2}) \to E_{p_1 + p_2}$. Moreover,

$$\omega_1 \wedge \omega_2 = (-1)^{p_1 p_2} \omega_2 \wedge \omega_1 \quad \forall \omega_1 \in E_{p_1}, \omega_2 \in E_{p_2}.$$

Remarks (2.4.29)

1. E_p is a linear subspace of $\mathcal{T}_p^0$ of dimension $\binom{m}{p}$.
2. With the natural basis on a chart, a p-form ω may be expressed as

$$\omega = \frac{1}{p!} \sum_{(j)} c_{j_1, \ldots, j_p} \, dq^{j_1} \wedge \cdots \wedge dq^{j_p}, \qquad c_{(j)} \in C^\infty(U).$$

3. $\omega(v_1, \ldots, v_p)$ is an antisymmetric mapping $(v_i) \to \mathbb{R}$, linear in the v_i. If a basis is chosen so that only the first p components $(v_i)^j, j = 1, \ldots, p$, are different from zero, then $\omega(v_1, \ldots, v_p)$ must be proportional to $\mathrm{Det}((v_i)^j)$, and hence proportional to the volume of the p-dimensional parallelotope spanned by the v_i. We shall return to this aspect of p-forms in our discussion of integration theory, where they replace concepts like surface tensors and tensor densities.
4. There is a canonical n-form on $\mathbb{R}^n$,

$$dx^1 \wedge dx^2 \cdots \wedge dx^n = \varepsilon_{i_1, \ldots, i_n} \, dx^{i_1} \otimes \cdots \otimes dx^{i_n},$$

where $\varepsilon_{i_1, \ldots, i_n}$ are the components of the totally antisymmetric tensor for which $\varepsilon_{1, 2, \ldots, n} = 1$. This tensor establishes a canonical isomorphism $*: E_p \to E_{n-p}, p = 0, 1, \ldots, n$ (cf. Problem 2), that assigns the $n - p$-form†

$$*\omega = \frac{1}{p!(n-p)!} \omega_{i_1}, \ldots, {}_{i_p} \varepsilon_{i_1}, \ldots, {}_{i_n} \, dx^{i_{p+1}} \wedge \cdots \wedge dx^{i_n}$$

to the p-form

$$\omega = \frac{1}{p!} \omega_{i_1, \ldots, i_p} \, dx^{i_1} \wedge \cdots \wedge dx^{i_p}.$$

Note that $* \circ * = (-1)^{p(n-p)} \cdot \mathbf{1}$.

Example (2.4.30)

$n = 3$, $* \circ * = \mathbf{1}$. To the exterior product of any two vectors the $*$-mapping assigns a new vector, the vector product (or cross-product). Here we denote it by $[v \times w]$, using the brackets because it is not associative. In this case the definition of the $*$-mapping reduces to:

$p = 0, 3:$ $\quad \mathbf{1} \overset{*}{\leftrightarrow} dx^1 \wedge dx^2 \wedge dx^3$

$p = 1, 2:$ $\quad (dx^1, dx^2, dx^3) \overset{*}{\leftrightarrow} (dx^2 \wedge dx^3, dx^3 \wedge dx^1, dx^1 \wedge dx^2).$

† Not to be confused with the $*$ of (2.4.19).

For the components of a p-form ω,

$p = 0$: $\dfrac{\omega}{3!}\varepsilon_{ijk} = (*\omega)_{ijk}$

$p = 1$: $\tfrac{1}{2}\omega_k\varepsilon_{kij} = (*\omega)_{ij}$;   and $(v \wedge w)_{ij} = (*[v \times w])_{ij}$

$p = 2$: $\omega_{kj}\varepsilon_{kji} = (*\omega)_i$;   and $[v \times w]_i = (*(v \wedge w))_i$

$p = 3$: $\omega_{ijk}\varepsilon_{ijk} = *\omega$;   and $*(*v \wedge w)) = (v \cdot w)$.

The scalar product of a vector of $T(M)$ with a vector of $T^*(M)$ generalizes to a mapping of tensor fields:

Definition (2.4.31)

A **contraction** of a tensor field is a linear mapping $V_k^\ell : \mathcal{T}_s^r(M) \to \mathcal{T}_{s-1}^{r-1}(M)$:

$$V_k^\ell(\partial_{i_1} \otimes \cdots \otimes \partial_{i_r} \otimes dq^{j_1} \otimes \cdots \otimes dq^{j_s} c_{j_1, \ldots, j_s}^{i_1, \ldots, i_r})$$
$$= \partial_{i_1} \otimes \cdots \otimes \partial_{i_{r-1}} \otimes dq^{j_1} \otimes \cdots \otimes dq^{j_{s-1}} c_{j_1, \ldots, j_{k-1}, j, j_k, \ldots, j_{s-1}}^{i_1, \ldots, i_{\ell-1}, j, i_\ell, \ldots, i_{r-1}}$$
$$\ell = 1, 2, \ldots, r, \qquad k = 1, 2, \ldots, s.$$

Remarks (2.4.32)

1. As with the scalar product, for a diffeomorphism Φ, $V_k^\ell\Phi^* = \Phi^*V_k^\ell$. In particular, the definition is chart-independent.
2. V's with different ℓ or k produce different tensor fields unless the contracted tensor field has some special symmetry properties.

Combining a contraction with a tensor product of a vector field and a form leads to

Definition (2.4.33)

The **interior product** of a vector field $X \in \mathcal{T}_0^1(M)$ is the linear mapping

$$i_X : E_p(M) \to E_{p-1}(M) : i_X(\omega) = \sum_{j=1}^{p} \frac{1}{p}(-1)^{j+1} V_j^1(X \otimes \omega).$$

Example (2.4.34)

For $p = 1$, $i_X(\omega) = (\omega \mid X) \in C^\infty(M)$.

Problems (2.4.35)

1. Show that the transpose $L \to L^t$ introduced at the beginning of the section is bijective.

2. Show that $* : E_p \to E_{n-p}$ is bijective.

3. Calculate the explicit form of $\Phi^*\omega$, $\omega \in \mathcal{T}_1^0$, in local coordinates. How is this related to the elementary transformation of differentials of coordinates (or of the gradient)?

4. Calculate the components of the gradient df in spherical and cylindrical coordinates in $\mathbb{R}^3$.

5. Show that $T(M_1 \times M_2) = T^*(M_1) \times T^*(M_2)$.

6. Show

(a) $i_X\omega(X_1, \ldots, X_{p-1}) = \omega(X, X_1, \ldots, X_{p-1})$, $\omega \in E_p(M)$;
(b) $i_{fX}\omega = f i_X\omega$, $f \in E_0(M)$, $\omega \in E_p(M)$;
(c) $i_X(\omega \wedge v) = (i_X\omega) \wedge v + (-1)^p\omega \wedge (i_X v)$, $\omega \in E_p(M)$, $v \in E_q(M)$.

7. What is the transformation law for an m-form in m dimensions?

Solutions (2.4.36)

1. $v \to (w^*|Lv)$ is a linear functional on E, and can therefore be written $(L'w^*|v)$, where L' is a linear mapping $F^* \to E^*$. The association $L \to L'$ is injective, for

$$(w^*|Lv) = (L_1'w^*|v) = (L_2'w^*|v), \quad \forall v \in E, w \in F^*$$
$$\Rightarrow (L_1' - L_2')w^* = 0, \quad \forall w^* \in F^* \Rightarrow L_1' = L_2'.$$

$E^{**} = E$ and $F^{**} = F$, and therefore to each L' there corresponds exactly one $L'': E \to F$, which must equal L. Hence the association is also surjective.

2. A linear mapping $\omega \to {}^*\omega$.

$$({}^*\omega)_{i_{p+1}, \ldots, i_n} = \frac{1}{p!(n-p)!}\, \omega_{i_1, \ldots, i_p} \varepsilon_{i_1, \ldots, i_n},$$

was defined in remark (2.4.29; 4). It is injective because $\omega \neq 0 \Rightarrow {}^*\omega \neq 0$. But for linear mappings of finite-dimensional spaces of equal dimension, injectiveness is equivalent to surjectiveness.

3. For $\Phi: q^i \to \bar{q}^i(q)$, we have $(T(\Phi))_{ij} = \partial\bar{q}^i/\partial q^j$. Therefore $(T(\Phi^{-1}))_{ij}' = (T^*(\Phi))_{ij} = \partial q^j/\partial\bar{q}^i$. Let $\omega: q \to (q, \omega_i(q)dq^i)$. Then $\Phi^*\omega: \bar{q} \to (\bar{q}, (\partial q^j/\partial\bar{q}^i)\omega_j(\Phi^{-1}(\bar{q}))d\bar{q}^i)$. The covariant components transform in the same way as the basis ∂_i of $T(M)$, and hence as the gradient $(\partial/\partial\bar{q}^i)f(q(\bar{q})) = (\partial q^j/\partial\bar{q}^i)(\partial f/\partial q^j)$. On the other hand, one can rewrite the differential using $dq^i = (\partial q^i/\partial\bar{q}^j)d\bar{q}^j$ and leave the components unchanged.

4. With a change of charts, $(x, y, z) = (r \sin \vartheta \cos \varphi, r \sin \vartheta \sin \varphi, r \cos \vartheta)$ (or respectively $(\rho \cos \varphi, \rho \sin \varphi, z)$), the metric $dx^2 + dy^2 + dz^{2\dagger}$ becomes

$$dr^2 + r^2\, d\vartheta^2 + r^2 \sin^2 \vartheta\, d\varphi^2$$

(or $d\rho^2 + \rho^2\, d\varphi^2 + dz^2$). Following remark (2.4.16), from the covariant components $(f_{,r}, f_{,\vartheta}, f_{,\varphi})$ (or $(f_{,\rho}, f_{,\varphi}, f_{,z})$) we obtain the components $v_i/(g_{ii})^{1/2}$:

$$\left(f_{,r}, \frac{1}{r}f_{,\vartheta}, \left(\frac{1}{r \sin \vartheta}\right)f_{,\varphi}\right) \quad \left(\text{or } \left(f_{,\rho}, \frac{1}{\rho}f_{,\varphi}, f_{,z}\right)\right).$$

5. Use a product chart; everything factors out.

---

† Pedantically, dx^2 should be written $dx \otimes dx$, etc.

6. (a) On a chart containing q,

$$X \otimes \omega = X^i \omega_{j_1, \ldots, j_p} \partial_i \otimes dq^{j_1} \otimes \cdots \otimes dq^{j_p},$$

and consequently

$$(i_X \omega)(X_1, \ldots, X_{p-1}) = \sum_k \frac{1}{p} (-1)^{k+1} V^1_k (X \otimes \omega)(X_1, \ldots, X_{p-1})$$

$$= \sum_k \frac{1}{p} (-1)^{k+1} X^i \omega_{j_1, \ldots, j_k, \ldots, j_p} \delta^{j_k}_i X^{j_1}_1 \cdots$$

$$\times X^{j_{k-1}}_{k-1} X^{j_{k+1}}_k \cdots X^{j_p}_{p-1}.$$

If the index j_k is brought to the first position by a permutation, each term in the above expression is multiplied by $(-1)^{k+1}$, because of the antisymmetry of ω; one then sums the same expression p times to find

$$(i_X \omega)(X_1, \ldots, X_{p-1}) = \omega_{i, j_1, \ldots, j_{p-1}} X^i X^{j_1}_1, \ldots, X^{j_p}_{p-1} = p\omega(X, X_1, \ldots, X_{p-1}).$$

(b) This follows from (a), since $\omega(fX, X_1, \ldots, X_{p-1}) = f\omega(X, X_1, \ldots, X_{p-1})$

(c) Let $\omega = \dfrac{1}{p!} \sum_{(i)} \omega_{i_1, \ldots, i_p} dx^{i_1} \wedge \cdots \wedge dx^{i_p}$, and

$$v = \frac{1}{q!} \sum_{(i)} v_{i_{p+1}, \ldots, i_{p+q}} dx^{i_{p+1}} \wedge \cdots \wedge dx^{i_{p+q}}. \text{ Then}$$

$$i_X(\omega \wedge v) = \frac{1}{(p+q)!} \sum_{j=1}^{p+q} (-1)^{j+1} X^i (\omega \wedge v)_{i_1, \ldots, i_{j-1}, i, i_{j+1}, \ldots, i_{p+q}} dx^{i_1} \wedge \cdots$$

$$\wedge dx^{i_j-1} \wedge dx^{i_j+1} \wedge \cdots \wedge dx^{i_{p+q}}$$

$$= \frac{1}{p!q!} \sum_{j=1}^{p} (-1)^{j+1} X^i \omega_{i_1, \ldots, i_{j-1}, i, i_{j+1}, i_p} v_{i_{p+1}, \ldots, i_{p+q}} dx^{i_1} \wedge$$

$$\cdots \wedge dx^{i_j-1} \wedge dx^{i_j+1} \wedge \cdots \wedge dx^{i_{p+q}}$$

$$+ (-1)^p \frac{1}{p!q!} \sum_{j=1}^{q} (-1)^{j+1} \omega_{i_1, \ldots, i_p}$$

$$\times X^i v_{i_{p+1}, \ldots, i_{p+j-1}, i, i_{p+j+1}, \ldots, i_{p+q}} dx^{i_1} \wedge \cdots$$

$$\wedge dx^{i_p+j-1} \wedge dx^{i_p+j+1} \wedge \cdots \wedge dx^{i_{p+q}}$$

$$= i_X \omega \wedge v + (-1)^p \omega \wedge i_X v.$$

7. Let $\omega = \omega_{1, \ldots, m} dx^1 \wedge \cdots \wedge dx^m$

$$= \omega_{1, \ldots, m} \frac{\partial x^1}{\partial \bar{x}^{j_1}} \cdots \frac{\partial x^m}{\partial \bar{x}^{j_m}} d\bar{x}^{j_1} \wedge \cdots \wedge d\bar{x}^{j_m} = \omega_{1, \ldots, m} \text{ Det} \left| \frac{\partial x^i}{\partial \bar{x}^j} \right| d\bar{x}^1 \wedge \cdots \wedge d\bar{x}^m.$$

2.5   Differentiation

The only generalization of the elementary operation of differentiation for a manifold with no additional structure is the exterior differential of a form. If a local flow is given by some vector field, then it defines the Lie derivative of an arbitrary tensor field.

The differential d (2.4.3) generalizes to a mapping $d: E_p(M) \to E_{p+1}(M)$, which contains the differentiation operations of elementary vector calculus as special cases.

Definition (2.5.1)

Let ω be a p-form, which is written as

$$\omega = \frac{1}{p!} \sum_{(i)} c_{(i)} \, dq^{i_1} \wedge \cdots \wedge dq^{i_p}, \qquad c_{(i)} \in C^\infty(M)$$

on some chart. Then the $p + 1$-form,

$$d\omega = \frac{1}{p!} \sum_{(i)} dc_{(i)} \wedge dq^{i_1} \wedge \cdots \wedge dq^{i_p},$$

is known as its **exterior differential**.

From the definition follow the

Rules of exterior differentiation (2.5.2)

(a) $d(\omega_1 + \omega_2) = d(\omega_1) + d(\omega_2)$, $\omega_i \in E_p(M)$,
(b) $d(\omega_1 \wedge \omega_2) = (d\omega_1) \wedge \omega_2 + (-1)^p \omega_1 \wedge d\omega_2$, $\omega_1 \in E_p$, $\omega_2 \in E_q$,
(c) $d(d\omega) = 0$, $\omega \in E_p$, $p = 0, 1, \ldots, m$.

Rules (a) and (b) are obvious. Rule (c) follows from the symmetry of the partial derivative:

$$d(d\omega) = \sum_{(i)} \sum_{k,j} \frac{1}{p!} \frac{\partial^2 c_{(i)}}{\partial q^k \partial q^j} \, dq^k \wedge dq^j \wedge dq^{i_1} \wedge \cdots \wedge dq^{i_p} = 0.$$

Remarks (2.5.3)

1. Since we wish definition (2.5.1) to be independent of the coordinate system, it is essential for d to be natural with respect to diffeomorphisms, a phrase which means that for a diffeomorphism $\Phi: M_1 \to M_2$,

$$\Phi^* \, d\omega = d\Phi^* \omega;$$

or equivalently that the diagram

$$
\begin{array}{ccc}
E_p(M_1) & \xrightarrow{\ \Phi^*\ } & E_p(M_2) \\
\Big\downarrow{\scriptstyle d} & & \Big\downarrow{\scriptstyle d} \\
E_{p+1}(M_1) & \xrightarrow{\ \Phi^*\ } & E_{p+1}(M_2)
\end{array}
$$

is permutable. This follows from the special case proved above (2.4.20; 3), by which

$$\Phi^*\omega = \sum_{(i)} \Phi^*(c_{(i)})\Phi^*(dq^{i_1}) \wedge \cdots \wedge \Phi^*(dq^{i_p})$$

$$= \sum_{(i)} c_{(i)} \circ \Phi^{-1} \, d(q^{i_1} \circ \Phi^{-1}) \wedge \cdots \wedge d(q^{i_p} \circ \Phi^{-1})$$

and

$$\Phi^*(d\omega) = \sum_{(i)} d(c_{(i)} \circ \Phi^{-1}) \wedge d(q^{i_1} \circ \Phi^{-1}) \wedge \cdots \wedge d(q^{i_p} \circ \Phi^{-1}).$$

If in particular Φ is the diffeomorphism of a change of charts, then $d\omega$ is constructed in the new coordinate system exactly as in the old one, except that everything is expressed in the new coordinates.

2. The relationship $\Phi^* \, d\omega = d\Phi^*\omega$ does not hold only for diffeomorphisms, but for all inverse images of forms (2.4.18). In that derivation and the following steps (2.4.20; 3), only the existence of the mapping $\Phi^{-1}: M_2 \to M_1$ was used. Given any mapping $\Psi: M_2 \to M_1$, we can carry forms over from M_1 to M_2 with $(\Psi^{-1})^*$. If M_2 happens to be a submanifold of M_1 and Φ^{-1} is the natural injection, and thus Φ^* is the restriction to M_2, then the relationship merely means that the differential of the restriction is the restriction of the differential.

Examples (2.5.4)

Let $M = \mathbb{R}^3$. As in (2.4.30) we identify E_0 with E_3 and E_1 with E_2. The connection between our notation and that of vector calculus is: $(df)_i = (\nabla f)_i = (\text{grad } f)_i$; $*(dv)_i = (\nabla \times v)_i = (\text{curl } v)_i$; and $*(*dv) = \nabla \cdot v = \text{div } v$. The rules (2.5.2) contain the following special cases:

1(b) $p = q = 0$: $\nabla(f \cdot g) = f \nabla g + g \nabla f$;
2(b) $p = 0, q = 1$: $\nabla \times (f \cdot v) = [\nabla f \times v] + f \nabla \times v$;
3(b) $p = q = 1$: $\nabla \cdot [v \times w] = *(d(v \wedge w)) = *(dv \wedge w) - *(v \wedge dw)$
$\qquad\qquad\qquad\qquad = (w \cdot \nabla \times v) - (v \cdot \nabla \times w)$;
4(c) $p = 0$; $\nabla \times \nabla f = 0$;
5(c) $p = 1$: $\nabla \cdot (\nabla \times v) = 0$;
6(b) and (c) $\nabla \cdot (f \cdot \nabla \times v) = (\nabla f \cdot \nabla \times v)$.

In vector calculus one learns that curl-free vectors ($\nabla \times v = 0$ everywhere) can be written as gradients, and divergence-free vectors can be written as curls. In order to state the analogous fact for manifolds, we make use of

Definition (2.5.5)

A p-form ω is said to be **closed** iff $d\omega = 0$ and **exact** iff $\omega = dv$ for some $v \in E_{p-1}(M)$.

Remarks (2.5.6)

1. By (2.5.2(c)), exact $\Rightarrow$ closed, and the exact forms are a linear subspace of the closed forms.
2. The exact forms are in general a proper subspace. Consider on $M = \mathbb{R}^2 \backslash \{0\}$ the 1-forms

$$\omega_i = \frac{-y\,dx + x\,dy}{x^2 + y^2} = \operatorname{Im}\frac{dz}{z}, \qquad z = x + iy,$$

and

$$\omega_r = \frac{x\,dx + y\,dy}{x^2 + y^2} = \operatorname{Re}\frac{dz}{z}.$$

Certainly $d\omega_i = d\omega_r = 0$, and locally $\omega_r + i\omega_i = d\ln z$. But since $\ln z$ is not defined continuously on M, the forms are not exact. Here it is crucial that we have removed the origin of $\mathbb{R}^2$, at which point the forms are singular and their differentials by no means zero.
3. If M is a starlike† open set in $\mathbb{R}^n$ (see Figure 13), then there exists a mapping $A: E_p \to E_{p-1}$ such that $A \circ d + d \circ A = 1$ (Problem 7). It follows that $d\omega = 0$ implies $\omega = d(A\omega)$ (Poincaré's lemma). Since in $\mathbb{R}^n$ every neighborhood contains a convex set, closed $\Rightarrow$ exact on small enough subsets. That is, locally (2.5.6; 1) holds the other way around.

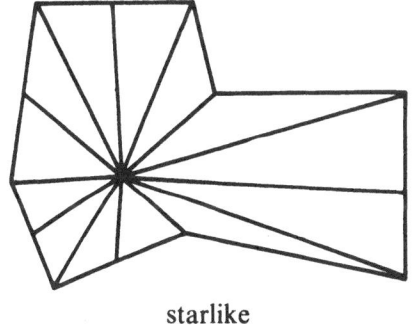
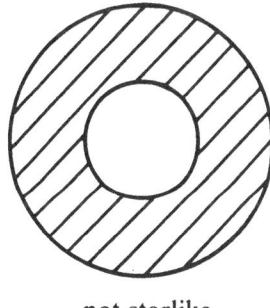

starlike   not starlike

Figure 13   Starlikeness in the plane.

4. Since p-forms with $p < 0$ are identically zero by definition, it would seem that $df = 0$ implies $f = 0$. But this is the degenerate case, and in fact it only implies that f is locally constant.

It is not always possible to attribute a coordinate-independent sense to the derivative of a tensor field $\mathcal{T}$. One would have to compare $\mathcal{T}(q)$ and

---

† A set $S \subset \mathbb{R}^n$ is starlike with respect to a point P iff the line connecting any point of S with P lies wholly within S. A convex set is starlike with respect to all its points.

$\mathcal{T}(q + \delta q)$, but the relative orientation of the tangent spaces depends on the coordinate system (2.2.10). Taking as an example $X^* \in \mathcal{T}_1^0$, the derivative $X_{i;k}^*$ does not transform as a tensor of degree two, though the unwanted terms cancel out in the transformation of the combination $X_{i;k}^* - X_{k;i}^*$ that comes from the exterior differential. Yet if a vector field X is given on M, it induces a local flow Φ_t^X, and in order to define the derivative of another vector field t at the point q, one could map the tangent vectors along the path through q, $q(t) = \Phi_t^X(q)$, back into $T_q(M)$ by using $T_{q(t)}(\Phi_{-t}^X)$. Then both vectors, $t(q)$, which is the value of the vector field at q, and the vector generated from $t(q(t))$ by running time backwards, can be compared at the same point q. The second of these vectors can also be written as $(\Phi_{-t}^X{}^* t)(q)$ (cf. (2.2.21) and Figure 14).

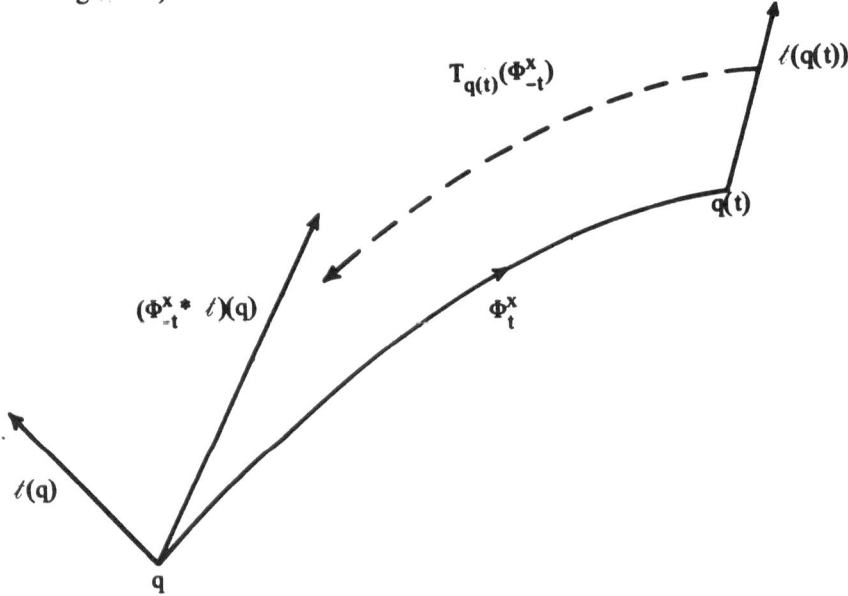

Figure 14   The Lie derivative.

The corresponding derivative

$$\frac{d}{dt}(\Phi_{-t}^X{}^* t)(q)$$

is independent of the coordinate system used, since it involves vectors in a single tangent space $T_q(M)$, and because differentiation of vectors commutes with linear transformations. This line of reasoning applies as well to arbitrary tensor fields, leading us to make (cf. (2.5.3))

Definition (2.5.7)

The **Lie derivative** $L_X : \mathcal{T}_s^r \to \mathcal{T}_s^r$ is defined by

$$L_X t = \frac{d}{dt} \Phi_{-t}^X{}^* t_{|t=0}, \qquad t \in \mathcal{T}_s^r.$$

Remarks (2.5.8)

1. It is important to have a vector field X and not merely a vector of $T_q(M)$. Our expressions for $L_X \ell(q)$ will contain not only the values of the components of X at the point q, but also the values of their derivatives.
2. If X induces an analytic flow, then we may generalize (2.3.11; 2) to

$$\Phi^X_{-t}{}^* \ell = e^{tL_X}\ell, \qquad \ell \in \mathcal{T}^r_s.$$

3. L_X measures the change due to an infinitesimal chart transformation $q^i \to q^i + tX^i(q)$.

Properties of the Lie derivative (2.5.9)

1. The mapping Φ^* that a diffeomorphism Φ induces on the tensor fields preserves their algebraic structure. For $\ell_i \in \mathcal{T}^r_s(M)$,

(a) $\Phi^X_{-t}{}^*(\ell_1 + \ell_2) = \Phi^X_{-t}{}^* \ell_1 + \Phi^X_{-t}{}^* \ell_2$;

(b) $\Phi^X_{-t}{}^*(\ell_1 \otimes \ell_2) = (\Phi^X_{-t}{}^* \ell_1) \otimes (\Phi^X_{-t}{}^* \ell_2)$;

(c) $\Phi^X_{-t}{}^* V\ell = V\Phi^X_{-t}{}^* \ell$, for V a contraction.

For infinitesimal t it follows that

(a) $L_X(\ell_1 + \ell_2) = L_X\ell_1 + L_X\ell_2$,

(b) $L_X(\ell_1 \otimes \ell_2) = (L_X \ell_1) \otimes \ell_2 + \ell_1 \otimes L_X \ell_2$,

(c) $L_X V\ell = VL_X \ell$.

2. The permutability of the diagram

$$
\begin{array}{ccc}
\mathcal{T}^r_s(M_1) & \xrightarrow{\ \Psi^*\ } & \mathcal{T}^r_s(M_2) \\
\Big\downarrow{\scriptstyle \Phi^X_{-t}{}^*} & & \Big\downarrow{\scriptstyle (\Phi^{\Psi^* X}_{-t})^*} \\
\mathcal{T}^r_s(M_1) & \xrightarrow{\ \Psi^*\ } & \mathcal{T}^r_s(M_2)
\end{array}
$$

implies the permutability of its infinitesimal version

$$
\begin{array}{ccc}
\mathcal{T}^r_s(M_1) & \xrightarrow{\ \Psi^*\ } & \mathcal{T}^r_s(M_2) \\
\Big\downarrow{\scriptstyle L_X} & & \Big\downarrow{\scriptstyle L_{\Psi^* X}} \\
\mathcal{T}^r_s(M_1) & \xrightarrow{\ \Psi^*\ } & \mathcal{T}^r_s(M_2)
\end{array},
$$

that is, $L_{\Psi^* X} \Psi^* \ell = \Psi^* L_X \ell$. This means that the flow on the transformed system is determined by the transformed vector field, so that the image with Ψ^* of the Lie derivative of a tensor field is the same as the Lie derivative with respect to the image of the vector field of the image of the tensor field. This naturalness of L_X with respect to diffeomorphisms is a consequence of the chart-independence of its definition.

3. Since d is natural with respect to diffeomorphisms (2.5.3), it commutes with L_X. Formally it is like this: $(\omega \in E_p)$

$$L_X \, d\omega = \frac{d}{dt} \Phi_{-t}^{X} {}^{*} \, d\omega_{|t=0} = \frac{d}{dt} d\Phi_{-t}^{X} {}^{*} \, \omega_{|t=0}$$

$$= d \frac{d}{dt} \Phi_{-t}^{X} {}^{*} \, \omega_{|t=0} = dL_X \omega.$$

Expressed diagrammatically:

$$
\begin{array}{ccc}
E_p(M) & \xrightarrow{\;\;L_X\;\;} & E_p(M) \\
\downarrow{\scriptstyle d} & & \downarrow{\scriptstyle d} \\
E_{p+1}(M) & \xrightarrow{\;\;L_X\;\;} & E_{p+1}(M)
\end{array}
$$

commutes.

4. On E_p, L_X may be expressed in terms of d and the interior product i_X (2.4.33),

$$L_X = i_X \circ d + d \circ i_X.$$

For the proof, see Problem 6. This also shows that $dL_X = d \circ i_X \circ d = L_X d$.

5. The Lie derivative L_X is consistent with linearity of X:

(a) $L_{X_1 + X_2} = L_{X_1} + L_{X_2}$,
(b) $L_{cX} = cL_X$, c any constant.

As for the module structure of the vector fields, it follows from Property 4 that

$$L_{fX} = fL_X + df \wedge i_X \quad \text{on } E_p,$$

since

$$i_{fX} \, d\omega + di_{fX}\omega = fi_X \, d\omega + d(fi_X\omega)$$
$$= f(i_X d + di_X)\omega + df \wedge i_X\omega, \quad \forall \omega \in E_p.$$

The extra term containing df reflects the presence of the derivatives of X in L_X.

6. According to 1(a) and Property 3, $(df \,|\, L_X Y) = L_X(df \,|\, Y) - (dL_X f \,|\, Y)$ $= (L_X L_Y - L_Y L_X)f$. Let us denote the vector field $L_X Y$ by the Lie bracket $[X, Y]$. The calculation just done implies

$$L_{L_X Y} = L_X L_Y - L_Y L_X \quad \text{on } \mathcal{T}_0^0,$$

from which it follows that $L_X Y = -L_Y X$, because a vector field is completely characterized by its action on $\mathcal{T}_0^0$. The relationship

$$L_{[X, Y]} = L_X L_Y - L_Y L_X$$

can be extended to all $\mathcal{T}^r_s$. If $L_X L_Y - L_Y L_X$ is applied to $t \in \mathcal{T}^0_s$:

$$t = \sum_{(i)} c_{(i)} \, dq^{i_1} \otimes \cdots \otimes dq^{i_s},$$

using the rules for sums and tensor products, only the terms in which one factor is differentiated twice remain. The others cancel out because of antisymmetry. For the remaining terms, the relationship in question holds by Property 3. Then by Property 1 it must hold for all $\mathcal{T}^r_s$.

Examples (2.5.10)

1. $r = s = 0$. $\Phi^X_{-t}{}^* f = f \circ \Phi^X_t = e^{tL_X} f = \tau^X_t f$. In this case (2.5.7) coincides with the earlier definition (2.3.11; 2). If on some chart X is $X^i \partial_i$, (2.5.7) yields

$$L_X f = X^i f_{,i}.$$

Observe that Φ^*_{-t} induces the automorphism τ_t. The reversal of the signs arises from the definition $\Phi^*_t f(q(t)) = f(q)$.

2. $r = 0, s = 1$: $\omega = \omega_i \, dq^i$. The rules imply that

$$L_X \omega = (L_X \omega_i) dq^i + \omega_i \, d(L_X q^i)$$
$$= (X^k \omega_{i,k} + \omega_k X^k_{,i}) dq^i.$$

3. $r = 1, s = 0$. For $\omega \in \mathcal{T}^0_1$, $Y = Y^i \partial_i \in \mathcal{T}^1_0$, we calculate

$$L_X(\omega | Y) = \omega_i Y^i_{,k} X^k + \omega_{i,k} Y^i X^k$$
$$= \omega_i (Y^i_{,k} X^k - X^i_{,k} Y^k) + (\omega_{i,k} X^k + \omega_k X^k_{,i}) Y^i$$
$$= (\omega | L_X Y) + (L_X \omega | Y).$$

Thus the i-th component of the Lie derivative of Y is

$$Y^i_{,k} X^k - X^i_{,k} Y^k.$$

4. $X = \partial_i$ and $Y = \partial_j$. Then $[X, Y] = 0$. The vanishing of the Lie bracket of the natural basis vectors means that the partial derivatives commute.

The Lie bracket provides the vector fields with an additional algebraic structure. It distributes over addition, and instead of the associative law it satisfies

Jacobi's identity (2.5.11)

$$[X, [Y, Z]] + [Y, [Z, X]] + [Z, [X, Y]] = 0.$$

Proof

Follows from the identity

$$L_{[X,[Y,Z]]} + L_{[Y,[Z,X]]} + L_{[Z,[X,Y]]} = L_X(L_Y L_Z - L_Z L_Y)$$
$$- (L_Y L_Z - L_Z L_Y)L_X + L_Y(L_Z L_X - L_X L_Z) - (L_Z L_X - L_X L_Z)L_Y$$
$$+ L_Z(L_X L_Y - L_Y L_X) - (L_X L_Y - L_Y L_X) L_Z = 0,$$

because $L_X = 0$ (even if only on $\mathcal{T}_0^0(M)$) implies $X = 0$.   □

Problems (2.5.12)

1. Why is it not possible to define d on all of $\mathcal{T}_s^0$ independently of the charts?

2. How are the covariant components of $d\omega$ written in the notation of (2.5.1)?

3. Show explicitly that $\Phi^* d\omega = d\Phi^*\omega$ for $M = \mathbb{R}^n$, $p = 1$.

4. Show directly for $f \in C(M)$ that $L_X df = d(L_X f)$

5. Calculate the components of $L_X Y$, X and $Y \in \mathcal{T}_0^1$, and of $L_X \alpha$, $\alpha \in \mathcal{T}_1^0$. Prove that $L_X(\alpha | Y) = (L_X \alpha | Y) + (\alpha | L_X Y)$.

6. Show that $L_X = i_X \circ d + d \circ i_X$ on $E_p(2.5.9; 4)$.

7. Define the mapping $A : E_p \to E_{p-1}$ of Remark (2.5.6; 3) as follows: Let U be starlike with respect to the origin and $h : (0, 1) \times U \to U$ be the mapping $(t, \bar{x}) \to t\bar{x}$. For $\omega \in E_p(U)$ we may decompose the inverse image under h into one part with dt and another without dt:

$$(h^{-1})^*\omega = \omega_0 + dt \wedge \omega_M, \omega_0 \in E_p((0, 1) \times U), \omega_M \in E_{p-1}((0, 1) \times U).$$

Then

$$A\omega \equiv \int_0^1 dt\, \omega_M \in E_{p-1}(U).$$

Show that $A \circ d + d \circ A = 1$, and calculate $A\omega$ in $\mathbb{R}^3$ for $\omega \in E_1$ and E_2.

8. Find an example of a vector field $\mathbf{E}$ that is divergence-free on $\mathbb{R}^3 \setminus \{0\}$, but which can not be written $\mathbf{E} = \nabla \times A$.

9. Show that for $\omega \in E_1$,

$$(d\omega | X \otimes Y) = L_X(\omega | Y) - L_Y(\omega | X) - (\omega | [X, Y]).$$

Solutions (2.5.13)

1. Antisymmetry is the key to the proof that $d(d\omega) = 0$.

2. $(d\omega)_{i_1, \ldots, i_{p+1}} = \sum_{\ell=1}^{p+1} \frac{\partial}{\partial x_\ell} (\omega)_{i_1, \ldots, i_{\ell-1}, i_{\ell+1}, \ldots, i_{p+1}} \cdot (-1)^\ell$

3.
$$\omega = c_k(x)dx^k, \qquad d\omega = c_{k,i}\, dx^i \wedge dx^k,$$

$$\Phi^*\omega = c_k(x(\bar{x}))\frac{\partial x^k}{\partial \bar{x}^j}\, d\bar{x}^j,$$

$$d\Phi^*\omega = \left(c_{k,i}\frac{\partial x^i}{\partial \bar{x}^r}\frac{\partial x^k}{\partial \bar{x}^j} + c_k\frac{\partial^2 x^k}{\partial \bar{x}^r \partial \bar{x}^j}\right)d\bar{x}^r \wedge d\bar{x}^j$$

$$= c_{k,i}\frac{\partial x^i}{\partial \bar{x}^r}\frac{\partial x^k}{\partial \bar{x}^j}\, d\bar{x}^r \wedge d\bar{x}^j \equiv \Phi^*\, d\omega.$$

4. $L_X L_Y f = L_X(df\,|\,Y) = (df\,|\,L_X\,Y) + (L_X\,df\,|\,Y) = L_{[X,\,Y]}f + (L_X\,df\,|\,Y)$,
   so $(L_X\,df\,|\,Y) = L_Y L_X\,f = (d(L_X\,f)\,|\,Y)\,\forall Y \in \mathcal{T}_0^1(M) \Rightarrow L_X\,df = d(L_X\,f)$, i.e., $dL_X = d_{i_X}\,d = L_X d$ on f.

5. For $X: q \rightarrow (q_i, X^i)$, etc.,

$$(L_X\alpha\,|\,Y) + (\alpha\,|\,L_X\,Y) = (L_X\alpha)_i\,Y^i + \alpha_i(L_X\,Y)^i = (\alpha_{i,k}X^k + \alpha_k X^k_{,i})Y^i$$
$$+ \alpha_i(Y^i_{,k}X^k - X^i_{,k}Y^k) = \alpha_i\,Y^i_{,k}X^k + \alpha_{i,k}\,Y^iX^k = L_X(\alpha\,|\,Y).$$

(And similarly.)

6. Proof by induction: for $p = 0$, $i_X\,f = 0$ by definition, and $i_X\,df = (df\,|\,X) = L_X\,f$. Every $p + 1$-form may be written as

$$\sum_i df_i \wedge \omega_i, \qquad \omega_i \in E_p,\, f \in C^\infty.$$

Now,

$$(i_X \circ d + d \circ i_X)df \wedge \omega = i_X \circ (-df \wedge d\omega) + d \circ ((i_X\,df)\omega - df \wedge i_X\omega)$$
$$= -(i_X\,df) \wedge d\omega + df \wedge (i_X\,d\omega) + (d(i_X\,df)) \wedge \omega$$
$$+ (i_X\,df) \wedge d\omega + df \wedge d(i_X\omega)$$
$$= df \wedge L_X\omega + (L_X\,df) \wedge \omega = L_X(df \wedge \omega).$$

Since both sides of the equation in (2.5.9; 4) are linear operators, this relationship also holds for $\sum_i df_i \wedge \omega_i$, and consequently on $E_{p+1}(M)$.

7. For

$$\omega = \frac{1}{p!}\,\omega_{(i)}(x)dx^{i_1} \wedge \cdots \wedge dx_{i_p},$$

we find

$$(h^{-1})^*\omega = \omega_{(i)}(xt)(t\,dx^{i_1} + x^{i_1}\,dt) \wedge \cdots \wedge (t\,dx^{i_p} + x^{i_p}\,dt)\frac{1}{p!}$$

$$= \omega_0 + dt \wedge \omega_M.$$

Let us designate the exterior derivative with t held constant by d'; then

$$dA\omega = \int_0^1 dt\, d'\omega_M, \qquad A\, d\omega = \int_0^1 dt\left(\frac{\partial\omega_0}{\partial t} - d'\omega_M\right).$$

As defined above, $\dot{\omega}_{0|t=1} = \omega$ and $\omega_{0|t=0} = 0$, and so $dA\omega + A\,d\omega = \omega$.

$$p = 1: \quad A\omega = \int_0^1 dt\, x_i v_i(xt) = \int_0^x ds \cdot v;$$

$$p = 2: \quad \omega_{ij} = \varepsilon_{ijk}B_k, \qquad (A\omega)_i = \int_0^1 dt\, tB_k x_j \varepsilon_{kji}.$$

8. $\mathbf{E} = \mathbf{x}/|\mathbf{x}|^3$. It is impossible that $\mathbf{E} = \mathbf{V} \times \mathbf{A}$, for then we would have

$$4\pi = \int_{S^2} d\mathbf{S} \cdot \mathbf{E} = \int_{S^2} d\mathbf{S} \cdot \mathbf{V} \times \mathbf{A} = \int_{\partial S^2} d\mathbf{s} \cdot \mathbf{A} = 0,$$

because $\partial S^1 = \varnothing$.

9. $L_X(\omega|Y) = ((i_X \circ d + d \circ i_X)\omega|Y) + (\omega|L_X Y) = (d\omega|X \otimes Y) + L_Y(\omega|X)$
    $+ (\omega|[X, Y])$.

2.6   Integration

An m-form defines a measure on a manifold. An integral is an inverse of the exterior derivative in the sense that integration by parts can be generalized as Stokes's theorem.

A differential volume element in $\mathbb{R}^m$ is written dV or $dx^1 dx^2 \cdots dx^m$, both of which are convenient but rather poor notation. With a chart transformation it transforms by being multiplied by the Jacobian, that is, it is like an exterior product rather than the ordinary product of the dx^i. This property makes it an m-form (cf. 2.4.29; 3), and as such its exterior derivative is zero†. Hence on $\mathbb{R}^m$ it can be written dV, $V \in E_{m-1}(\mathbb{R}^m)$. However, this is not always possible for manifolds with holes (2.5.6). If we intend to use m-forms to measure volumes, we are faced with a question of the sign. This is because if we change the sign of one coordinate, an m-form changes its sign; but a volume should always be positive. Let us remove this two-valuedness at this point.

Definition (2.6.1)

An m-dimensional manifold is **orientable** iff there exists a nowhere-vanishing m-form Ω on it.

Examples (2.6.2)

1. An open subset of $\mathbb{R}^m$ is orientable, because the m-form $dx^1 \wedge \cdots \wedge dx^m$ generated by $\varepsilon_{i_1,\ldots,i_m}$ never vanishes.
2. The Möbius strip (2.2.16; 3) is not orientable; the 2-form $d\varphi \wedge dx$ defined locally can not be continuously defined at all points.
3. The product of two orientable manifolds is orientable, since one can take $\Omega = \Omega_1 \times \Omega_2$.
4. The cotangent bundle $T^*(M)$ is orientable, even when M is not. We shall soon learn of a nowhere-vanishing $2m$-form on it.

---

† The exterior derivative vanishes because it is an $m + 1$-form.

Remarks (2.6.3)

1. Since Ω does not vanish, every m-form can be written $f\Omega$, $f \in C^\infty$.
2. Definition (2.6.1) is equivalent to the existence of an atlas for which the Jacobian $\text{Det}(\bar{\Phi} \circ \Phi^{-1})$ is always positive on the overlap of two charts (Problem 4).
3. Parallelizability is sufficient but not necessary for orientability.
4. In case of doubt we will assume orientability from now on.

Choosing Ω as positive, we can define an integral over m-forms on an orientable manifold. Of course, it will be necessary to check invariance under chart-transformations and the convergence of the integral.

On a chart (U, Φ) the image of $\Omega_{|U}$ will be of the form $w(x)dx^1 \wedge \cdots \wedge dx^m \equiv w\, d^m x$, and we can choose the chart so that $w > 0$. If an m-form has compact support in U, then we define its integral as

$$\int \Omega f \equiv \int \Phi^*(\Omega f) = \int_{-\infty}^{\infty} dx^1 \int_{-\infty}^{\infty} dx^2 \cdots \int_{-\infty}^{\infty} dx^m\, w(x)(f \circ \Phi^{-1})(x)$$

$$(2.6.4)$$

The value of this integral does not change under a diffeomorphism, since by (2.4.17) w gets multiplied by $\text{Det}(\partial x^i / \partial \bar{x}^j)$, and

$$\int d^m \bar{x}\, \text{Det}\left(\frac{\partial x^i}{\partial \bar{x}^j}\right) f(x(\bar{x})) = \int d^m x\, f(x).$$

(See [(1), 16.22.1].)

More generally, if $N = \text{supp}\, f$ is compact†, then finitely many charts (U_i, Φ_i) of an atlas can be chosen such that $N \subset \bigcup_i U_i$. By the use of a partition of unity (see, e.g., [(1), 12.6.4]), f may be written as $f = \sum_i f_i$, where $\text{supp}\, f_i \subset U_i$. This enables us to make

Definition (2.6.5)

The **integral** of an m-form $f\Omega$ with compact support on an orientable manifold M is

$$\int_M \Omega f = \sum_i \int \Phi_i^*(\Omega f_i),$$

where $f = \sum_i f_i$, f_i is of compact support on the domain of the chart (U_i, Φ_i), and the integrals summed over are given by (2.6.4).

---

† Since infinite regions are diffeomorphic to finite ones, compact sets take on the role of sets of finite size.

Remarks (2.6.6)

1. Since the sum is finite and $w \in C^\infty$, there are no questions of convergence.
2. For all C_0^∞-functions f, $\int \Omega f$ is a linear functional bounded by $\sup|f| \cdot$ some constant depending only on supp f, and so defines a measure on M.
3. If ω is a p-form and N an orientable p-dimensional submanifold of M, then $\int_N \omega$ is defined by (2.6.5) with $\omega_{|N}$.
4. There is no meaning independent of the charts for an integral over other tensor fields.

If $M = (a, b)$ and ω is the 1-form df with supp $f \subset M$, then

$$\int df = \int_a^b dx \, \frac{\partial f}{\partial x} = 0,$$

because f vanishes at the boundary. Without the condition on the support of f, $\int df = f(b) - f(a)$. If we make the immediate extension of definition (2.6.5) to manifolds with boundaries, this rule generalizes to

Stokes's theorem (2.6.7)

Let M be an orientable m-dimensional manifold with a boundary and ω be an $m - 1$-form with compact support. Then

$$\int_M d\omega = \int_{\partial M} \omega.$$

Remarks (2.6.8)

1. It does not need to be assumed that ∂M is orientable, since the orientation of M induces one on ∂M. Indeed, it is a consequence of the proof of the theorem that if on some chart of the form (2.1.20) the orientation of M is given by $w(x) \, dx^1 \wedge dx^2 \wedge \cdots \wedge dx^m$, $w > 0$, then we ascribe the orientation $-dx^2 \wedge \cdots \wedge dx^m$ to ∂M. The sign is important, for, if it were reversed, (2.6.7) would be false: for $M = [0, \infty)$,

$$\int_0^\infty \frac{df}{dx} \, dx = -f(0).$$

2. The requirement of a compact support is necessary even if M is a finite part of $\mathbb{R}^n$. E.g., $M = (a, b)$, $\partial M = \varnothing$, $f = x$, and

$$\int_a^b df = b - a \neq \int_{\partial M} f = 0.$$

3. Note that the rule $d \circ d = 0$ follows from the fact that a boundary has no boundary: Let V be a compact submanifold of M with a boundary. Then

$$\int_V d \circ d\omega = \int_{\partial V} d\omega = \int_{\partial \partial V} \omega = 0.$$

It is easy to convince oneself that an m-form vanishes if its integral over every compact submanifold with a boundary vanishes, and hence that $d \circ d = 0$.

Proof

Let us again write $\int d\omega = \sum_i \int d\omega_i$, where each ω_i has compact support in the domain U_i of a chart of the form of (2.1.20); then it suffices to show that $\int_M d\omega_i = \int_{\partial M} \omega_i$. On a chart,

$$\Phi_i^* \omega_i = \sum_{j=1}^{m} g_j \, dx^1 \wedge \cdots \wedge (dx^j)^{\natural} \wedge \cdots \wedge dx^m,\dagger$$

and we choose $dx^1 \wedge dx^2 \wedge \cdots \wedge dx^m$ as the orientation. Then

$$\int_M d\omega_i = \sum_{j=1}^{m} (-1)^{j+1} \int_0^{\infty} dx^1 \int_{-\infty}^{\infty} dx^2 \cdots \int_{-\infty}^{\infty} dx^m \frac{\partial g_j}{\partial x^j}$$

$$= - \int_{-\infty}^{\infty} dx^2 \cdots \int_{-\infty}^{\infty} dx^m \, g_1(0, x^2, \ldots, x^m).$$

On the other hand we know (cf. (2.6.8; 1)) that

$$\int_{\partial M} \omega_i = - \int_{-\infty}^{\infty} dx^2 \cdots \int_{-\infty}^{\infty} dx^m \, g_1(0, x^2, \ldots, x^m),$$

because the restriction of dx^1 to ∂M vanishes, so that

$$\omega_{i|\partial M} = g_1 \, dx^2 \wedge \cdots \wedge dx^m. \qquad \qquad \square$$

Examples (2.6.9)

1. $M = \{(x, y) \in \mathbb{R}^2, \frac{1}{2} \leq x^2 + y^2 \leq 1\}$, $\omega = \dfrac{-y \, dx + x \, dy}{x^2 + y^2}$, $d\omega = 0$,

(cf. (2.5.6; 2))

$$0 = \int_{x^2+y^2=1} \omega - \int_{x^2+y^2=1/2} \omega = 2\pi - 2\pi.$$

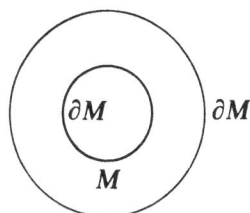

∂M ∂M M

---

† The symbol $(dx^j)^{\natural}$ indicates that the j-th differential is missing.

It is again apparent that the compact support of ω is essential, as otherwise one could take ω on $M = \{(x, y) \in \mathbb{R}^2, 0 < x^2 + y^2 \leq 1\}$, $\partial M = S^1$, and get the contradiction $0 = 2\pi$. We also see that ω can not be exact, as $\omega = dv$ would imply

$$2\pi = \int_{S^1} \omega = \int_{S^1} dv = \int_{\partial S^1} v = 0, \quad \text{since } \partial S^1 = \varnothing.$$

2. $C =$ any one-dimensional submanifold with a boundary in $\mathbb{R}^3$, $\partial C = \{a, b\}$,

$$\int_C df = \int_{\partial C} f, \quad \text{or} \quad \int_C ds \cdot \nabla f = f(b) - f(a).$$

3. $M =$ a two-dimensional submanifold of $\mathbb{R}^3$ (not necessarily a part of a plane), and ω is the 1-form $\mathbf{w}$. In vector notation (2.6.7) reads

$$\int_M d\mathbf{S} \cdot \nabla \times \mathbf{w} = \int_{\partial M} d\mathbf{s} \cdot \mathbf{w}.$$

4. $M =$ a three-dimensional manifold with a boundary of $\mathbb{R}^3$, and ω is the 2-form $*\mathbf{w}$. We obtain Gauss's theorem:

$$\int_M dV \nabla \cdot \mathbf{w} = \int_{\partial M} d\mathbf{f} \cdot \mathbf{w}.$$

In order to discover the relationship of the Lie derivative to integration, recall that the integral is invariant under diffeomorphisms:

$$M_1 \overset{\Phi}{\rightarrow} M_2 : \int_{M_1} \omega = \int_{M_2} \Phi^* \omega. \tag{2.6.10}$$

If Φ is specifically a flow on $M = M_1 = M_2$, then the infinitesimal version of (2.6.10) is

$$\int_M L_X \omega = 0, \quad X \in \mathcal{T}_0^1, \quad \omega \in E_m(M). \tag{2.6.11}$$

These facts have physically interesting formulations when we consider an m-form Ω which is invariant under the flow. This is the case for the Hamiltonian flows we shall be interested in, as they leave the Liouville measure $dq_1 \cdots dq_m \, dp_1 \cdots dp_m$ of phase space invariant (see § 3.2).

Incompressibility of the Flow (2.6.12)

Let Φ_t be a flow on M and Ω an m-form such that $\Phi_t^* \Omega = \Omega$. Then $\forall f \in C^\infty(M)$,

$$\int \Omega \cdot f = \int \Omega \cdot f \circ \Phi_t.$$

Proof

Use (2.6.10) and the fact that

$$\Phi_t^*(\Omega \cdot f) = \Phi_t^*\Omega \cdot \Phi_t^*f = \Omega \cdot f \circ \Phi_t.$$

This holds for all measurable functions. If f is the characteristic function χ_A of a set A, then the equation states that the volume of the set, as measured by Ω, stays unchanged during the time-evolution. The motion thus resembles that of an incompressible fluid.   □

Poincaré's Recurrence Theorem (2.6.13)

Let $A \subset M$, $\Phi_t(A) \subset A \; \forall t \in \mathbb{R}$, and $\Omega(A) \equiv \int \Omega \chi_A < \infty$. If $\Phi_t^*\Omega = \Omega$, then for almost every† point p of A, the trajectory through p returns infinitely often to each of its neighborhoods.

Proof

Let $B \subset A$ be an arbitrary measurable set, $\Omega(B) > 0$, and let $\tau \in \mathbb{R}^+$ be a unit of time. $K_n = \bigcup_{j=n}^{\infty} \Phi_{-j\tau}(B)$, j and $n \in \mathbb{Z}^+$, is the set of points that enter B after n or more time units (and possibly earlier as well). We clearly have the inclusions $B \subset K_0 \supset K_1 \supset \cdots \supset K_{n-1} \supset K_n$. The set of points of B that return after arbitrarily long times is $B \cap (\bigcap_{n \geq 0} K_n)$. This is disjoint from the set of points which do not return infinitely often, but, instead, are in B for a last time, and never come back. We want to show that the measure of the first set equals the measure of B. By assumption,

$$\Omega(K_n) = \Omega(\Phi_\tau K_n) = \Omega(K_{n-1}) < \infty,$$

because of the successive ordering by inclusion of the K_n's, and

$$\Omega\left(B \cap \left(\bigcap_{n \geq 0} K_n\right)\right) = \Omega(B \cap K_0) - \sum_{j=1}^{\infty} \Omega(B \cap (K_{i-1} \backslash K_i)) = \Omega(B),$$

since $B \cap K_0 = B$, and $K_{n-1} \supset K_n$ and $\Omega(K_n) = \Omega(K_{n-1}) \Rightarrow \Omega(K_{n-1} \backslash K_n) = 0$. Hence the measure of the arbitrary measurable set B equals that of the set of its points that return to B infinitely often.   □

Under the right circumstances conservation of energy provides a time-invariant submanifold of finite volume in phase space, for which the theorem applies. However, invariant regions of finite measure for unbounded forces (1.1.2) and more than two particles are not known, as the trajectories for which particles escape to infinity fill up a large portion of phase space. The strongest theorem is as follows.

---

† With respect to Ω.

Schwarzschild's Capture Theorem (2.6.14)

Assume again that $\Phi_t^*\Omega = \Omega$ and let $A \subset \Omega$ be measurable, $\Omega(A) < \infty$. Then for almost every† point $p \in A$, if the trajectory through p will always remain in A in the future, it must always have been in A in the past.

Proof

Let $A_\pm = \bigcap_{\tau \geqslant 0} \Phi_\tau(A)$, the set of points which will remain in A forever, or respectively which have always been in A. Then

$$\Omega(A_+) = \Omega(\Phi_{-t} A_+) = \Omega\left(\bigcap_{\tau > -t} \Phi_\tau(A)\right)$$

$$= \Omega\left(\bigcap_{-\infty < \tau < \infty} \Phi_\tau(A)\right) = \Omega(A_+ \cap A_-) = \Omega(A_-);$$

hence

$$\Omega(A_+ \setminus A_+ \cap A_-) = \Omega(A_- \setminus A_+ \cap A_-) = 0.$$

Trajectories that come from infinity and get bound in A, or, conversely, those that leave A forever, having formerly always been in A, can thus compose at most a subset of A of measure zero. Of course, the system could be unstable and $\Omega(A_+) = 0$. $\square$

Remarks (2.6.15)

1. The invariance of Ω is the basis of what is called ergodic theory, which is concerned with establishing the existence of the time-average

$$f_\infty = \lim_{T \to \infty}\left(\frac{1}{T}\right) \int_0^T dt\, \tau_t f$$

for functions f in general‡ (the ergodic theorems of Birkhoff and von Neumann). Because the spectral theory of operators in Hilbert space is used in the proof, we defer it until a later volume (*Quantum Mechanics of Large Systems*), where it appears as a special case.
2. The utility of this theorem for physics is limited, in that it does not state how long it will be before a trajectory returns or before a time-average approaches its asymptotic value.
3. Ω serves only to measure the probabilities of various configurations. This interpretation of Ω seems well justified for the Liouville measure.

---

† With respect to Ω.

‡ The function f need not even be C^∞. Measurability is sufficient for the almost everywhere existence of the time average.

We sum up by collecting the important formulas for differentiation and integration:

$$\int_M d\omega = \int_{\partial M} \omega$$

$$d(\alpha_i \omega_i) = \alpha_i \, d\omega_i, \qquad \alpha_i \in \mathbb{R}, \; \omega_i \in E_m, \qquad \int \alpha_i \omega_i = \alpha_i \int \omega_i,$$

$$d(\omega_1 \wedge \omega_3) = d\omega_1 \wedge \omega_2 + (-1)^p \omega_1 \wedge d\omega_2,$$

$$d \circ d\omega = 0, \qquad d\Phi^* \omega = \Phi^* \, d\omega,$$

$$\partial(M_1 \times M_2) = \partial M_1 \times M_2 \cup M_1 \times \partial M_2,$$

$$\partial \partial M = \varnothing, \qquad \int_{\Phi M} \Phi^* \omega = \int_M \omega$$

$$\int L_X \omega = 0, \qquad \text{if} \quad \Phi_t^X M = M. \tag{2.6.16}$$

Problems (2.6.17)

1. Show that the Liouville measure $d^m q \, d^m p \equiv dq_1 \wedge \cdots \wedge dq_m \wedge dp_1 \wedge \cdots \wedge dp_m$ is invariant under a point transformation $q \to \bar{q}$.

2. Work through Example (2.6.9; 2), following the steps of the definition of the integral.

3. Using (2.6.7), prove the following theorem: $\oint_C dz \, f(z) = 0$ if the path of integration C encircles a region in a neighborhood of which f is a meromorphic function without poles of the first order.

4. Show that orientability is equivalent to the existence of an atlas $\bigcup_i (U_i, \Phi_i)$ with $d_{ij} \equiv \mathrm{Det} \, D(\Phi_i \circ \Phi_j^{-1}) > 0 \; \forall i, j$ such that $U_i \cap U_j \neq \varnothing$.

5. Two C^∞-mappings f and $g: M \to N$ of two manifolds are said to be homotopic to each other iff there exists a C^∞-mapping $F: [0, 1] \times M \to N$, such that $f = F \circ i_0$ and $g = F \circ i_1$, where i_0 and i_1 are the imbeddings $i_0: M \to \{0\} \times M$ and $i_1: M \to \{1\} \times M$. Show that if M and N are orientable, compact, and n-dimensional, then $\forall \omega \in E_n(N)$,

$$\int_M f^* \omega = \int_M g^* \omega,$$

if f and g are homotopic. (First show that $\varphi \in E_n$ and closed $\Rightarrow g^* \varphi - f^* \varphi$ is exact.)

6. Use Problem 5 to prove the theorem that you can't comb a hedgehog: If n is even, then every C^∞-vector field X on S^n has at least one point where it vanishes.

7. The hydrodynamic equations, $\dot{v}_i + v_k v_{i|k} = -p_{,i}$ are written as $\dot{v} + L_v v = d(v^2/2 - p)$ (cf. (2.5.12; 5)), if we construe v as a 1-form on $\mathbb{R}^3$ and denote the Lie derivative with respect to v by L_v, by making the covariant vector field contravariant with the metric $g_{ik} = \delta_{ik}$. Let C_t be a closed curve that follows the flow of v. Show that

$$\frac{d}{dt} \int_{C_t} v = 0. \quad \text{(Thompson's theorem)}$$

8. For a divergence-free vector field $(E \in \mathcal{T}_0^1(\mathbb{R}^3), g_{ik} = \delta_{ik}, d*E = 0)$, show that the field strength is proportional to the density of the lines of force in the following sense: Lines of force are the trajectories of Φ_t^E, so that the same number of lines of force pass through both N and $\Phi_t^E N$. For $E \| df$,

$$\int_N {}^*E = \int_N \mathbf{E} \cdot d\mathbf{S} \propto \frac{\text{field strength}}{\text{density of lines of force}},$$

because (surface area) $\times$ (density of lines of force) is constant. Prove the invariance of this relationship under Φ_t^E.

Solutions (2.6.18)

1.
$$d^m q = d^m \bar{q} \, \text{Det}\left(\frac{\partial q_i}{\partial \bar{q}_j}\right), \qquad d^m p = d^m \bar{p} \, \text{Det}\left(\frac{\partial \bar{q}_i}{\partial q_j}\right).$$

2. The 1-form df should be integrated along the curve $u: I = (a, b) \to \mathbb{R}^3$. To do this we use the chart $(u(I), u^{-1})$ on this one-dimensional submanifold. Then

$$\int_{u(I)} df = \int_I (u^{-1})^* \, df = \int_I d((u^{-1})^* f) = \int_I d(f \circ u) = \int_{\partial I} (f \circ u)$$

$$= f(u(b)) - f(u(a)).$$

3. It follows from the assumptions that $f(z) = (\partial/\partial z)F(z)$, so

$$\int_C dz \, f(z) = \int_C dF = \int_{\partial C} F = 0,$$

since $\partial C = \emptyset$ for a closed path C.

4. $\Rightarrow$: Choose Φ_i such that in the coordinate system $(x_1, \ldots, x_m) = \Phi_i(x)$, $\Omega = g_i dx_1 \wedge \cdots \wedge dx_m, g_i > 0$. Then $d_{ij} = g_j/g_i > 0$.
$\Leftarrow$: Let $\omega_i = dx_1 \wedge \cdots \wedge dx_m$ on U_i, and let x_k be as above. Construct $\sum_i \omega_i f_i$, where f_i is a partition of unity as in (2.6.5). Given $x \in M$, let I be the set of i such that $f_i(x) \neq 0$, and fix $i_0 \in I$. Then since f_i and $d_{ii_0}(x) > 0$ by assumption,

$$\omega(x) = \sum_{i \in I} [f_i(x) d_{ii_0}(x)] \omega_{i_0}(x) \neq 0.$$

5. In analogy with (2.5.12; 7), define a mapping $K: E_{p+1}(I \times M) \to E_p(M)$ such that $\omega_0 + dt \wedge \omega_M \to \int_0^1 dt \, \omega_M$. This mapping satisfies $d \circ K + K \circ d = i_1^* - i_0^*$. Hence $d \circ K \circ F^* + K \circ F^* \circ d = (d \circ K + K \circ d) \circ F^* = (i_1^* - i_0^*) \circ F^* = (F \circ i_1)^* - (F \circ i_0)^* = g^* - f^*$. If $d\varphi = 0$, then $g^* \varphi - f^* \varphi = d \circ K \circ F^* \varphi$ is exact. But then $\int_M (g^* \omega - f^* \omega) = 0$, because ω, as an n-form, is closed, and M has no boundary.

6. Imagine S^n imbedded in $\mathbb{R}^{n+1}$ and $T_q(S^n)$ equipped with the corresponding Riemannian structure, and let $X(x) \neq 0 \ \forall x \in S^n$. If we replace X with $X/\|X\|$, then we can treat it as a mapping from S^n to S^n. Let $F: [0, 1] \times S^n \to S^n$ be given by $(t, x) \to x \cos \pi t + X(x) \sin \pi t$ (note that $X(x) \perp x$); this furnishes a homotopy between $x \to x$ and

$x \to -x$ (the antipodal mapping a). The homotopy, however, reverses the orientation of S^n for even n, so that by Problem 5 we would conclude that

$$\int_{S^n} \omega = \int_{S^n} a^* \omega = - \int_{S^n} \omega, \quad \forall \omega \in E_n(S^n),$$

which is absurd.

7. Since C_t follows the flow of v, $\int_{C_{t+\tau}} v = \int_{C_t} e^{\tau L_v} v$, which contributes the term $\int_{C_t} L_v v$ to the derivative:

$$\frac{d}{dt} \int_{C_t} v = \int_{C_t} (\dot{v} + L_v v) = \int_{C_t} d\left(\frac{v^2}{2} - p\right) = 0.$$

8. Let M be the cylinder spanned by N and $\Phi_t^E N$ (see figure). $\partial M = N \cup \Phi_t^E N \cup$ the outer surface. Since $E \perp df$ on the outer surface,

$$0 = \int_M d^* E = \int_{\Phi_t^E N} {}^* E - \int_N {}^* E.$$

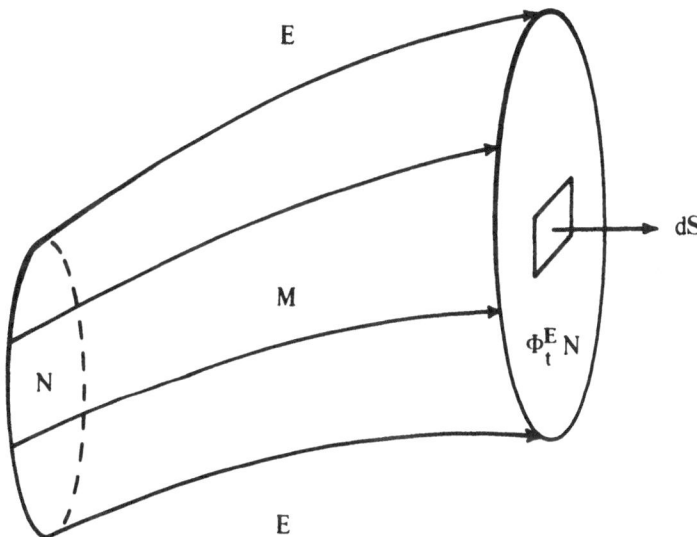

The cylinder spanned by the lines of force.

3 Hamiltonian Systems

3.1 Canonical Transformations

A 2-form is canonically defined on the cotangent bundle of a manifold. Diffeomorphisms leaving this 2-form invariant are called canonical transformations.

The differential equations of the mechanics of point particles are of second order, and, as explained in §2.3, they generate a local flow on the tangent bundle. We are now going to pass from this Lagrangian form of the equations of motion to the Hamiltonian form, which determines the corresponding flow on the cotangent bundle. This flow has the special property of leaving the canonical 2-form (defined below) invariant.

We start by constructing a 1-form on $T^*(M)$ out of its projection Π, without requiring any additional structure. We have already written[†]

$$\Pi : (q, p) \in T^*(M) \to q \in M.$$

So

$$T(\Pi) : \left(q, p; \frac{\partial}{\partial q}, \frac{\partial}{\partial p} \right) \in T(T^*(M)) \to \left(q, \frac{\partial}{\partial q} \right) \in T(M).$$

On the fibers, $T_{(q, p)}(\Pi)$ is simply the linear transformation having the matrix

---

† Dropping the indices for coordinates and vectors.

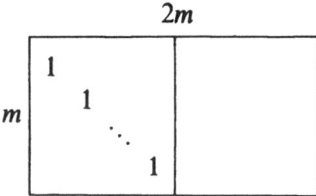

The transposed matrix $T^t_{(q, p)}(\Pi)$:

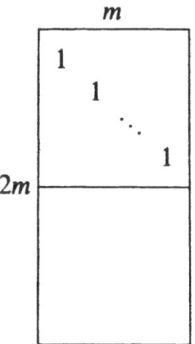

maps the dual spaces into each other by $p \in T^*_q(M) \rightarrow (p, 0) \in T^*_{(q, p)}(T^*(M))$. On the basis of a chart this last vector is $\sum p_i \, dq^i$. Since the bundles inherit this map from the fibers, we make the following

Definition (3.1.1)

The 1-form $\Theta: (q, p) \in T^*(M) \rightarrow (q, p; p_i \, dq^i) \in T^*(T^*(M))$ and its differential $\omega \equiv -d\Theta$ are called the **canonical forms** on $T^*(M)$.

Remarks (3.1.2)

1. In the sense of Example (2.4.3; 2) one often writes $\Theta = p_i \, dq^i$ and $\omega = dq^i \wedge dp_i$.
2. The forms are chart-independent. More generally, under diffeomorphisms $\Phi: M_1 \rightarrow M_2$, for which $\Psi \equiv T^*(\Phi): T^*(M_1) \rightarrow T^*(M_2)$, the canonical forms $\Theta_{1, 2}$ and $\omega_{1, 2}$ are transformed into each other: $\Psi^*\Theta_1 = \Theta_2$, and $\Psi^*\omega_1 = \omega_2$ (Problem 8).
3. The 1-form Θ vanishes at the origin of $T^*(M)$, but ω is always nonzero. The m-fold exterior product,

$$\Omega \equiv \frac{(-1)^{(m-1)m/2}}{m!} \underbrace{\omega \wedge \omega \wedge \cdots \wedge \omega}_{m \text{ times}}$$

is a $2m$-form which likewise never vanishes, since on a chart it is $dq_1 \wedge dq_2 \wedge \cdots \wedge dq_m \wedge dp_1 \wedge \cdots \wedge dp_m$ (Liouville measure).

As we have seen, diffeomorphisms of a manifold, also known as point transformations, leave the canonical forms invariant. However, we can ask about more general transformations of $T^*(M)$ in which the new q depends on the old q and p, but which leave ω invariant.

Definition (3.1.3)

A diffeomorphism $T^*(M_1) \supset U_1 \xrightarrow{\Psi} U_2 \subset T^*(M_2)$ that takes the canonical 2-form $\omega_{|U_1}$ to $\omega_{|U_2}$ is called a **local canonical transformation**. If $U_1 = U_2 = T^*(M)$, then Ψ is called a **canonical transformation**.

Remarks (3.1.4)

1. Point transformations are canonical.
2. An example of a canonical transformation that is not a point transformation is the interchange $(q, p) \rightarrow (p, -q)$ on $M = \mathbb{R}$, where $T^*(M) = \mathbb{R} \times \mathbb{R}$ and $\omega = dq \wedge dp$.
3. The interchange $(q, p) \rightarrow (p, q)$ changes the sign of ω; thus not every linear transformation is canonical.
4. Because $(\Psi_1 \circ \Psi_2)^* = \Psi_1^* \circ \Psi_2^*$, the canonical transformations form a group.
5. Since $\Psi^*(\omega_1 \wedge \omega_2) = (\Psi^*\omega_1) \wedge (\Psi^*\omega_2)$, canonical transformations also leave the $2m$-form Ω invariant.

In general the 1-form Θ will change under a canonical transformation. However, it is still true that $\Psi^* \, d\Theta - d\Theta = d(\Psi^*\Theta - \Theta) = 0$, so that, at least in a neighborhood U, $\Psi^*\Theta = \Theta + d\tilde{f}$, for some $\tilde{f} \in C^\infty(U)$. Letting $\Psi: (\bar{q}, \bar{p}) \rightarrow (q, p)$, on some chart, this formula may be written

$$p_i \, dq^i = \bar{p}_i \, d\bar{q}^i + d\tilde{f} = -\bar{q}^i \, d\bar{p}_i + df,$$
$$f(\bar{q}, \bar{p}) = \tilde{f}(\bar{q}, \bar{p}) + \bar{q}^i \bar{p}_i. \tag{3.1.5}$$

If we want an explicit expression for Ψ, we must evaluate (3.1.5) on a basis. If

$$\mathrm{Det}\left(\frac{\partial q^i}{\partial \bar{q}^j}\right)_{|\bar{p}\,\text{constant}} \neq 0,$$

which holds, for instance, for a point transformation or for Ψ sufficiently close to 1, it will suffice to express everything in terms of dq^i and $d\bar{p}_j$ and to equate coefficients. Then we may write $\bar{q}$ locally as a function of q and $\bar{p}$ by inverting $q(\bar{p}, \bar{q})$. If we also call $f(\bar{q}(q, \bar{p}), \bar{p})$ simply $f(q, \bar{p})$ and plug it into (3.1.5), we obtain

Lemma (3.1.6)

A local canonical transformation $\Psi: (\bar{q}, \bar{p}) \rightarrow (q, p)$ *with*

$$\mathrm{Det}\left(\frac{\partial q^i}{\partial \bar{q}^j}\right)_{|\bar{p}\,\text{constant}} \neq 0$$

may be written locally as

$$p_i = \frac{\partial f}{\partial q^i}, \qquad \bar{q}^i = \frac{\partial f}{\partial \bar{p}_i}, \qquad f(q, \bar{p}) \in C^\infty.$$

The function f is known as the local generator. Conversely, if $f(q, \bar{p}) \in C^\infty(U)$ such that $\mathrm{Det}(\partial^2 f / \partial q^i \, \partial \bar{p}_j) \neq 0$ is given, then the above equations define a local canonical transformation.

Remarks (3.1.7)

1. The canonical transformation $q^i = \bar{p}_i$, $p_i = -\bar{q}^i$ on $T^*(M) = \mathbb{R}^{2m}$ is not induced in this way. The construction fails because $\partial q^i / \partial \bar{q}^j|_{\bar{p}} = 0$.
2. If $\mathrm{Det}(\partial q^i / \partial \bar{p}_j) \neq 0$, then $\bar{p}(q, \bar{q})$ may be calculated locally. Substituting into $\bar{f}$, we obtain the alternative form

$$p_i = \frac{\partial \bar{f}}{\partial q^i}, \qquad \bar{p}_i = -\frac{\partial \bar{f}}{\partial \bar{q}^i}.$$

Point transformations can not be written like this.
3. We learn from integration theory (§2.6) that integrals over ω and Ω are left invariant by canonical transformations, where of course the new integral is taken over the image of the original integration region:

$$\int_N \omega = \int_{\Psi N} \omega, \qquad \int_U \Omega = \int_{\Psi U} \Omega,$$

in which N and U are respectively 2-dimensional and $2m$-dimensional submanifolds of M. Nothing like this is necessarily true for Θ. However, if C is a one-dimensional submanifold without a boundary, contained in some neighborhood in which Equation (3.1.5) holds, that is, C is a sufficiently small closed curve, then

$$\int_{\Psi^{-1} C} \Theta = \int_C \Theta + \int_{\partial C} \bar{f} = \int_C \Theta.^\dagger$$

It is easy to go astray for arbitrary closed curves (Problem 6).

Examples (3.1.8)

1. $p = \sqrt{2\omega\bar{p}} \cos \bar{q}, q = \sqrt{2\bar{p}/\omega} \sin \bar{q}, \omega \in \mathbb{R}^+$, is a local canonical transformation from $\mathbb{R}^+ \times S^1 \subset T^*(S^1)$ to $\mathbb{R}^2 \setminus \{0\} \subset T^*(\mathbb{R})$. Calculating

$$dp = \tfrac{1}{2} d\bar{p} \cos \bar{q} \sqrt{2\omega/\bar{p}} - d\bar{q}\sqrt{2\omega\bar{p}} \sin \bar{q},$$
$$dq = \tfrac{1}{2} d\bar{p}\sqrt{2/\omega\bar{p}} \sin \bar{q} + d\bar{q}\sqrt{2\bar{p}/\omega} \cos \bar{q},$$

we see

$$dq \wedge dp = d\bar{q} \wedge d\bar{p}.$$

This can obviously not be continued to a canonical transformation.

---

† The forms ω and Ω are called **integral invariants**, and Θ is called a **relative integral invariant**.

2. We would like to determine when a linear transformation on $T^*(\mathbb{R}^m) = \mathbb{R}^{2m}$ is canonical. Let us treat $(q^1, \ldots, q^m, p_1, \ldots, p_m)$ as a single vector $(x_1, \ldots, x_{2m})$ and write

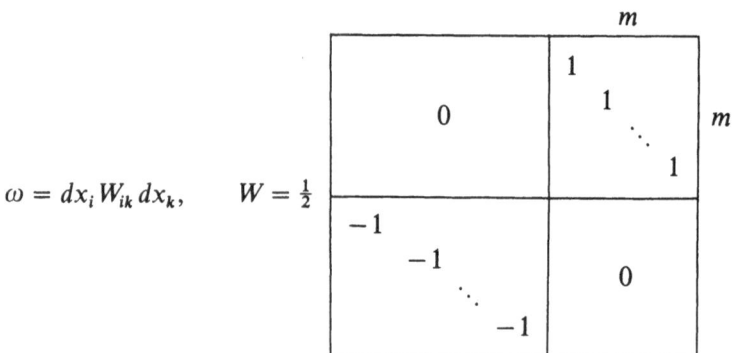

$$\omega = dx_i \, W_{ik} \, dx_k, \qquad W = \tfrac{1}{2}$$

A linear transformation $x_k = L_{kj} \bar{x}_j$ is canonical in case $L'WL = W$, that is, L is a symplectic matrix. Symplectic matrices have the following properties (see Problem 4):

(a) Det L must be either $+1$ or -1.
(b) If λ is an eigenvalue, then so are $1/\lambda$, λ^*, and $1/\lambda^*$.

The canonical 2-form is given everywhere as the invertible matrix W. Hence ω, just like g from (2.4.13; 2), defines a nondegenerate bilinear form, and consequently a bijection from $\mathcal{T}_1^0(T^*(M))$ to $\mathcal{T}_0^1(T^*(M))$.

Definition (3.1.9)

The bijection produced by ω assigns to $v^* \in T^*_{(q, p)}(T^*(M))$ the vector $bv^* \in T_{(q, p)}(T^*(M))$ such that for all $w \in T_{(q, p)}(T^*(M))$, the equation $(v^* | w) = \omega_{(q, p)}(bv^*, w)$ holds. The corresponding bijection from $\mathcal{T}_1^0(T^*(M))$ to $\mathcal{T}_0^1(T^*(M))$ will also be given the symbol $b: b(q \to (q, v^*(q))) = q \to (q, bv^*(q))$. The vector fields $X_H \equiv b(dH)$, $H \in C^\infty(T^*(M))$ are said to be **Hamiltonian**. If H can only be defined locally, X_H is said to be **locally Hamiltonian**.

Remarks (3.1.10)

1. The action of b can be expressed schematically, in that the diagram

$$
\begin{array}{ccc}
\mathcal{T}_0^1(T^*(M)) \times \mathcal{T}_0^1(T^*(M)) & \xrightarrow{\omega} & \\
{\scriptstyle b \times 1} \uparrow & & C^\infty(T^*(M)) \\
\mathcal{T}_1^0(T^*(M)) \times \mathcal{T}_0^1(T^*(M)) & \xrightarrow{(|)} &
\end{array}
$$

commutes.

2. On a chart, $b(v_{q^i}^* \, dq^i + v_{p_i}^* \, dp_i) = v_{p_i}^* \partial q^i - v_{q^i}^* \partial p_i$, and

$$X_H : (q, p) \rightarrow \left(q, p; \left(\frac{\partial H}{\partial p_i} \right) \partial q^i, \; -\left(\frac{\partial H}{\partial q^i} \right) \partial p_i \right).$$

Note that X_H is exactly of the form of the vector field that came up in (2.3.25).

3. The distinction made in Definition (3.1.9) between Hamiltonian and locally Hamiltonian vector fields is easy to illustrate. Let $M = T^1$, and $T^*(M) = T^1 \times \mathbb{R} \ni (\varphi, p)$, and let X be the vector field $(\varphi, p) \rightarrow (\varphi, p; 0, 1)$. It is locally Hamiltonian with $H = -\varphi$, but not Hamiltonian, because φ is not defined globally.

4. If Ψ is a canonical transformation, then $\Psi^* \circ b = b \circ \Psi^*$, because $\Psi^* \omega = \omega$. Hence $\Psi^* X_H = b \Psi^* \, dH = bd(H \circ \Psi^{-1}) = X_{H \circ \Psi^{-1}}$. For practical purposes this means that Hamilton's equations in the new system are obtained simply by substituting into H.

Still another mapping from $C^\infty(T^*(M)) \times C^\infty(T^*(M))$ to $C^\infty(T^*(M))$ can be defined using ω, by applying it to the vector fields associated with two functions.

Definition (3.1.11)

The **Poisson bracket** of two functions F and G in $C^\infty(T^*(M))$ is defined by

$$\{G, F\} \equiv \omega(X_G, X_F) = i_{X_F} i_{X_G} \omega = i_{X_F} \, dG = L_{X_F} G = -L_{X_G} F.$$

Remarks (3.1.12)

1. In the natural basis,

$$i_{X_G} \omega = (X_G)_{q^i} \, dp_i - (X_G)_{p_i} \, dq_i,$$

from which the relationship used above, $i_{X_G} \omega = dG$, follows. The last equality of (3.1.11) uses the antisymmetry of ω, that is, $\{G, F\} = -\{F, G\}$.

2. On a chart,

$$\{G, F\} = \sum_{i=1}^\infty \left(\frac{\partial G}{\partial q^i} \frac{\partial F}{\partial p_i} - \frac{\partial F}{\partial q^i} \frac{\partial G}{\partial p_i} \right),$$

and in particular, $\{q^i, q^j\} = \{p_i, p_j\} = 0$, and $\{q^i, p_j\} = \delta_{ij}$.

3. Poisson brackets are invariant under canonical transformations Ψ in the sense that in the new system they just become the Poisson brackets of the new functions: By (3.1.10; 4), $\{F, G\} \circ \Psi = \{F \circ \Psi, G \circ \Psi\}$. Conversely, if Ψ is a diffeomorphism of $T^*(M)$ that satisfies this equation for all functions F and G, then Ψ is canonical. To see this, choose functions that equal q^i and p_i on the domain of a chart; then the transformed coordinates are $\bar{q}^i \equiv q^i \circ \Psi$ and $\bar{p}_i \equiv p_i \circ \Psi$, and they also satisfy $\{\bar{q}^i, \bar{p}_j\} = \delta_{ij}$ and $\{\bar{q}^i, \bar{q}^j\} = \{\bar{p}_i, \bar{p}_j\} = 0$. Hence the transformed $\omega = d\bar{q}^i \wedge d\bar{p}_i$ and Ψ are canonical.

4. The rules (2.2.24) for the Lie derivative have the consequences that
$\{F + G, H\} = \{F, H\} + \{G, H\}$ and $\{F \cdot G, H\} = G\{F, H\} + F\{G, H\}$.

Problems (3.1.13)

1. Find the generator $\tilde{f}$ of the local canonical transformation $q = \sqrt{2\bar{p}/\omega} \sin \bar{q}$, $p = \sqrt{2\omega\bar{p}} \cos \bar{q}$.

2. Calculate $\{e^{\alpha q}, e^{\beta p}\}$.

3. Let us write the matrix L of the linear canonical transformation (3.1.8; 2) in block form

$$L = \begin{array}{|c|c|} \hline A & B \\ \hline C & D \\ \hline \end{array}.$$

What are the conditions for L to be symplectic, and what is the generator $\tilde{f}$ of the transformation?

4. Let λ be an eigenvalue of the symplectic matrix L. Show that $1/\lambda$ (and hence also λ^* and $1/\lambda^*$) is an eigenvalue. If L is an element of a one-parameter group generated by a function F, $L = e^{tF}$, then what does this imply about the eigenvalues of F?

5. Consider the flow on $T^*(M)$ generated by the canonical vector field $b\Theta$. Is it canonical?

6. Construct a canonical transformation Ψ for which $\Psi^*\Theta \neq \Theta + d\tilde{f}$, and hence $\int_C \Theta \neq \int_{\Psi C} \Theta$ for a closed curve C.

7. What is the form of the generator f (cf. (3.1.5)) of a point transformation $q \to \bar{q}(q)$?

8. Show that $\Psi^*\Theta = \Theta$ (and hence $\Psi^*\omega = \omega$ and $\Psi^*\Omega = \Omega$) for the point transformation $\Psi = T^*(\Phi)$, where Φ is a diffeomorphism of neighborhoods in M. Is this the most general local canonical transformation that leaves Θ invariant?

Solutions (3.1.14)

1. $\tilde{f} = \bar{p} \cos \bar{q} \sin \bar{q}$, because

$$p \, dq = \sqrt{2\omega\bar{p}} \cos \bar{q} \left(d\bar{p} \sqrt{\frac{1}{2\bar{p}\omega}} \sin \bar{q} + d\bar{q} \sqrt{\frac{2\bar{p}}{\omega}} \cos \bar{q} \right)$$

$$= \bar{p} \, d\bar{q} + d(\bar{p} \cos \bar{q} \sin \bar{q}).$$

Note that $\tilde{f}$ is defined globally on $T(S^1)$, but even so it generates only a local canonical transformation.

2.
$$\{e^{\alpha q}, p\} = \sum_{n=0}^{\infty} \frac{\alpha^n}{n!} \{q^n, p\} = \alpha \sum_{n=0}^{\infty} \frac{\alpha^n}{n!} q^n = \alpha e^{\alpha q},$$

$$\{e^{\alpha q}, p^m\} = p^{m-1}\{e^{\alpha q}, p\} + \{e^{\alpha q}, p^{m-1}\}p = m\alpha p^{m-1} e^{\alpha q},$$

$$\{e^{\alpha q}, e^{\beta p}\} = \sum_{m=0}^{\infty} \frac{\beta^m}{m!} \{e^{\alpha q}, p^m\} = \sum_{m=0}^{\infty} \frac{\beta^m}{m!} \alpha \beta e^{\alpha q} p^m = \alpha \beta e^{\alpha q + \beta p}.$$

3. The matrices $A^t C$ and $B^t D$ must be symmetric, and $A^t D - C^t B = 1$.

$$\tilde{f} = \bar{p} B^t C \bar{q} + \tfrac{1}{2}(\bar{q} C^t A \bar{q} + \bar{p} D^t B \bar{p}).$$

4. (a) Take the determinant of $L'WL = W$, to conclude that $(\text{Det } L)^2 = 1$.
   (b) $\text{Det}(L^{-1} - \lambda) = \text{Det}(W^{-1}L'W - \lambda) = \text{Det}(L' - \lambda) = \text{Det}(L - \lambda) = 0$, which implies that λ is an eigenvalue of L^{-1}. Therefore if φ is an eigenvalue of F, then so are $-\varphi$, φ^*, and $-\varphi^*$, since L is real.

5. $b\Theta = p_i \partial / \partial p_i$ generates the flow $q^i(t) = q^i(0)$, $p_i(t) = e^t p_i(0)$. No, it is not canonical.

6. $M = S^1$, $T^*(M) = S^1 \times \mathbb{R}$, and $\Psi : (\varphi, p) \to (\varphi, p + a)$, $a \in \mathbb{R}$, for both charts (2.1.7; 2) on S^1. Locally, $\tilde{f} = a\varphi$, but $\varphi \notin C^\infty(T^*(M))$. So for $C : t \in [0, 2\pi) \to (t, p) \in T^*(M)$,

$$\int_C \Theta = 2\pi p \neq \int_{\Psi C} \Theta = 2\pi(p + a).$$

Note that C is not the boundary of a surface. If S^1 were imbedded in $\mathbb{R}^2$, then C would be the boundary of a circular disc D, and one might be led to the false conclusion that

$$\int_C \Theta = -\int_D \omega = -\int_{\Psi D} \omega = \int_{\Psi C = \partial \Psi D} \Theta.$$

7. $f(\bar{p}, q) = \bar{p}_i \bar{q}^i(q)$.

8. By Problem 7, $f = \bar{p}\bar{q}(q)$, and so $\tilde{f} = 0$ and $\Psi^*\Theta = \Theta$. If on the other hand Θ is invariant, then $f = \bar{p}\bar{q}(q)$, and Ψ is a point transformation.

3.2 Hamilton's Equations

Hamiltonian vector fields generate local flows that leave ω invariant. These flows are determined by Hamilton's equations.

A local flow Ψ_t on $T^*(M)$ that leaves ω invariant, $\Psi_t^*\omega = \omega$, has a vector field X which generates it and satisfies $L_X \omega = 0$. It turns out that Hamiltonian vector fields (3.1.9) X_H, $H \in C^\infty(T^*(M))$, have this very property. To see that this is so, recall that the inverse of the bijection from (3.1.9), $b^{-1} : \mathcal{T}_0^1(T^*(M)) \to \mathcal{T}_1^0(T^*(M))$ is the mapping $X \to i_X \omega$ (2.4.33), and consequently one can write $dF = i_{X_F} \omega$ (see (3.1.12; 1)). Then using Property (2.5.9; 4) and the relationship $d\omega = -d \, d\Theta = 0$,

$$L_{X_F} \omega = d i_{X_F} \omega = d \, dF = 0. \tag{3.2.1}$$

Moreover, locally the argument goes the other way, too, since $0 = L_X \omega = d i_X \omega$, which implies $i_X \omega = dF$, for $F \in C^\infty(T^*(M))$, which then implies $X = X_F$. This proves

Theorem (3.2.2)

Every locally Hamiltonian vector field generates a local flow of canonical transformations. Conversely, any local flow of canonical transformations has a locally Hamiltonian generator.

Remark (3.2.3)

The above statement is false if the qualification "local" is dropped. A Hamiltonian vector field need not be complete even if H is defined globally. And conversely, the flow $\Psi_t: (\varphi, p) \to (\varphi, p + t)$ on $T^*(S^1) = S^1 \times \mathbb{R}$ has the generator φ, which is only defined locally (cf. (3.1.10: 3)).

Let us consolidate the foregoing results.

Definition (3.2.4)

For all $H \in C^\infty(T^*(M))$, **Hamilton's equations,**

$$\dot{u} = X_H \circ u,$$

define the **local canonical flow** Φ_t.

Remarks (3.2.5)

1. The variation in time of an observable is determined by its Poisson bracket with H:

$$\frac{d}{dt} F \circ \Phi_t = \{F \circ \Phi_t, H\}.$$

2. According to Remark (3.1.10; 4), Hamilton's equations are invariant under canonical transformations in the sense that in the new system it is necessary merely to use the transformed H. More explicitly, let $\Psi: (\bar{q}, \bar{p}) \to (q(\bar{q}, \bar{p}), p(\bar{q}, \bar{p}))$, and let us call $H_T(q, p) = H(\bar{q}(q, p), \bar{p}(q, p))$. Then the pairs of equations $\dot{\bar{q}}^i = \partial H/\partial \bar{p}_i, \dot{\bar{p}}_i = -\partial H/\partial \bar{q}^i$ and $\dot{q}^i = \partial H_T/\partial p_i, \dot{p}_i = -\partial H_T/\partial q^i$ are equivalent.
3. Because of the antisymmetry of the Poisson bracket, H is constant in time. For the same reason, any quantity that generates a local canonical flow that does not change H is constant in time.
4. Time-evolution leaves invariant not only ω, but also, as in Remark (3.1.4; 5), the phase-space volume Ω. (This is Liouville's theorem.[†])
5. If H and F are analytic, and X_H is complete, then the canonical flow may be written more explicitly following Remark (2.3.11; 2) as

$$F \circ \Phi_t = \sum_{n=0}^\infty \frac{t^n}{n!} \underbrace{\{\{\cdots \{\{F, H\}, H\} \cdots \}, H\}}_{n \text{ times}}.$$

---

† In the framework of classical mechanics the proof of this theorem requires some effort. But modern concepts are so formulated that there is really nothing to prove.

Examples (3.2.6)

1. The canonical transformation of Example (3.1.8; 1) transforms the Hamiltonian of the harmonic oscillator,

$$H = \tfrac{1}{2}(p^2 + \omega^2 q^2),$$

into

$$H_T = \bar{p}\omega.$$

With these coordinates the time-evolution is $\bar{p} = $ constant, $\bar{q}(t) = \bar{q}(0) + \omega t$. With the old coordinates, $q(t) = \sqrt{2\omega\bar{p}}\cos(\bar{q}(0) + \omega t)$ and $p(t) = \sqrt{2\bar{p}/\omega}\sin(\bar{q}(0) + \omega t)$, which is precisely the solution of the equations for p and q using H. Note that (i) Time-evolution is a rotation in phase space, (q, p), and therefore it leaves phase-space volumes invariant. (ii) This canonical transformation can be used even though it is only local, because it maps time-invariant regions into each other.

2. $M = \mathbb{R}^3$, $T^*(M) = \mathbb{R}^3 \times \mathbb{R}^3$, $H = |\mathbf{p}|^2$, and $G_1 = p_1$. The vector field X_{G_1} generates the flow $(q_1, q_2, q_3; p_1, p_2, p_3) \to (q_1 + \lambda, q_2, q_3; p_1, p_2, p_3)$. This leaves H invariant, which is equivalent to $G_1 = $ constant (conservation of momentum for a free particle).

3. M and H as in Example 2, but $G_2 = p_1 q_2 - q_1 p_2$. The vector field X_{G_2} generates the flow $(q_1, q_2, q_3; p_1, p_2, p_3) \to (q_1 \cos \lambda + q_2 \sin \lambda, - q_1 \sin \lambda + q_2 \cos \lambda, q_3; p_1 \cos \lambda + p_2 \sin \lambda, -p_1 \sin \lambda + p_2 \cos \lambda, p_3)$. The flow leaves H invariant, which is equivalent to $G_2 = $ constant (conservation of angular momentum for a free particle).

The fact that Hamiltonian vector fields generate (locally) canonical transformations establishes a connection between the Lie and Poisson brackets.

Theorem (3.2.7)

The Lie bracket (2.5.9; 6) *of two Hamiltonian vector fields is the Hamiltonian vector field of their Poisson bracket*: $[X_H, X_G] = X_{\{G, H\}}$.

Proof

Suppose that X_H generates the local flow Ψ_t. Differentiating $\Psi^*_{-t} X_G = X_{G \circ \Psi_t}$ by time at $t = 0$ and using (2.5.7), we obtain $L_{X_H} X_G = X_{L_{X_H} G}$. But $L_{X_H} X_G = [X_H, X_G]$, while $L_{X_H} G = \{G, H\}$. ☐

Remarks (3.2.8)

1. This can also be expressed as the commutativity of a diagram:

$$
\begin{array}{ccc}
\mathscr{T}^1_0(T^*(M)) \times \mathscr{T}^1_0(T^*(M)) & \xrightarrow{\;[\ \ ]\;} & \mathscr{T}^1_0(T^*(M)) \\[1em]
\Big\uparrow{\scriptstyle b \circ d \times b \circ d} & & \Big\uparrow{\scriptstyle b \circ d} \\[1em]
C^\infty(T^*(M)) \times C^\infty(T^*(M)) & \xrightarrow{\;-\{\ \}\;} & C^\infty(T^*(M))
\end{array}
$$

2. As with the Lie bracket, the Poisson bracket is not associative, but instead Jacobi's identity,

$$\{F, \{G, H\}\} + \{G, \{H, F\}\} + \{H, \{F, G\}\} = L_{X_F} L_{X_G} H$$
$$- L_{X_G} L_{X_F} H + L_{X_{\{F, G\}}} H$$
$$= (L_{[X_F, X_G]} + L_{X_{\{F, G\}}})H = 0$$

holds. Cf. (2.5.9; 6).

3. A consequence of Remark 2 is that the Poisson bracket of two constants of motion is itself a constant, as $\{G, H\} = \{F, H\} = 0 \Rightarrow \{H, \{F, G\}\} = 0$. E.g., in Examples (3.2.6; 2 and 3), the momentum $p_2 = \{G_1, G_2\}$ is a constant.

The canonical flow does not leave Θ invariant; the function that gives its rate of change, the $\tilde{f}$ of (3.1.5), turns out to be the same as the action W introduced in equation (2.3.20).

Theorem (3.2.9)

Locally,

$$\Phi^*_{-t} \Theta = \Theta + d\tilde{f}_t,$$

where in their explicit forms, $\Phi^*_{-t}: (q(0), p(0)) \rightarrow (q(t), p(t))$ and

$$\tilde{f}_t = W(q(0), p(0), t) = \int_0^t dt' \, L(q(t'), p(t')).$$

Explanatory Comment

Here we consider the Lagrangian as a function on $T^*(M)$, which is made possible by the diffeomorphism $T^*(M) \leftrightarrow T(M)$ (2.4.15; 3):

$$L = \sum_i p_i \frac{\partial H}{\partial p_i} - H.$$

Of course L can just as well be expressed in terms of q and $\dot{q}$; in either case the integration is along the trajectory that passes through $(q(0), p(0))$. As for the sign of t, compare (2.5.10; 1).

Proof

If the equation in question is differentiated by time, then with (2.5.9; 4) and the equation $dH = i_{X_H} \omega = -i_{X_H} d\Theta$, there results

$$L_{X_H} \Theta = (i_{X_H} \circ d + d \circ i_{X_H})\Theta = -d(H - i_{X_H} \Theta) = d \frac{\partial \tilde{f}}{\partial t}.$$

Up to a constant (which is still arbitrary), it follows that

$$\frac{\partial \tilde{f}}{\partial t} = -H + i_{X_H} \Theta.$$

On the standard chart,

$$i_{X_H} \Theta = \sum_i p_i (dq^i | X_H) = \sum_i p_i \frac{\partial H}{\partial p_i}.$$

Thus $\partial \bar{f}_t / \partial t = L$, and the theorem follows by integration. $\qquad\square$

Remarks (3.2.10)

1. This $\bar{f}_t$ is a time-dependent system of generators of the transformation $(q(0), p(0)) \to (q(t), p(t))$, for brevity written $(\bar{q}, \bar{p}) \to (q, p)$. In writing this we have treated $\bar{f}_t$ as a function of $\bar{q}$ and $\bar{p}$, but, as in (3.1.7; 2), it is more convenient to use the variables $\bar{q}$ and q. Since by Theorem (2.3.18) $\mathrm{Det}(\partial q^i / \partial \bar{p}_j)_{|\bar{q}} \neq 0$, we can consider the initial momentum as a function $\bar{p}(\bar{q}, q, t)$. If we define the action

$$W(\bar{q}, q, t) = \bar{f}(\bar{q}, \bar{p}(\bar{q}, q, t), t),$$

then by (3.1.7; 2),

$$p_i = \frac{\partial W}{\partial q^i}, \quad \text{and} \quad \bar{p}_i = -\frac{\partial W}{\partial \bar{q}^i}.$$

The time-dependence is affected in the following way: Taking a partial derivative, with $\bar{q}$ and q fixed, yields

$$\frac{\partial W}{\partial t} = \frac{\partial \bar{f}}{\partial t} + \frac{\partial \bar{f}}{\partial \bar{p}_i} \frac{\partial \bar{p}_i}{\partial t},$$

where $\bar{p}_i$ stands for $\bar{p}_i(\bar{q}, q, t)$. But since the initial conditions do not depend on time,

$$\frac{\partial \bar{p}_i}{\partial t} + \frac{\partial \bar{p}_i}{\partial q^j} \frac{\partial q^j}{\partial t} = 0.$$

Finally, because

$$\frac{\partial W}{\partial q^j} = \frac{\partial \bar{f}}{\partial \bar{p}_i} \frac{\partial \bar{p}_i}{\partial q^j}$$

we conclude that

$$\frac{\partial W}{\partial t} = L - p_j \frac{dq^j}{dt} = -H.$$

When written out explicitly, the action satisfies the Hamilton–Jacobi partial differential equation with $H(q^i, p_j)$,

$$\frac{\partial}{\partial t} W(\bar{q}, q, t) + H\left(q^i, \frac{\partial}{\partial q^j} W(\bar{q}, q, t)\right) = 0. \qquad (3.2.11)$$

2. This whole treatment is as yet only local; in particular, it shows that local solutions of the Hamilton–Jacobi equations exist. Whether there exist global solutions is a problem of a much higher level of difficulty.

It is often convenient to introduce time as a dependent variable. The formalism is then changed as follows.

Definition (3.2.12)

We shall call $M_e \equiv M \times \mathbb{R}$ **extended configuration space** and $T^*(M_e)$ **extended phase space**. Let t and $-E$ be the coordinates of the final Cartesian factor, so that the canonical 1-form becomes $\Theta_e = p_i \, dq^i - E \, dt$ on $T^*(M_e)$. Now

$$\mathcal{H} \equiv H(p, q; t) - E \in C^\infty(T^*(M_e))$$

generates a local canonical flow (with the parameter s), for which Hamilton's equations are as in (3.2.4), and

$$\frac{dt}{ds} = 1, \quad \text{and} \quad \frac{dE}{ds} = \frac{\partial H}{\partial t}. \tag{3.2.13}$$

Remarks (3.2.14)

1. As always, $\mathcal{H}$ is a constant (since we do not consider the case where it depends explicitly on s), and we may restrict ourselves to the submanifold $\mathcal{H} = 0$, where $E = H$, i.e., the actual energy.
2. It is possible for H to depend explicitly on t, in which case (3.2.13) says that energy is conserved iff H is invariant under the transformation $t \to t + c$.
3. The invariance of the equations of motion under displacements in time is irrelevant for conservation of energy. For example, the equation of the damped oscillator, $\ddot{x} = -\mu\dot{x} - \omega^2 x$, is invariant under time-displacements, although its energy is not conserved, because its Hamiltonian,

$$H = e^{-\mu t}\frac{p^2}{2} + \omega^2 \frac{x^2}{2}e^{\mu t},$$

depends explicitly on t (Problem 3).
4. If a potential is turned on, so that $H = H_0 + V(q)e^{\alpha t}$, it causes a change in the energy between $t = -\infty$ and $t = 0$:

$$\delta E = \int_{-\infty}^0 \alpha \, ds \, e^{\alpha s} V(q(s)),$$

which is the Cesàro average of V.
5. From Equation (3.2.13), t and s are equal up to a constant; but time-dependent coordinate transformations, such as the passage to an accelerated frame of reference, are point transformations on M_e.

6. For many purposes it is desirable to choose $t \neq s$ (cf. (1.1.6) and (1.1.4)). E.g., the Hamiltonian $\mathcal{H} = f(q, p)(H(q, p) - E)$ with f positive yields the equations

$$\frac{dq}{ds} = fH_{,p} + f_{,p}(H - E), \qquad \frac{dp}{ds} = -fH_{,q} - f_{,q}(H - E),$$

$$\frac{dt}{ds} = f.$$

These equations are equivalent to (2.3.25) on the invariant surface $\mathcal{H} = 0$. Thus if the canonical equations can be solved after a factor has been separated off from H, then the above equations solve the problem with another parameter in place of t — their solution gives the trajectories directly, and it only remains to integrate $dt/ds = f(q(s), p(s))$ in order to calculate the time-evolution.

Examples (3.2.15)

1. Constant acceleration. $M = \mathbb{R}$, $M_e = \mathbb{R}^2$, and $\mathcal{H} = p^2/2 + gx - E$.

   (a) One might at first consider changing to a co-moving coordinate system. The transformation

   $$\Phi : x = \bar{x} - \frac{g}{2}\bar{t}^2, \qquad t = \bar{t},$$

   is a point transformation on M_e, and with

   $$T_{(\bar{x}, \bar{t})}(\Phi) = \begin{vmatrix} 1 & -g\bar{t} \\ 0 & 1 \end{vmatrix}, \qquad T_{(\bar{x}, \bar{t})}(\Phi)^{\prime - 1} = \begin{vmatrix} 1 & 0 \\ g\bar{t} & 1 \end{vmatrix},$$

   it induces the canonical transformation $T^*(\Phi)$:

   $$(x, t; p, -E) = \left(\bar{x} - \frac{g}{2}\bar{t}^2, \bar{t}; \bar{p}, -\bar{E} + g\bar{t}\bar{p} \right):$$

   $$\mathcal{H} = \frac{\bar{p}^2}{2} + g\left(\bar{x} + \bar{p}t - \frac{g}{2}\bar{t}^2 \right) - \bar{E}.$$

   This indeed produces an equivalent set of equations of motion, but makes no real advance.

   (b) If we wish to separate off the influence of the gravitational field g, we can make use of a canonical transformation that agrees with the above Φ on M, but acts differently on the fibers $(p, -E)$:

   $$(x, t; p, -E) = \left(\bar{x} - \frac{g}{2}\bar{t}^2, \bar{t}; \bar{p} - g\bar{t}, -\bar{E} + g(\bar{t}\bar{p} - \bar{x}) \right):$$

   $$\mathcal{H} = \frac{\bar{p}^2}{2} - \bar{E}.$$

With these coordinates the motion is the same as free motion without gravitation.

(c) It is always possible to transform the system canonically to equilibrium, so that everything but t becomes a constant:

$$(x, t; p, -E) = \left(\bar{x} + \bar{p}\bar{t} - \frac{g}{2}\bar{t}^2, \bar{t}; \bar{p} - g\bar{t}, -\bar{E} - g\bar{x} - \frac{\bar{p}^2}{2} \right):$$

$$\mathcal{H} = -\bar{E}.$$

Then $\dot{\bar{x}} = \dot{\bar{p}} = \dot{\bar{E}} = 0, s = \bar{t} = t$ (cf. (2.3.12)).

2. A rotating system. $M = \mathbb{R}^2$, $M_e = \mathbb{R}^3$, and $\mathcal{H} = |\mathbf{p}|^2/2 + V(|\mathbf{x}|) - E$. With the canonical transformation

$$x = \bar{x}\cos \omega\bar{t} + \bar{y}\sin \omega\bar{t}, \qquad p_x = \bar{p}_x \cos \omega\bar{t} + \bar{p}_y \sin \omega\bar{t},$$

$$y = -\bar{x}\sin \omega\bar{t} + \bar{y}\cos \omega\bar{t}, \qquad p_y = -\bar{p}_x \sin \omega\bar{t} + \bar{p}_y \cos \omega\bar{t},$$

$$t = \bar{t}, \qquad\qquad E = \bar{E} + \omega(\bar{x}\bar{p}_y - \bar{y}\bar{p}_x),$$

(x and p are transformed the same way by orthogonal transformations) there results

$$\mathcal{H} = \frac{|\bar{\mathbf{p}}|^2}{2} - \omega(\bar{x}\bar{p}_y - \bar{y}\bar{p}_x) + V(|\bar{x}|) - \bar{E}.$$

The extra term contains the Coriolis ($\sim \omega$) and centrifugal ($\sim \omega^2$) forces,

$$\ddot{\bar{x}} = -V_{,x} + 2\omega\dot{\bar{y}} + \omega^2\bar{x}.$$

3. The situation of Example 1 (c) can be formulated generally. Let a canonical transformation on $T^*(M_e)$ be given by

$$\sum_i p_i \, dq^i - E \, dt = \sum_i \bar{p}_i \, d\bar{q}^i - \bar{E} \, d\bar{t} + d(f(q, \bar{p}, t) + \bar{E}(\bar{t} - t) - \sum_i \bar{p}_i\bar{q}^i).$$

This implies that

$$\bar{t} = t, E = \bar{E} - \frac{\partial f}{\partial t}, p_i = \frac{\partial f}{\partial q^i}, \quad \text{and} \quad \bar{q}^i = \frac{\partial f}{\partial \bar{p}_i}.$$

Now if f satisfies the Hamilton–Jacobi equation

$$\frac{\partial f}{\partial t} + H\left(q, \frac{\partial f}{\partial q}, t \right) = 0, \tag{3.2.16}$$

then it follows that $\mathcal{H} = -\bar{E}$, and thus $t = s$ and $\bar{q}, \bar{p}$, and $\bar{E}$ are constant. Hence Equation (3.2.16) determines the generator of the transformation on $T^*(M_e)$ that always keeps the system in equilibrium.

Problems (3.2.17)

1. Using (2.3.27), calculate L in the rotating system (3.2.15; 2).

2. Check whether the transformations given in (3.2.15; 1) are canonical on extended phase space.

3. Verify Hamilton's equations for

$$H = e^{-\mu t} \frac{p^2}{2} + e^{\mu t} \omega^2 \frac{x^2}{2}.$$

4. The transformation $t \to t - a$, $x \to xe^{\mu a/2}$, $p \to pe^{-\mu a/2}$ leaves $\mathcal{H} = H - E$ invariant, where H is as in Problem 3. Show that its generator is constant.

5. Show that the function f introduced in Equation (3.1.5) also satisfies the Hamilton–Jacobi equation (3.2.11) with suitable variables.

6. Derive the equations $p = \partial W/\partial q$ and $\bar{p} = -\partial W/\partial \bar{q}$ with a variational argument, where $q \to q + \delta q$, by using Equation (2.3.20).

7. Verify the calculations of (3.2.10; 1) explicitly for the action W of the harmonic oscillator, $H = (p^2 + q^2)/2$. Verify the same equations for the f defined in (3.1.5).

8. Show that $dF = i_{X_F}\omega$ by using the expression for ω on a bundle chart.

Solutions (3.2.18)

1. L can be written as

$$\tfrac{1}{2}(|\dot{\mathbf{x}}| - [\omega \cdot \mathbf{x}])^2 - V(|\mathbf{x}|),$$

(where ω points in the z-direction), which also follows from direct substitution into $L(\dot{x}, x)$.

2. Calculate the Poisson brackets.

3.

$$-e^{\mu t}\omega^2 x = \dot{p} = \frac{d}{ds}(\dot{x}e^{\mu t}) = e^{\mu t}(\ddot{x} + \dot{x}\mu).$$

4. The generator is $E + (\mu/2)xp$, and

$$\dot{E} = \dot{H} = -\mu\left(e^{-\mu t}\frac{p^2}{2} - e^{\mu t}\frac{\omega^2 x^2}{2}\right) = -\frac{d}{ds}\frac{\mu}{2}xp.$$

5. With $f(\bar{p}, q, t) = W(\bar{q}(\bar{p}, q, t), q, t) + \sum_i \bar{q}^i(\bar{p}, q, t)\bar{p}_i$, we find that

$$\frac{\partial f}{\partial q^i} = p_i, \qquad \frac{\partial f}{\partial \bar{p}_i} = \bar{q}^i, \qquad \frac{\partial f}{\partial t} = -H(q, p(\bar{p}, q, t)),$$

so

$$\frac{\partial}{\partial t} f(\bar{p}, q, t) + H\left(q, \frac{\partial}{\partial q} f(\bar{p}, q, t)\right) = 0.$$

6. $W = \int_{t_0}^{t_1} dt\, L(q, \dot{q})$, and for

$$q^i(t) \to q^i(t) + \delta q^i(t),\ \dot{q}^i(t) \to \dot{q}^i(t) + \frac{d}{dt}\,\delta q^i(t),$$

$$\delta W = \delta q^i \frac{\partial L}{\partial \dot{q}^i}\Big|_{t_0}^{t} + \int_{t_0}^{t} dt\, \delta q^i\left(\frac{\partial L}{\partial q^i} - \frac{d}{dt}\frac{\partial L}{\partial \dot{q}^i}\right).$$

Along the trajectory, $(\cdots) = 0$, and $q^i(t_0) = \bar{q}^i$ and $q^i(t) = q^i$, and so $dW = p_i\, dq^i - \bar{p}_i\, d\bar{q}^i$.

7. $q(t) = \bar{q} \cos t + \bar{p} \sin t,\ p(t) = -\bar{q} \sin t + \bar{p} \cos t$,

$$\bar{f}(\bar{q}, \bar{p}, t) = -\bar{q}\bar{p} \sin^2 t + \frac{\sin t \cos t}{2}(\bar{p}^2 - \bar{q}^2),$$

$$\frac{\partial \bar{f}}{\partial t} = \frac{1}{2}(p(t)^2 - q(t)^2) = L,$$

$$W(\bar{q}, q, t) = -\bar{q}q \sin t + \frac{\sin t \cos t}{2}\bar{q}^2 + \frac{\cos t}{2 \sin t}(q - \bar{q}\cos t)^2,$$

$$\frac{\partial W}{\partial q} = -\bar{q}\sin t + \frac{q - \bar{q}\cos t}{\sin t}\cos t = p,$$

$$\frac{\partial W}{\partial \bar{q}} = -q \sin t + \sin t \cos t\bar{q} - \frac{\cos^2 t}{\sin t}(q - \bar{q}\cos t) = -\bar{p},$$

$$\frac{\partial W}{\partial t} = -\frac{1}{2}\left(\bar{q}^2 + \left(\frac{\bar{q}\cos t - q}{\sin t}\right)^2\right) = -H,$$

$$f(q, \bar{p}, t) = q\bar{p} \cos t - \frac{\bar{p}^2}{2}\sin t \cos t - \frac{\sin t}{2 \cos t}(q - \bar{p}\sin t)^2,$$

$$\frac{\partial f}{\partial q} = \bar{p} \cos t - \frac{q - \bar{p}\sin t}{\cos t}\sin t = p,$$

$$\frac{\partial f}{\partial \bar{p}} = q \cos t - \bar{p} \sin t \cos t + \frac{\sin^2 t}{\cos t}(q - \bar{p}\sin t) = \bar{q},$$

$$\frac{\partial f}{\partial t} = -\frac{1}{2}\left(\frac{q - \bar{p}\sin t}{\cos t}\right)^2 + \bar{p}^2 = -H.$$

8. $i_{X_F}\omega = (i_{F,p\,\partial_q} - i_{F,q\,\partial_p})dq \wedge dp = F_{,p}\, dp + F_{,q}\, dq = dF$.

3.3 Constants of Motion

Constants of motion divide phase space into time-invariant submanifolds. A Hamiltonian system always has at least one constant of motion. A trajectory is completely determined by $2m - 1$ constants; yet m constants are often sufficient for a solution of the problem.

As we saw with Theorem (2.3.12), all flows are locally diffeomorphic to linear fields of motion, except at points of equilibrium. This leaves two kinds of questions open, dealing on the one hand with the behavior in the vicinity of an equilibrium position, and on the other with global characteristics of the trajectories. In this section we shall be concerned with the latter questions, and leave the former to §3.4.

We know that a trajectory can not fill up all of M; on the contrary, it always remains on the energy surface $H = $ const., which, as long as $dH \neq 0$, is a $2m - 1$-dimensional submanifold. More generally, if r independent constants, $K_1 = H$, and $K_2, \ldots, K_r$, are known, then the motion must take place only on $N: K_i = $ constant for all i, which is a $2m - r$-dimensional submanifold closed in M. By "independent" we mean that for all q and p in N, the dK_i are independent vectors in $T^*_{(q,\, p)}(T^*(M))$ (cf. (2.1.10; 3)). If it happens that $f(K) = 0$ for some differentiable f, then they are dependent, because $0 = df = \sum_i dK_i \partial f / \partial K_i$.

The K_i reduce the problem by allowing the motion to be determined by the restriction of X_H to N. It was noted above (2.4.21) that it is not always possible to speak of the restriction of a vector field. It is possible in the present case, because the values of X_H lie within $T(N)$. Vectors of $T_q(N)$ can be characterized as being perpendicular to the dK_i, because the derivatives of the K_i in the direction of $T(N)$ all vanish, as the K_i are constant on N. That is,

$$(dK_i | X_H) = L_{X_H} K_i = \{K_i, H\} = \frac{d}{dt} K_i = 0.$$

If $r = 2m - 1$, then N is precisely the trajectory. When $m = 1$, it suffices simply to write $H = p^2 + V(q)$ in order to solve for the trajectory without further integration.

Examples (3.3.1)

1. $M = \mathbb{R}$, $T^*(M) = \mathbb{R} \times \mathbb{R}$, and $V = q^2$.

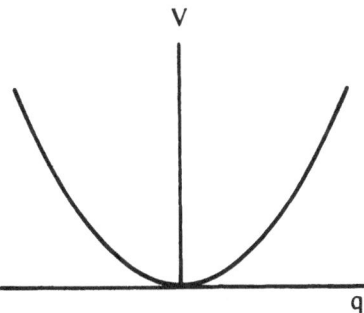

Figure 15   Harmonic   oscillator potential.

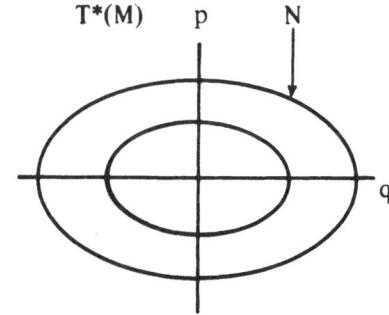

Figure 16   The trajectories of the harmonic oscillator.

2. $M = \mathbb{R}$, $T^*(M) = \mathbb{R} \times \mathbb{R}$, and $V = -q^2 + q^4$.

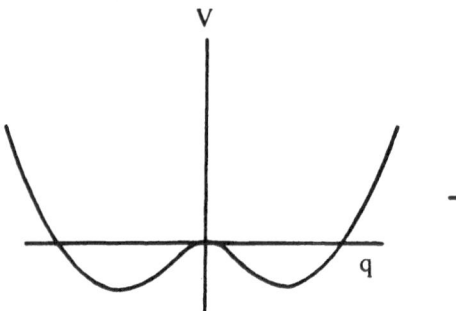

Figure 17   Potential with two wells.   Figure 18   The trajectories of the
                                                two-well potential

3. $M = \mathbb{R}^+$, $T^*(M) = \mathbb{R}^+ \times \mathbb{R}$, and $V = 1/q$.

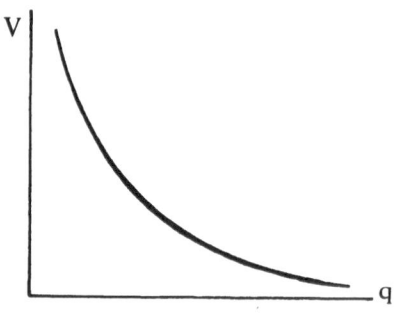
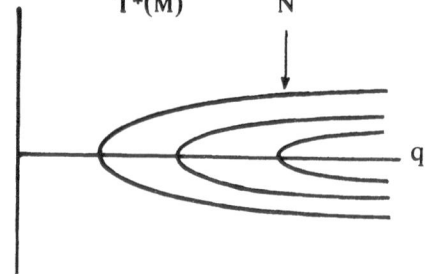

Figure 19   A $1/q$ potential   Figure 20   Trajectories of the $1/q$ po-
                                        tential.

4. $M = \mathbb{R}^+$, $T^*(M) = \mathbb{R}^+ \times \mathbb{R}$, and $V = -1/q + 1/q^2$.

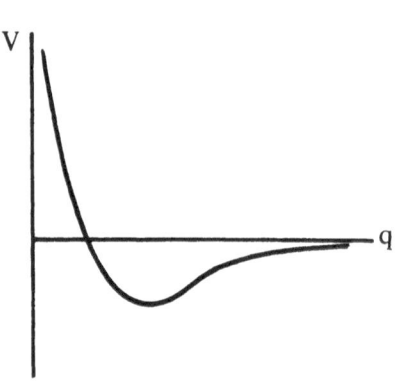
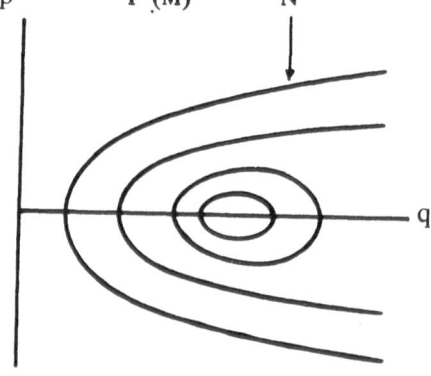

Figure 21   Potential of the form $-1/q +$ Figure 22   Trajectories of the potential
$1/q^2$. $-1/q + 1/q^2$.

Remarks (3.3.2)

1. The equilibrium positions (where $dH = 0$) are 0-dimensional trajectories. In Examples 1, 2, 3, and 4 there are respectively 1, 3, 0, and 1 such points.
2. The 1-dimensional trajectories of Example 1 are all diffeomorphic to T^1, and those of Example 3 are diffeomorphic to $\mathbb{R}$. In Example 2 there are two trajectories diffeomorphic to $\mathbb{R}$, namely the ones where $H = 0$, and the rest are diffeomorphic to T^1. In Example 4 there are infinitely many of both kinds.
3. In these cases $(dH | X_H) = 0$. The restriction of X_H is not simply $b \cdot$ (the restriction of dH), which would make it zero. Thus X_H is not a Hamiltonian vector field on N. If $T^*(\mathbb{R}) = \mathbb{R}^2$, then dH and X_H are as in Figure 23.

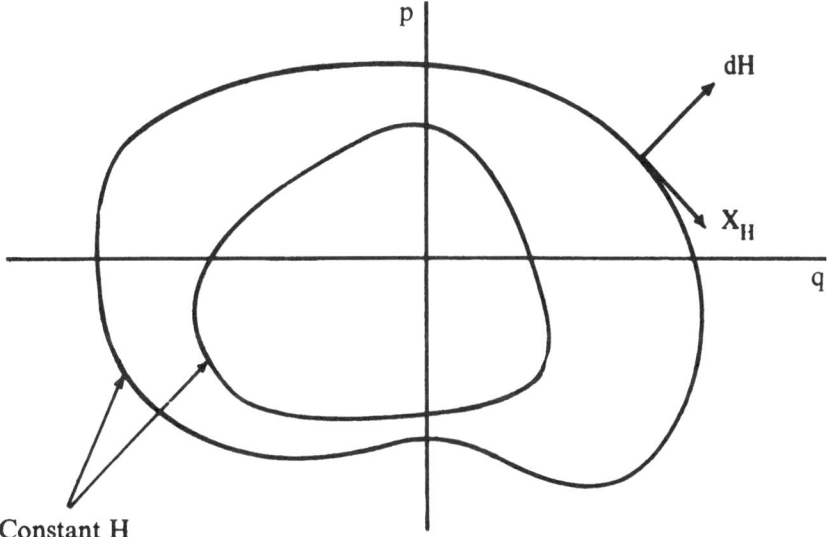

Figure 23   The vectors dH and X_H.

It should be emphasized again that it is essential that the K_i be defined on all of M, as locally it is always possible to find $2m - 1$ time-independent constants of motion (2.3.13). For most problems, the local constants can not be extended continuously to all of M and do not define a closed $2m - 1$-dimensional submanifold.† It then happens that the trajectory is dense in a submanifold of dimension > 1, as the following rather typical case shows.

Lemma (3.3.3)

Let Φ_t be the flow generated by $H = \frac{1}{2}(p_1^2 + p_2^2 + \omega_1^2 q_1^2 + \omega_2^2 q_2^2)$ on

$$(T^*(\mathbb{R}) \backslash \{0\}) \times (T^*(\mathbb{R}) \backslash \{0\}).$$

---

† On compact manifolds Hamiltonian systems with additional global constants are exceptional [18].

The functions

$$K_i \equiv \frac{\omega_i^2}{2} q_i^2 + \frac{p_i^2}{2}, \qquad i = 1, 2,$$

are constant, i.e., $\Phi_t^* K_i = K_i$, and are independent on this manifold. If the frequencies ω_i have a rational ratio, then all trajectories are submanifolds diffeomorphic to T^1. If their ratio is irrational, then every trajectory is dense in some 2-dimensional submanifold defined by the K_i.

Proof

Map $(\mathbb{R}^2 \setminus \{0\}) \times (\mathbb{R}^2 \setminus \{0\})$ onto $\mathbb{R}^+ \times \mathbb{R}^+ \times T^1 \times T^1$ with the transformation $(q_i, p_i) = (\sqrt{K_i} \sin \varphi_i, \sqrt{K_i} \cos \varphi_i)$, $i = 1, 2$ (cf. (3.1.8; 1)). On this chart the time-evolution is given by

$$\Phi_t : (K_1, K_2, \varphi_1, \varphi_2) \to (K_1, K_2, \varphi_1 + \omega_1 t, \varphi_2 + \omega_2 t)$$

(cf. (3.2.6; 1)). Let $\Psi_n \equiv \Phi_{2\pi n/\omega_1}$, $n \in \mathbb{Z}$, and consider its restriction to the last T^1 factor; on the other factors, $\Psi_n = 1$.

1. Suppose $\omega_1/\omega_2 = g_1/g_2$, where $g_i \in \mathbb{Z}$. Then $\Psi_{g_1} = 1$, and to each value of φ_1 on the trajectory there correspond g_2 values of φ_2. (See figure.)

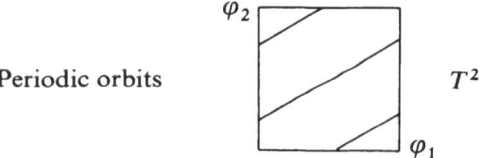

Periodic orbits

cf. (2.3.14). Trajectories like these, which return to their initial points (in $T^*(M)$), are called **periodic**, or **closed**, **orbits**. They are closed submanifolds diffeomorphic to T^1.

2. Suppose ω_1/ω_2 is irrational. Then there is no value g_1 other than 0 for which $\Psi_{g_1} = 1$. Since T^1 is compact, there must be a point of accumulation; i.e., $\forall \varepsilon > 0$, there exist integers g_1 and g_2 such that

$$|\Psi_{g_1}(\varphi_2) - \Psi_{g_2}(\varphi_2)| < \varepsilon,$$

and thus $\Psi_{g_1 - g_2}(\varphi_2) = \varphi_2 + \eta$, where $|\eta| < \varepsilon$. Therefore the set

$$\{\Psi_{g(g_1 - g_2)}(\varphi_2), g \in \mathbb{Z}\}$$

fills T^1 with points that are only some small ε apart. This means that for all φ_1 the points attained by the trajectory are dense in the second factor. Since the trajectory obviously takes on every value of φ_1, it is dense in T^2. Such trajectories are called **almost-periodic orbits**. $\square$

Remarks (3.3.4)

1. This lemma can be generalized to the case of the n-dimensional harmonic oscillator by iterating the proof.
2. Projected onto configuration space (q_1, q_2), the trajectory is dense in a rectangle; it is a Lissajou figure.

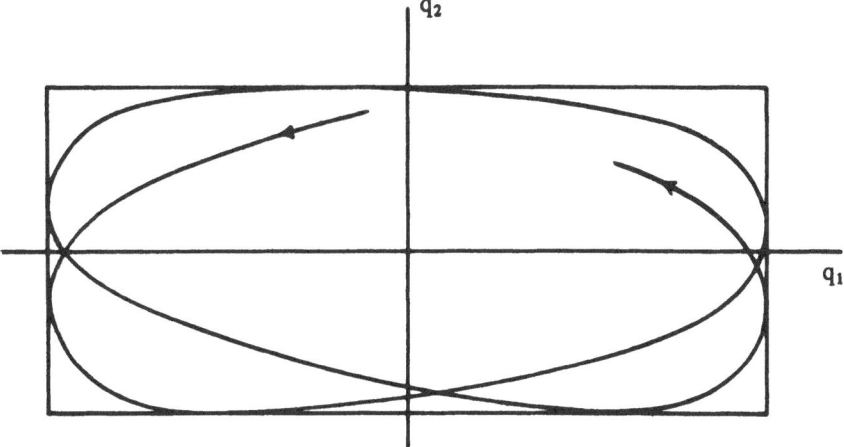

Figure 24   Almost-periodic orbits.

3. There exist curves, known as Peano curves, which completely fill up higher-dimensional manifolds. Differentiable curves can at most be dense in them.
4. When the constants are not independent in the sense considered above, they can still restrict the trajectories, but it is not possible to say anything in general about the dimension of $N = \{(q, p) \in T^*(M): K_i(q, p) = \alpha_i \in \mathbb{R}\}$, or even whether it is a manifold. Recall Example (3.3.1; 1). If $H = 0$, then $dH = 0$ also, and N is a point; thus a single constant reduces the dimension by 2. In (3.3.1; 2) the energy surface $H = E < 0$ divides into two pieces. By choosing $f(q) \in C(T^*(M))$ such that f equals different constants on the two pieces, and then multiplying by $g(H) \in C(T^*(M))$ such that $g = 0$ for $H > E$, we can produce a constant that is not a function of H alone and that forces the trajectory to stay in one part of the energy surface. The constant is not independent of H in our sense, because its differential is proportional to dH.[†] In this case two constants only reduce the dimension by one. And lastly, in Example (3.3.1; 2) the set where $H = 0$ is not a manifold at all; near the origin it has the structure of Example (2.1.7; 6).
5. As mentioned in Example (2.3.14), when $\omega_1/\omega_2 = g_1/g_2$, where the g_i are integers, there exists a constant, $\sin(2\pi(\varphi_1 g_2 - \varphi_2 g_1))$. One might

---

[†] Of course, this is only on an energy surface $H = E < 0$.

suspect that whenever the trajectories remain restricted to a submanifold there are always additional constants. That this is not generally true is shown by the following non-Hamiltonian example: $M = \mathbb{R}$, $X \in \mathcal{T}_0^1(M)$: $x \rightarrow (x, x)$; for the time-evolution, $\dot{x} = x \Rightarrow x(t) = x(0)e^t$. There are three trajectories, $\mathbb{R}^-$, 0, and $\mathbb{R}^+$, each of which is a submanifold of M, but none of which is dense. Yet there is no time-independent constant in $C(M)$; it would have to be constant on the trajectories, and therefore, as a continuous function, on all $\mathbb{R}$. Consequently it would have a vanishing differential.

6. Suppose that M is an open set of $\mathbb{R}^{2m}$ on which the $2m$ independent coordinate functions z_k are defined globally. If there exists a function $J \in C(M)$ that increases sufficiently fast with time (specifically, $\forall z \in M$, $\exists c > 0$: $(d/dt)J \circ \Phi_t(z) > c \; \forall t$), then the trajectories are one-dimensional submanifolds. The initial values of the coordinate functions are $C(M)$ and are obviously constants in time. In this case there are necessarily $2m - 1$ independent constants of motion (see Problem 6). Therefore, even an ergodic system (i.e., any trajectory is dense on the energy shell) can be considered as a subsystem of a system with one more degree of freedom and the maximal number of constants of motion. For instance, this is the situation in extended phase space with $dt/ds = 1$, and every system with M_e open in $\mathbb{R}^m$ has $2m - 1$ independent constants in $T^*(M_e)$. Hence the existence of constants does not imply that the motion is simple. The projection of a trajectory onto a subsystem may be complicated.

The existence of the time-average

$$f_\infty \equiv \lim_{T \to \infty} \frac{1}{T} \int_0^T dt \; \tau_t f$$

was alluded to in (2.6.15). Then

$$\tau_t f_\infty = \lim_{T \to \infty} \frac{1}{T} \int_t^{t+T} dt' \tau_{t'} f = f_\infty \; \forall f \in C^\infty(T^*(M)),$$

and thus f_∞ is constant in time. If it is also spatially constant for all f, that means that with the passage of time the points of any arbitrarily small neighborhood travel throughout the whole manifold. One might hope to always find a nonconstant function f_∞ on the energy shell N whenever the trajectory does not fill N densely. This line of reasoning fails, however, because there is no guarantee that f_∞ is continuous, even for $f \in C^\infty(T^*(M))$.

Example (3.3.5)

$$M = T^2, \; T^*(M) = T^2 \times \mathbb{R}^2, \quad \text{and} \quad H = p_1^2/2 + p_2^2/2.$$

The time-evolution is $(\varphi_1, \varphi_2; p_1, p_2) \rightarrow (\varphi_1 + p_1 t, \varphi_2 + p_2 t; p_1, p_2)$. If

$$f = g(\varphi_1) \cdot g(\varphi_2), \quad g(\varphi) = \begin{cases} e^{-1/\varphi - 1/(\varphi - 2\pi)} & \text{for } \varphi \neq 0 \\ 0 & \text{for } \varphi = 0, \end{cases}$$

then

$$f_\infty(\varphi_1, \varphi_2) = \lim_{T \to \infty} \frac{1}{T} \int_0^T dt\, f(\varphi_1 + p_1 t, \varphi_2 + p_2 t)$$

$$= \frac{1}{4\pi^2} \int_0^{2\pi} d\varphi_1\, d\varphi_2\, g(\varphi_1) g(\varphi_2),$$

when p_1/p_2 is irrational; but on the other hand

$$f_\infty = \frac{1}{2\pi} \int_0^{2\pi} d\varphi\, |g(\varphi)|^2$$

when $p_1 = p_2$ and $\varphi_1 = \varphi_2$. Hence f_∞ is quite discontinuous. Constants of this sort can always be found; it is only necessary to assign arbitrary numbers to the trajectories. However, these constants are uninteresting because they do not define manifolds.

Global generators of groups of canonical transformations that leave H invariant are constants of the motion. To each parameter of the group there corresponds one generator, yet the generators are not generally independent of H. For example, in (3.2.6; 2), p_1, p_2, and p_3 are constant, and $H = |\mathbf{p}|^2$. The generators themselves are certainly linearly independent, but may depend on each other algebraically.

Example (3.3.6)

The m-dimensional oscillator: $M = \mathbb{R}^m$, $m > 1$, and $H = \frac{1}{2} \sum_{i=1}^m (p_i^2 + x_i^2)$. The functions $M_{ik} = p_i p_k + x_k x_i$ and $L_{ik} = p_i x_k - p_k x_i$, where $i, k = 1, \ldots m$, are constant, as is easily verified. $\{M_{ik}, H\} = \{L_{ik}, H\} = 0$. They provide complete vector fields, which generate a group of canonical transformations that is isomorphic to U_m. There are $(m(m + 1)/2)$ M's and $(m(m - 1)/2)$ L's, a total of m^2 generators. There is no way that they could all be algebraically independent, as phase space has only $2m$ dimensions. For example, if $m = 2$, then $2H = M_{11} + M_{22}$, and $M_{12}^2 + L_{12}^2 = M_{11} \cdot M_{22}$.

Remarks (3.3.7)

1. The group mentioned in this example is far from the largest group that leaves H invariant. The largest such group is generated by the functions $K \in C^\infty(T^*(M))$ for which $\{K, H\} = 0$ and X_K is complete. It does not depend on only a finite number of parameters, and consequently it is not even locally compact. Even in the trivial example, $M = \mathbb{R}$, $H = p^2$, the functions $f(p)$, where $f \in C^\infty(\mathbb{R})$, generate the groups $(x, p) \to (x + \lambda f'(p), p)$, which are different unless the f's differ only by a constant. All together, the largest group that leaves H invariant has infinitely many generators, which of course are not all independent.

2. We can just as well pose the opposite question, of what group gives the

greatest number of constants with the fewest parameters. Although there can be no more than $2m - 1$ constants, under certain circumstances, any group that gives all the constants must have more than $2m - 1$ parameters. It can also happen that the minimal group is not unique. In the above example all the minimal groups have one parameter, and the groups generated by $f(p) = (p^2 + c^2)^{1/2}$, $c \in \mathbb{R}$, are equally good for all $c \neq 0$.

It is not often that one is lucky enough to find $2m - 1$ constants, but it frequently suffices to find m of them. This situation occurs frequently enough that it is given a name of its own:

Definition (3.3.8)

A Hamiltonian system is said to be **integrable**† iff there exist $m \equiv \dim M$ functions K_i on a time-invariant neighborhood $U \subset T^*(M)$ such that

(a) $\{K_i, H\} = 0$,
(b) $\{K_i, K_j\} = 0$,
(c) the dK_i are linearly independent.

Remark (3.3.9)

It is very common to find treatments of integrable systems; in fact most books on mechanics, including this one, are basically catalogs of them. This can lead to the wrong opinion that most systems are integrable on some U that is dense in $T^*(M)$. It is in fact exceptional for such cases to occur, and they are only popular because they are soluble.

The first interesting fact about integrable systems is that the K_i can be used as new coordinates:

Theorem (3.3.10) (Liouville)

Consider an integrable system. For all $(q, p) \in U$,

(a) *There exist $U_1 \subset U$ and $\varphi_i \in C(U_1)$, $i = 1, \ldots, m$, such that $\{K_i, K_j\} = \{\varphi_i, \varphi_j\} = 0$ and $\{K_i, \varphi_j\} = \delta_{ij}$ on U_1; and*
(b) *all other sets of variables that satisfy the relationships in (a) are of the form*

$$\bar{\varphi}_i = \varphi_i + \frac{\partial \mathscr{X}(K)}{\partial K_i}, \qquad \mathscr{X} \in C^\infty(\mathbb{R}^m).$$

Proof

Let $N_\alpha = \{(q, p) \in U : K_i(q, p) = \alpha_i, \ \alpha_i \in K_i(U) \subset \mathbb{R}\}$. Then the X_{K_i} constitute a basis for $T(N_\alpha)$; for $\{K_i, K_j\} = L_{X_{K_j}} K_i = (dK_i | X_{K_j}) = 0$ implies

---

† More precisely we should say "integrable on U."

that they lie in $T(N_\alpha)$, and since b (3.1.9) is bijective, they are linearly independent just like the dK_i. Because $\omega(X_{K_i}, X_{K_j}) = \{K_i, K_j\} = 0$, the restriction (2.4.24) of $\omega = -d\Theta$ vanishes. According to Remark (2.5.6; 3), $\exists U_1 \subset U$ such that $\Theta = d\tilde{f}$ on $U_1 \cap N_\alpha$. If q_i, $i = 1, \ldots, m$, are local coordinates on $U_1 \cap N_\alpha$, then the dq^i are a basis for $T^*(T^*(U_1 \cap N_\alpha))$, and the restriction of Θ to $U_1 \cap N_\alpha$ can be written as

$$\Theta_{|U_1 \cap N_\alpha} = \sum_i dq^i \, p_i(q, \alpha) = d\tilde{f}(q, \alpha)_{|U_1 \cap N_\alpha}.$$

The p_i and $\tilde{f}$ depend on α.† If we define $\varphi_i \equiv -(\partial/\partial K_i)\tilde{f}(q, K)$, then we can write

$$\sum_i dq^i \, p_i = \sum_i \varphi_i \, dK_i + d\tilde{f}$$

on all of U_1. According to (3.1.6), $\tilde{f}$ establishes a canonical transformation $(q^i, p_j) \to (K_i, \varphi_j)$ on U_1 (cf. Problem 6). With (3.1.7; 2), this proves (a). Because $p_i = \partial \tilde{f}/\partial q^i$, $\tilde{f}$ is determined up to a function of the K's alone, and (b) follows. $\qquad\qquad\square$

Corollary (3.3.11)

By (b), *the canonical transformation corresponding to time-evolution must be of the form*

$$(K_i, \varphi_j) \to \left(K_i, \varphi_j + \frac{\partial}{\partial K_j} \mathcal{X}(t, K)\right).$$

The group property (2.3.7) *implies that*

$$\frac{\partial}{\partial K_j} \mathcal{X}(t_1 + t_2, K) = \frac{\partial}{\partial K_j} \mathcal{X}(t_1, K) + \frac{\partial}{\partial K_j} \mathcal{X}(t_2, K).$$

Since $\mathcal{X}$ depends continuously on t and any contribution to $\mathcal{X}$ that is independent of the K_i vanishes, $\mathcal{X}$ is of the form $tH(K)$, and the time-evolution is

$$(K_i(t), \varphi_j(t)) = \left(K_i(0), \varphi_j(0) + t \frac{\partial}{\partial K_j} H(K)\right).$$

Because of this, there is again locally a linear field of motion (2.3.5) on N_α. However, among other problems the time-independent constants $\varphi_i \partial H/\partial K_j - \varphi_j \partial H/\partial K_i$ can not generally be extended to all of N_α and they do not restrict the motion to a manifold of dimension less than m. Until now all the statements we have made about the motion have only been local, and they contain no information that might answer global questions. With fairly harmless additional assumptions, though, some light is cast on the global structure by

---

† These q^i and p_i do not necessarily bear the standard relationship to Π (2.2.15).

Theorem (3.3.12) (Arnold)

Suppose that for an integrable system on $T^*(M)$

$$N_\alpha = \{(q, p) \in U, K_i(q, p) = \alpha_i \in K_i(U) \subset \mathbb{R}\}$$

is compact and connected. Then N_α is diffeomorphic to the torus T^m.

Proof

As we have already seen, on N_α all the $X_{K_i} \in \mathcal{T}_0^1(N_\alpha)$. By assumption,

$$L_{X_{K_i}} L_{X_{K_j}} = L_{X_{K_j}} L_{X_{K_i}},$$

and the X_{K_i} are complete since N_α is compact (2.3.6; 2). Consequently,

$$\exp\left(\sum_j \tau_j L_{X_{K_j}}\right), \qquad (\tau_j) \in \mathbb{R}^m,$$

generates an m-parameter group of diffeomorphisms of N_α, and the mapping

$$\Phi : \mathbb{R}^m \to N_\alpha, (\tau) \to \exp\left(\sum_j \tau_j L_{X_{K_i}}\right) \cdot q, \qquad q \in N_\alpha \text{ constant,}$$

is locally a diffeomorphism. In the vicinity of $0 \in \mathbb{R}^m$, Φ is

$$\tau_j \to q_j + \tau_i(dq_j | X_{K_j}) + 0(\tau^2),\dagger$$

and the matrix of derivatives is thus composed of the components of the X_{K_i}, as a result of which it is nonsingular [see (1), 16.5.6]. The group property transfers this to all $(\tau) \in \mathbb{R}^m$; the image of Φ in N_α is both open and closed, and therefore all of N_α. However, Φ is not injective. Because $\Phi(\tau + \tau') = \Phi(\tau) \circ \Phi(\tau')$, the stabilizer $G \equiv \{\tau : \Phi(\tau)q = q\}$ is a subgroup of the additive group $\mathbb{R}^m$, and the mapping of the factor group $\mathbb{R}^m/G$ to N_α is a diffeomorphism [(1), 16.10.8]. $\mathbb{R}^m/G$ is diffeomorphic to $T^r \times \mathbb{R}^{m-r}$ for some r, $0 \leq r \leq m$, and since N_α was assumed to be compact, r must equal m. $\qquad\square$

Remarks (3.3.13)

1. Since the trajectory always remains in a connected component of N_α, it is no real restriction to consider only connected N_α's (cf. (3.3.1; 2) for $H < 0$).
2. If N_α is not compact, then it is necessary to add the requirement that all the X_i are complete. In that case, the above argument shows that N_α is diffeomorphic to some $T^r \times \mathbb{R}^{m-r}$, $0 \leq r \leq m$.

If N_α is compact, then it can be parametrized in accordance with the standard charts of T^1, and it is possible to define a normal form for the coordinates for (3.3.10).

---

† Understanding the q_i as coordinates on some chart.

Definition (3.3.14)

Let N_α be diffeomorphic to T^m and C_j a curve in N_α that encircles the j-th torus, and whose part on the other tori is continuously contractible to a point. Then we call

$$I_j(\alpha) \equiv \frac{1}{2\pi} \int_{C_j} \Theta$$

an **action variable**. If $\text{Det}(\partial I_j/\partial \alpha_k) \neq 0$, then the α's can be expressed locally in terms of the I's, and the generator

$$S \equiv \int_{q_0}^{q} \Theta$$

can be treated as a function of the I's and the q's used in the proof of (3.3.10). This transforms q^i and $p_i = \partial S/\partial q^i$ into I_j and $\varphi_j \equiv \partial S/\partial I_j$. The φ_j are known as the **angle variables** belonging to the I_j.

Remarks (3.3.15)

1. Since C_j is transformed by Φ_t, we have to determine whether the I_j remain invariant. If C_j is the boundary of a surface D_j contained in $T^*(M)$, then it is trivial that

$$I_j = -\frac{1}{2\pi} \int_{D_j} \omega = -\frac{1}{2\pi} \int_{\Phi_t D_j} \Phi_t \omega = -\frac{1}{2\pi} \int_{\Phi_t D_j} \omega$$

is constant in time. However, the assumption of a surface can be misleading (cf. (3.1.13; 6)). The general fact of constancy in time can be seen by another argument. The change in I_j between two times is

$$\frac{1}{2\pi} \int_F d\Theta,$$

where C_j and $\Phi_t C_j$ compose a surface F, which lies in N_α (see Figure 25). But $d\Theta = 0$ on N_α, and hence

$$I_j - \Phi_t I_j = \frac{1}{2\pi} \int_{\partial F} \Theta = \frac{1}{2\pi} \int_F d\Theta = 0.$$

2. Although S is not defined globally, dS is.
3. The term "angle variable" makes sense in that φ_j changes by 2π if the j-th torus is circled:†

$$\int_{C_j} d\varphi_i = \int_{C_j} d\frac{\partial S}{\partial I_i} = \int_{C_j} \frac{\partial}{\partial I_i} dq^k \, p_k(q, I) = \frac{\partial}{\partial I_i} \int_{C_j} dS$$

$$= \frac{\partial}{\partial I_i} \int_{C_j} \Theta = \frac{\partial I_j}{\partial I_i} 2\pi = 2\pi \delta_{ij}.$$

† Only the q's vary on C_j, so

$$\int_{C_j} \frac{\partial}{\partial I_i} = \frac{\partial}{\partial I_i} \int_{C_j}.$$

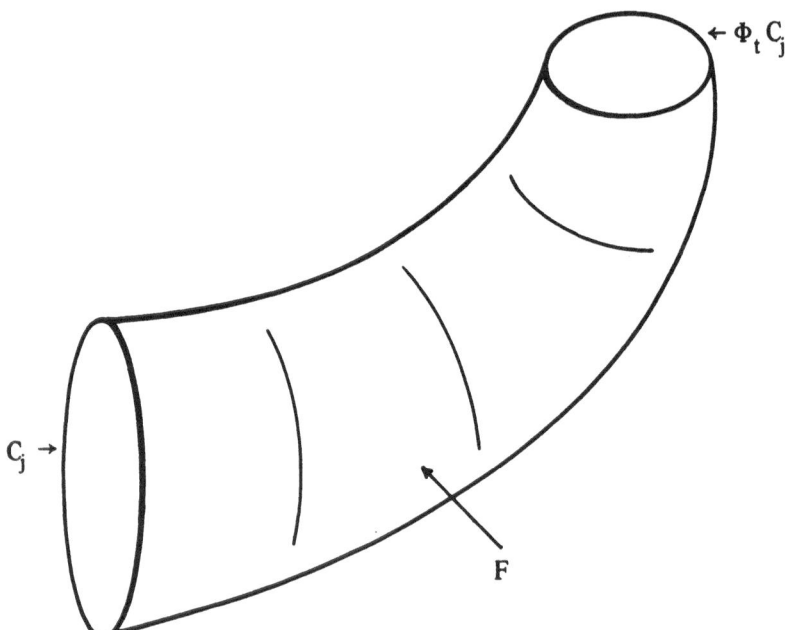

Figure 25 The paths of integration in the definition of I_j.

4. Since I and φ are canonical coordinates, the time-evolution is

$$(I_j(0), \varphi_j(0)) \to \left(I_j(0), \varphi_j(0) + t\,\frac{\partial H(I)}{\partial I_j}\right),$$

according to (3.3.11). The frequencies $\partial H(I)/\partial I_j$ depend continuously on the I's and as a rule do not always have rational ratios, which means that, usually, the trajectory covers N densely.

Examples (3.3.16)

1. The oscillator $M = \mathbb{R}$, and $H = (p^2 + \omega^2 q^2)/2$. One-dimensional systems are always integrable, if the points where $dH = 0$ are removed. Here that means the one point $p = q = 0$. We can then take $U = \{(q, p): H > 0\}$ and $N = \{(q, p): p^2 + \omega^2 q^2 = 2E \text{ (a constant)}\}$;

$$S = \int_0^q dq'\sqrt{2E - \omega^2 q'^2} = \frac{E}{\omega} \text{ arc sin} \frac{q\omega}{\sqrt{2E}} + \frac{q}{2}\sqrt{2E - q^2\omega^2}, \quad I = \frac{E}{\omega},$$

$$S(q, I) = I \text{ arc sin } q\sqrt{\frac{\omega}{2I}} + \frac{q}{2}\sqrt{2\omega I - q^2\omega^2},$$

$$\varphi = \frac{\partial S}{\partial I} = \text{arc sin } q\sqrt{\frac{\omega}{2I}}, \qquad p = \frac{\partial S}{\partial q} = \sqrt{2\omega I - q^2\omega^2},$$

and φ and I are the canonical variables of (3.1.8; 1).

2. The pendulum. $M = T^1$, and $H = p^2 - \lambda \cos \varphi$. If H does not equal λ or $-\lambda$, then $dH \neq 0$. In fact, if $H = -\lambda$, the trajectory is not a torus but a point (where the pendulum is at rest), and if $H = \lambda$, there are three trajectories: a point (in unstable equilibrium at the apex), and two trajectories diffeomorphic to $\mathbb{R}$ (asymptotically approaching the apex).

$$I(E) = \frac{1}{2\pi} \oint d\varphi \sqrt{E + \lambda \cos \varphi},$$

$$\frac{\partial I}{\partial E} = \frac{1}{\omega(E)} = \frac{1}{2\pi} \oint \frac{d\varphi}{2\sqrt{E + \lambda \cos \varphi}}.$$

If $E \gg \lambda$, the potential energy makes little difference, and the trajectories are roughly $p = \text{const}$. For $-\lambda < E < \lambda$, the trajectory returns to its starting point when $\cos \varphi_m = -E/\lambda$ (Figure 26), so the integral $\oint d\varphi$ runs only between $-\varphi_m$ and φ_m. $\omega(E)$ is an elliptic integral.

3. Small oscillations. $M = \mathbb{R}^m$, and

$$H = \sum_{i=1}^{m} \frac{p_i^2}{2m_i} + V(x).$$

Suppose that V has an equilibrium point, which we take as the origin of the coordinate system, so $dV(0) = 0$. Now replace V with the first three terms of its Taylor expansion,

$$V(x) \rightarrow V(0) + \tfrac{1}{2} x_i x_k V_{,ik}(0),$$

though as yet we can not tell how valid this replacement is (see §3.5). In any case, the kinetic and potential energy are turned into quadratic forms,

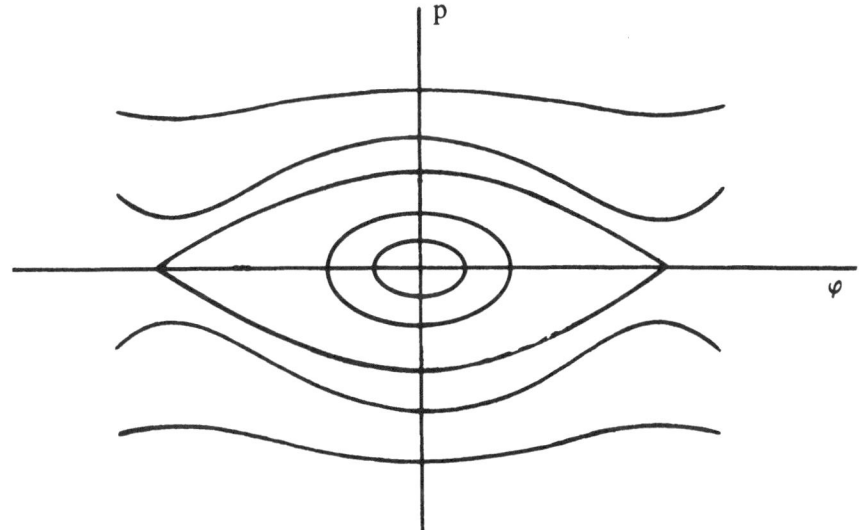

Figure 26   The trajectories of a pendulum.

and the system becomes integrable. The matrices of the quadratic forms do not commute, so they can not be simultaneously diagonalized by an orthogonal matrix. But we can put H in the form

$$H = \sum_{i=1}^{m} \frac{p_i^2}{2} + \frac{1}{2} \sum_{i,k} x_i x_k v_{ik},$$

$$v_{ik} = V_{,ik}(0)/\sqrt{m_i m_k},$$

with the point transformation $p_i \to p_i \sqrt{m_i}$, $x_i \to x_i/\sqrt{m_i}$, and then diagonalize $v: (M'vM)_{ik} = \delta_{ik} v_k$, where $M'M = 1$. In the canonical coordinates $(\bar{x}, \bar{p})$: $x_i = M_{ik}\bar{x}_k$, and $p_i = M_{ik}\bar{p}_k$,

$$H = \frac{1}{2} \sum_i (\bar{p}_i^2 + v_i \bar{q}^{i2}),$$

and the m constants we are looking for are $\bar{p}_i^2 + v_i(\bar{q}^i)^2$. The N_α are diffeomorphic to $T^r \times \mathbb{R}^{m-r}$, where r is the number of the v_i that are positive. For stable equilibrium ($r = m$ and $V_{,ik}$ is a positive matrix), there are action and angle variables which can be constructed as in Example 1.

4. The "Toda Molecule." In problems with several particles, a replacement of harmonic potentials with other functions generally destroys the integrability of the system. But sometimes a miracle occurs and one actually finds additional constants. The following model of a linear molecule with three identical particles:

$$H = \tfrac{1}{2}(p_1^2 + p_2^2 + p_3^2) + e^{q_1 - q_2} + e^{q_2 - q_3} + e^{q_3 - q_1}.$$

has, in addition to H, another constant, the momentum of the center of mass, $P = p_1 + p_2 + p_3$. This constant generates the transformation $q^i \to q^i + \lambda$, $p_i \to p_i$, which clearly leaves H invariant. Moreover,

$$\begin{aligned} K = \tfrac{1}{9}(p_1 + p_2 - 2p_3)(p_2 + p_3 - 2p_1)(p_3 + p_1 - 2p_2) \\ -(p_1 + p_2 - 2p_3)e^{q_1 - q_2} - (p_2 + p_3 - 2p_1)e^{q_2 - q_3} \\ -(p_3 + p_1 - 2p_2)e^{q_3 - q_1} \end{aligned}$$

is conserved, as can be verified from the equations of motion. Furthermore, K is invariant under the transformation generated by P; $\{P, K\} = \{H, K\} = 0$, and since dH, dP, and dK are independent on an open subset of $T^*(\mathbb{R}^3)$, the system is integrable.

Problems (3.3.17)

1. Suppose $M = \mathbb{R}^3$ and $H = |\mathbf{p}|^2$; $K_i = p_i$ for $i = 1, 2, 3$, and $K_{3+i} = [\mathbf{q} \times \mathbf{p}]_i$ (which is by definition also L_i). Are all the K_i independent constants?

2. Show that $\omega_{|N_\alpha} = 0$, with the help of a chart (see the proof of (3.3.10)).

3. Show that the harmonic oscillator with a periodic external force,

$$\mathcal{H} = \tfrac{1}{2}(p^2 + q^2) + \lambda q \cos \omega t - E, \qquad \omega \neq 1,$$

is an integrable system in extended phase space $\mathbb{R}^4$.

4. Show that K in (3.3.16; 4) is a constant.

5. Calculate the frequency of vibration of the H_2O molecule, in one dimension and linearized:

$$H = \frac{1}{2m}(p_1^2 + p_3^2) + \frac{1}{2M}p_2^2 + \frac{K}{2}((q_2 - q_1)^2 + (q_3 - q_2)^2).$$

What are the normal modes like?

6. Let the function $J \in C(M)$ be such that $\forall z \in M\ \exists c > 0$ with $(d/dt)J \circ \Phi_t(z) > c$. Suppose there exist $2m$ independent functions $\bar{z}_k$ in M (for example, the coordinate functions if M is an open set of $\mathbb{R}^{2m}$), and construct $2m - 1$ independent constants of the motion.

Solutions (3.3.18)

1. No; otherwise every trajectory would be a point. From $(\mathbf{L} \cdot \mathbf{p}) = 0$ it follows that $\mathbf{L} \cdot d\mathbf{p} + \mathbf{p} \cdot d\mathbf{L} = 0$.

2. Let A_q and A_p be the $m \times m$ matrices of partial derivatives of the K_i by the q^j and, respectively, the p_j. Choose the coordinates so that $\text{Det } A_p \neq 0$, and hence it is locally possible to write $p_j(q, K)$. Let P_q be the partial derivative of p_i by q^i, with K fixed. Then $A_p P_q + A_q = 0$. The vanishing of the Poisson bracket implies that $A_p A_q^t = A_q A_p^t$. Thus $A_p P_q A_p^t + A_p A_q^t = A_p(P_q A_p^t + A_q^t) = 0$. Since A_p^{-1} exists, we conclude that $A_p P_q^t + A_q = 0$, and so $P_q^t = P_q$. This is exactly the condition that $\sum p_i(q, \alpha)dq^i$ is closed on N_α.

3. In addition to $\mathcal{H}$ there is the constant

$$K = ((\omega^2 - 1)q - \lambda \cos \omega t)^2 + ((\omega^2 - 1)p + \lambda\omega \sin \omega t)^2;$$

$$\frac{dK}{ds} = 2[(\omega^2 - 1)p + \lambda\omega \sin \omega t][(\omega^2 - 1)q - \lambda \cos \omega t$$

$$- (\omega^2 - 1)(q + \lambda \cos \omega t) + \lambda\omega^2 \cos \omega t] = 0.$$

Here N is not compact, but instead is diffeomorphic to $T^1 \times \mathbb{R}$.

4. Periodic: $q_{i+2} = q_{i-1}$, etc., and

$$\frac{d}{dt}\frac{1}{9}\prod_{i=1}^{3}(p_{i+1} + p_{i-1} - 2p_i)$$

$$= \frac{1}{3}\sum_{i=1}^{3}(p_{i+1} + p_{i-1} - 2p_i)(p_{i-1} + p_i - 2p_{i+1})(e^{q_{i+1} - q_{i-1}} - e^{q_{i-1} - q_i})$$

$$= \frac{1}{3}\sum_{i=1}^{3}(p_{i+1} + p_{i-1} - 2p_i)(p_{i-1} - p_{i+1})e^{q_{i+1} - q_{i-1}}$$

$$= -\frac{1}{3}\frac{d}{dt}\sum_{i=1}^{3}(p_{i+1} + p_{i-1} - 2p_i)e^{q_{i+1} - q_{i-1}}.$$

5.

$$v_{ik} = K \begin{vmatrix} \dfrac{1}{m} & -\dfrac{1}{\sqrt{mM}} & 0 \\[2.2ex] -\dfrac{1}{\sqrt{mM}} & \dfrac{2}{m} & -\dfrac{1}{\sqrt{mM}} \\[2.2ex] 0 & -\dfrac{1}{\sqrt{mM}} & \dfrac{1}{m} \end{vmatrix}$$

$$\text{Det}(v - 1 \cdot \omega^2) = \omega^2 \left(\frac{K}{m} - \omega^2 \right) \left(\omega^2 - \frac{K}{m} - \frac{2K}{M} \right),$$

Frequency $\omega = 0$ $\omega = \sqrt{\dfrac{K}{m}}$ $\omega = \sqrt{\dfrac{K}{m} + \dfrac{2K}{M}}$

	H	O	H		H	O	H		H	O	H
mode	○	○	○		○	○	○		○	○	○
	→	→	→		←	○	→		→	←	→

6. The monotonic mapping $\mathbb{R} \to \mathbb{R}: t \to J \circ \Phi_t(z)$ is invertible for any z. The inverse image of, say, $0 \in \mathbb{R}$ under the mapping $\mathbb{R} \times M \to \mathbb{R}: (t, z) \to J \circ \Phi_t(z)$ assigns a value of t to each $z \in M$. Letting this function $M \to \mathbb{R}$ be called τ, we find that $\tau \circ \Phi_t(z) = \tau(z) - t \; \forall z \in M$, since $0 = J \circ \Phi_{\tau(z)}(z) = J \circ \Phi_{\tau(z)-t}(\Phi_t(z))$. Hence the mapping $M \to M: z \to \Phi_{\tau(z)}(z)$ is time-invariant: $\Phi_{\tau(\Phi_t(z))}(\Phi_t(z)) = \Phi_{\tau(z)} \circ \Phi_{-t} \circ \Phi_t(z) = \Phi_{\tau(z)}(z)$. Composing this with the coordinate functions $\bar{z}_k, k = 1, \dots, 2m$, produces the $2m$ constants $\bar{z}_k \circ \Phi_{\tau(\bar{z})}(\bar{z})$. Define $z(t, \bar{z}) \equiv \bar{z} \circ \Phi_t(\bar{z})$, $z_{i;k} \equiv \partial z_i / \partial \bar{z}_k$, and $\dot{z}_i \equiv \partial z_i / \partial t$; then the differentials of the constants are $d\bar{z}_i(z_{k;i} + \dot{z}_k \partial \tau / \partial \bar{z}_i)$. Since Φ_t is a diffeomorphism, the matrix $z_{k;i}$ has rank $2m$. The matrix $\dot{z}_k \partial \tau / \partial \bar{z}_i$ has rank 1, so the sum has at least rank $2m - 1$. The condition $J \circ \Phi_{\tau(z)}(z) = 0$ implies that the rank is in fact equal to $2m - 1$, which is then the number of independent constants.

3.4 The Limit $t \to \pm \infty$

Often the time-evolution of a system approaches that of an integrable system asymptotically. If so, its behavior after long times can be discovered.

Theoretical predictions usually become less precise for longer times, and the future of a system as $t \to \pm \infty$ may be wholly unknown. However, if the potential is of finite range, then particles that escape eventually act like free particles and their time-evolution becomes simple. As we shall see, on the part of phase space filled by the trajectories of escape, Φ_t is diffeomorphic to Φ_t^0, free time-evolution, and there are $2m - 1$ constants of motion.

We start by looking for quantities which are not necessarily constant, but approach limiting values.

Definition (3.4.1)

Let

$\mathcal{A} \equiv \{f \in C^\infty(T^*(M))$: the pointwise limit $\lim_{t \to \pm \infty} \tau_t f$ exists and
$\in C^\infty(T^*(M))\} \equiv \{$the **asymptotic constants** of the motion$\}$;
$\{H\}' \equiv \{f \in C^\infty(T^*(M))$: $\tau_t f = f\} \equiv \{$the **constants** of the motion$\}$;
and
$\tau_\pm : \mathcal{A} \to \{H\}'$, such that $f \to \lim_{t \to \pm \infty} \tau_t f$.

Remarks (3.4.2)

1. Since τ_t commutes with the algebraic operations, $\mathcal{A}$ and $\{H\}'$ are algebras, and $\tau_\pm$ are homomorphisms.
2. Timewise limits are constants in time, and since $\tau_{\pm |\{H\}'} = 1$, $\tau_\pm$ are mappings from $\mathcal{A}$ onto $\{H\}'$.
3. It is not necessary for $\tau_\pm$ to be injective; $\{H\}'$ may be a proper subset of $\mathcal{A}$.
4. $\mathcal{A}$ (or its quantum-mechanical generalization) is of especial interest in atomic physics, where only the asymptotic parts of the trajectories can be measured directly. The deflection angle of the particles is given by the difference between $\tau_- \mathbf{p}$ and $\tau_+ \mathbf{p}$.

Examples (3.4.3)

1. $M = T^1$ and $H = \omega p$ (an oscillator). Since the time-evolution is given by $(\varphi, p) \to (\varphi + \omega t, p)$, a function $f(\varphi, p) \in C^\infty(T^*(T^1))$ is an asymptotic constant of the motion iff f depends only on p: this case is trivial, as $\mathcal{A} = \{H\}'$ and $\tau_\pm = 1$.
2. $M = \mathbb{R}^+$ and $H = p^2/2 + \gamma/r^2$, $\gamma > 0$. This system is integrable, because $p = \dot{r} = \sqrt{2(E - \gamma/r^2)}$, and if

$$t = \int \frac{dr\, r}{\sqrt{2Er^2 - 2\gamma}} = \frac{1}{2E}\sqrt{2Er^2 - 2\gamma},$$

then

$$r = \sqrt{\frac{\gamma}{E} + 2Et^2},$$

and

$$p = t\sqrt{2E}\left(\sqrt{\frac{\gamma}{2E^2} + t^2}\right)^{-1}.$$

Thus $r(t)$ is a hyperbola, and the trajectories $H = $ constant are asymptotically horizontal in phase space (see figure). If we express E in terms of
the initial values, we obtain the time-evolution,

$$\Phi_t: \begin{vmatrix} r \\ p \end{vmatrix} \to \begin{vmatrix} \left[p^2 + \dfrac{2\gamma}{r^2}\right]^{-1/2}\left[2\gamma + \left(t\left(p^2 + \dfrac{2\gamma}{r^2}\right) + rp\right)^2\right]^{1/2} \\ \left[p^2 + \dfrac{2\gamma}{r^2}\right]^{1/2}\left[2\gamma + \left(t\left(p^2 + \dfrac{2\gamma}{r^2}\right) + rp\right)^2\right]^{-1/2}\left(t\left(p^2 + \dfrac{2\gamma}{r^2}\right) + rp\right) \end{vmatrix}.$$

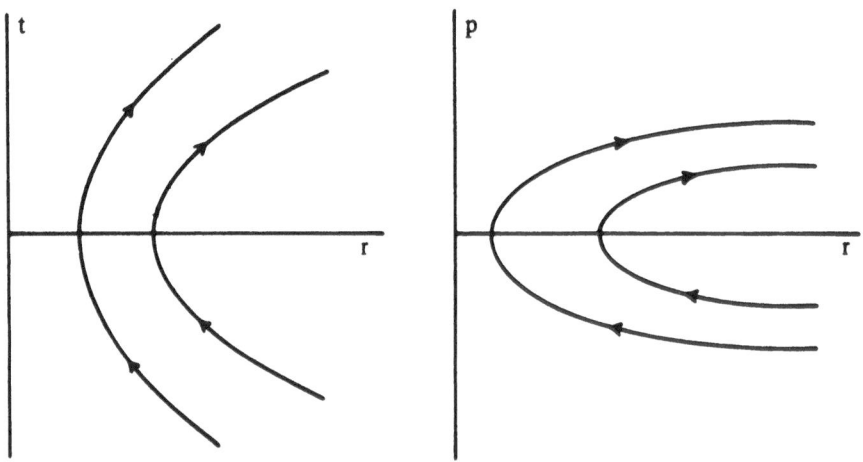

Trajectories for a $1/r^2$ potential.

Observe that

$$p \overset{t \to \pm\infty}{\to} \pm\sqrt{p^2 + \frac{2\gamma}{r^2}}, \quad \text{and} \quad \frac{1}{r} \to 0,$$

so

$$\mathcal{A} = \left\{ f\left(p, \frac{1}{r}\right), f \in C^\infty(\mathbb{R} \times (0, \infty)) \right\},$$

$$\tau_\pm : f\left(p, \frac{1}{r}\right) \to f\left(\pm\sqrt{p^2 + \frac{2\gamma}{r^2}}, 0\right).$$

This time $\{H\}'$ is just $\{f(H)\}$, and is a proper subset of $\mathcal{A}$.

3. $M = \mathbb{R}^2 \backslash \{0\}$ and $H = p^2/2 + \alpha/r^2$. The point transformation $(x, y) \to$
$(r \cos \varphi, r \sin \varphi)$ generates a canonical transformation $(x, y; p_x, p_y) \to$
$r, \varphi; p_r, L)$, by which H becomes $p_r^2/2 + \gamma/r^2$, where $\gamma = \alpha + L^2/2$.

Consequently the radial motion is as in Example 2, and the equation $\dot{\varphi} = \partial H/\partial L = L/r^2$ can be integrated by substituting $r(t)$ in, yielding

$$\Phi_t : (\varphi, L) \to \left(\varphi + \frac{L}{\sqrt{2\alpha + L^2}} \left[\arctan \frac{rp_r + t\left(p_r^2 + \dfrac{L^2 + 2\alpha}{r^2}\right)}{\sqrt{2\alpha + L^2}} \right.\right.$$
$$\left.\left. - \arctan \frac{rp_r}{\sqrt{2\alpha + L^2}}\right], L\right).$$

The functions $f(\varphi, L)$ now also belong to $\mathscr{A}$, because angular momentum is conserved, and the particle escapes at a definite angle:

$$\tau_\pm f(\varphi, L) = f\left(\varphi + \frac{L}{\sqrt{2\alpha + L^2}}\left[\pm \frac{\pi}{2} - \arctan \frac{rp_r}{\sqrt{2\alpha + L^2}}\right], L\right)$$

(see figure). Note that in this case a third constant independent of H and L also appears,

$$\tau_+ p_x = \sqrt{2H} \cos\left(\varphi + \frac{L}{\sqrt{2\alpha + L^2}}\left[\frac{\pi}{2} - \arctan \frac{rp_r}{\sqrt{2\alpha + L^2}}\right]\right).$$

In physics the connection between the observables as $t \to \pm \infty$ is quite important, and one would like to know what the mapping $\tau_+ \circ \tau_-^{-1}$ is. Unfortunately τ_-^{-1} is not uniquely defined, since τ_- is not injective. One can get around this problem by choosing a subalgebra of $\mathscr{A}$ on which $\tau_\pm$ are

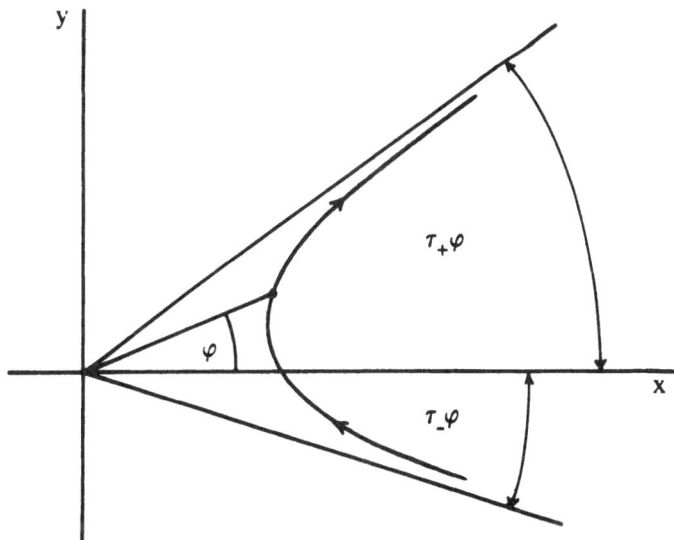

The trajectories in a plane.

injective. The mapping of the asymptotic quantities then depends on what subalgebra has been chosen; if the subalgebra is $\{H\}'$, for instance, then $\tau_+ \circ \tau_-^{-1} = 1$. If the time-evolution Φ_t asymptotically approaches that of a simple reference system Φ_t^0, then as in (2.3.11; 3) it is possible to construct a limiting diffeomorphism, which reproduces the action of the τ's on a subalgebra, and makes them invertible.

Definition (3.4.4)

Let H and H_0 be two Hamiltonians that generate the flows Φ_t and Φ_t^0 on $T^*(M)$. If

(a) $\lim_{t \to \pm\infty} \Phi_{-t} \circ \Phi_t^0 \equiv \Omega_{\pm}$ exist pointwise on some open sets $D_{\pm}$,
(b) $\Omega_{\pm}$ are local canonical transformations from $D_{\pm}$ onto neighborhoods $\mathscr{R}_{\pm}$, and
(c) $\lim_{t \to \pm\infty} \Phi_{-t}^0 \circ \Phi_t$ exist on $\mathscr{R}_{\pm}$ and equal $\Omega_{\pm}^{-1}$, then we say that the **Møller-transformations** $\Omega_{\pm}$ exist.

Remarks (3.4.5)

1. From (2.3.11; 3), $\Phi_t \circ \Omega_{\pm} = \Omega_{\pm} \circ \Phi_t^0$. According to (2.5.9; 2), $L_{\Omega_{\pm}^* X_{H_0}} = L_{X_H}$, and therefore $H_0 \circ \Omega_{\pm}^{-1} = H_{|\mathscr{R}_{\pm}}$.† What is more, $\forall t_0$, $\Phi_{-t} \circ \Phi_t^0$ also converges on $\Phi_{t_0}^0 D_{\pm}$, so we may assume that $D_{\pm}$ are τ_t^0-invariant, making $\mathscr{R}_{\pm}$ τ_t-invariant. Then the flow that H creates on $\mathscr{R}_{\pm}$ is diffeomorphic to the flow that H_0 creates on $D_{\pm}$.
2. For an observable f, $f \circ \Omega_{\pm}^{-1} \circ \Phi_t \circ \Omega_{\pm} = f \circ \Phi_t^0$ implies that the image under $\Omega_{\pm}^*$ of the time-evolution according to H_0 is the same as the time-evolution according to H of the image: $\tau_t(\Omega_{\pm}^* f) = \Omega_{\pm}^*(\tau_t^0 f)$. Thus $\{H_0\}'$ gets mapped into $\{H\}'$. If H_0 has $2m - 1$ independent constants (as for free motion), then so does H on $\mathscr{R}_{\pm}$. In particular, such a system is integrable on $\mathscr{R}_{\pm}$.
3. The transformations that $\Omega_{\pm}$ generate on $\{H_0\}'$ are exactly $\tau_{\pm}$, since $\forall f \in \{H_0\}'$, $f \circ \Phi_{-t}^0 \circ \Phi_t = f \circ \Phi_t \to f \circ \Omega_{\pm}^{-1}$ as $t \to \pm\infty$, i.e., $\tau_{\pm}(f) = \Omega_{\pm}^*(f)$. Hence $\mathscr{A}$ contains $\{H_0\}'$, and is usually larger than $\{H\}'$.
4. If H and H_0 are invariant under $T: (x, p) \to (x, -p)$ (reversal of the velocities, which is not a canonical transformation), then $\Phi_t \circ T = T \circ \Phi_{-t}$ and $\Phi_t^0 \circ T = T \circ \Phi_{-t}^0$; so the existence of Ω_+ on D_+ implies that of Ω_- on $T(D_+)$.

Examples (3.4.6)

1. $M = T^1$, and $H = \omega p$ (an oscillator). In this example $\Omega_{\pm}$ exist only for $H_0 = H$, and then $\Omega_{\pm}$ are trivially 1.

---

† Only up to a constant, of course.

2. $M = \mathbb{R}^+$, $H = p^2/2 + \gamma/r^2$, and $H_0 = p^2/2 + \gamma_0/r^2$. ($\gamma_0$ can not be set to 0, because then X_{H_0} would not be complete.) With the result of (3.4.3; 2), we calculate for $p \neq 0$ that

$$\lim_{t \to \pm \infty} \Phi_{-t} \Phi_t^0(r, p) = \left(r \sqrt{\frac{r^2 p^2 + 2\gamma}{r^2 p^2 + 2\gamma_0}}, \, p \sqrt{\frac{r^2 p^2 + 2\gamma_0}{r^2 p^2 + 2\gamma}} \right) \equiv (\bar{r}, \bar{p}).$$

In fact, $(r, p) \to (\bar{r}, \bar{p})$ is a canonical transformation (Problem 3), so that $H_0 = H(\bar{r}, \bar{p})$; that is, $H_0 = H \circ \Omega_\pm$. The domains of $\Omega_\pm$ and their ranges are $D_\pm = \mathcal{R}_\pm = T^*(\mathbb{R}^+)$. All H's of this form produce diffeomorphic flows for all $\gamma > 0$, and $\Phi_{-t}^0 \circ \Phi_t$ always converges to $\Omega_\pm^{-1}$. So in this example, the Møller transformation exists and is different from **1**.

3. $M = \mathbb{R}^2 \backslash \{0\}$, $H = |\mathbf{p}|^2/2 + \alpha/|\mathbf{x}|^2$, and $H_0 = |\mathbf{p}|^2/2$. Using polar coordinates as in (3.4.33; 3), the radial problem reduces to Example 2, and for the angles we find

$$\Omega_\mp(\varphi, L) = \left(\varphi + \arctan \frac{r p_r}{L} - \frac{L}{\sqrt{2\alpha + L^2}} \arctan \frac{r p_r}{\sqrt{2\alpha + L^2}} \right.$$

$$\left. \pm \frac{\pi}{2} \left(\frac{L}{\sqrt{2\alpha + L^2}} - 1 \right), L \right).$$

It is not hard to convince oneself that the $\Omega_\pm$ transform H_0 canonically into H. $D_\pm$ and $\mathcal{R}_\pm$ are $T^*(\mathbb{R}^2) \backslash \{\{0\} \times \mathbb{R}^2\}$, and the Møller transformations exist, and can be extended to all of $T^*(\mathbb{R}^2)$. $\mathbf{p} \in \{H_0\}'$, and in fact $\mathbf{p} \in \mathcal{A}$.

Some of the properties of the above example hold for a wider class of potentials:

Theorem (3.4.7)

Let $M = \mathbb{R}^m$, $H = |\mathbf{p}|^2/2 + V(x)$, where $V \in C_0^\infty(\mathbb{R}^m)$, and $H_0 = |\mathbf{p}|^2/2$. Then

(a) $\exists \Omega_\pm$, $D_\pm = T^*(\mathbb{R}^m) \backslash \{\mathbb{R}^m \times \{0\}\}$,

(b) $C\mathcal{R}_\pm = \bigcup_n \{(x, p) \in T^*(\mathbb{R}^m): \|\Phi_{\pm t} x\| < n \, \forall t > 0\}$,

(c) $\Omega(\mathcal{R}_+ \Delta \mathcal{R}_-) = 0.$†

Remarks (3.4.8)

1. The significance of (b) is that $\mathcal{R}_\pm$ are the complements of the trajectories that remain in compact sets for all $t \lessgtr 0$. Yet $\mathcal{R}_+$ need not be the same as $\mathcal{R}_-$. To see that, consider the following one-dimensional example:

---

† Δ is the symmetric difference and Ω is the Liouville measure.

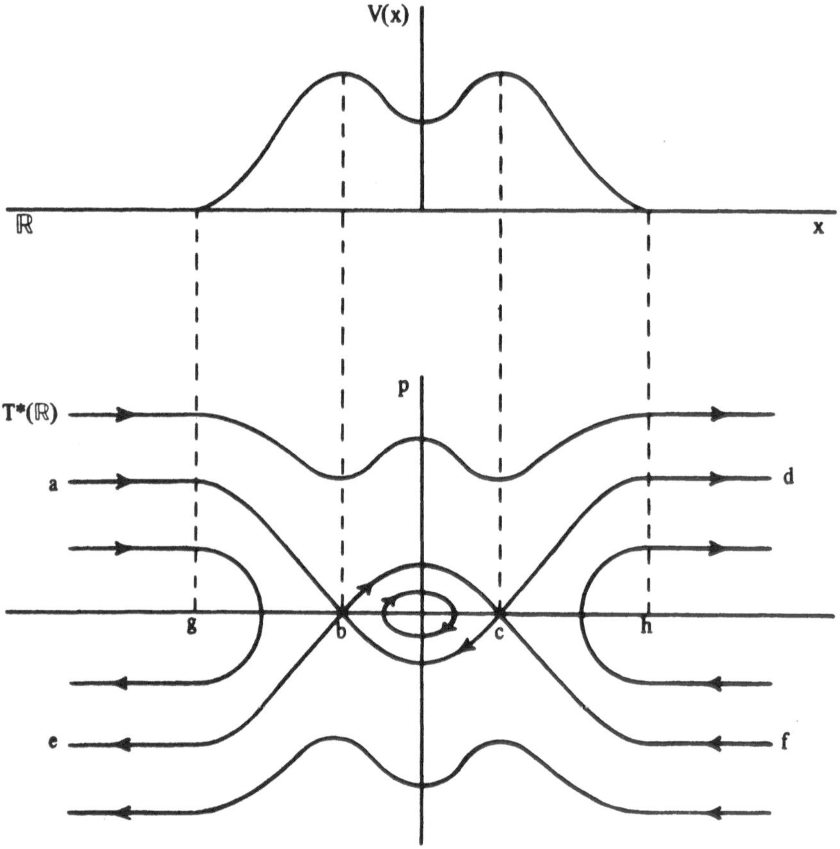

V(x)

ℝ   x

T*(ℝ)

p

a   d

g   b   c   h

e   f

Figure 27   Trajectories in phase space.

$C\mathscr{R}_+ \cap C\mathscr{R}_- =$ (a closed set bounded by the trajectories bc and cb) $\cup\{(-\infty, g) \times \{\mathbf{0}\}\} \cup \{(h, \infty) \times \{\mathbf{0}\}\}$. In addition, $C\mathscr{R}_+$ (respectively $C\mathscr{R}_-$) contains the trajectories ab and fc (respectively be and cd).

2. In the terminology of atomic physics, $\mathscr{R}_+ \cap \mathscr{R}_-$ is the set of scattering states, and $C\mathscr{R}_+ \cap C\mathscr{R}_-$ is the set of bound states:

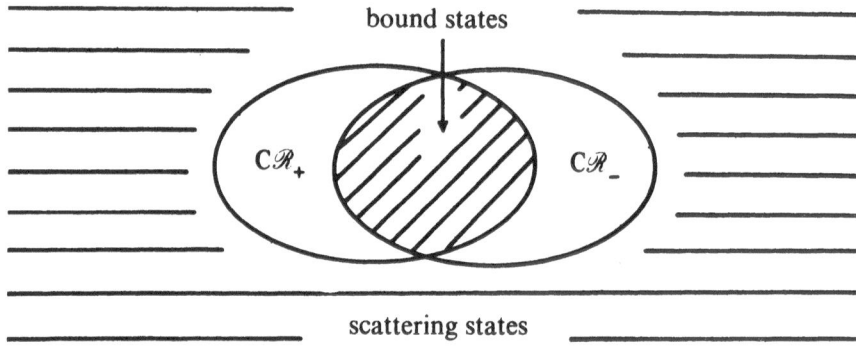

bound states

$C\mathscr{R}_+$ $C\mathscr{R}_-$

scattering states

Statement (c) expresses "asymptotic completeness": bound states and scattering states fill up almost all of phase space. The above example shows that it is not possible to drop the word "almost" here any more than in Theorems (2.6.13) and (2.6.14).

3. This theorem about C_0^∞-forces of finite range can be extended to cover forces that fall off fast enough at infinity. But in a system with many particles and potentials $V(\mathbf{x}_i - \mathbf{x}_k)$, the forces do not fall off along lines where $\mathbf{x}_i = \mathbf{x}_k + \text{constant}$, and a theory with many channels is required. We will develop such a theory in the context of quantum mechanics.†

4. The statement that, generically, there are no further constants beyond H depends on the topology used to define the word "generic." The theorem shows that with a topology for which the C_0^∞-functions are an open set, there is a neighborhood of the zero potential where there exist $2m - 1$ constants of motion. In such a topology, potentials that do not allow further constants can not be dense. Theorems about the absence of additional constants of motion generally refer to the situation where the trajectories are actually orbits, i.e., they remain in compact sets.‡

Proof

(a) $\Phi_t^0 : (x, p) \to (x + pt, p)$, and if $V(x) = 0 \, \forall \|x\| > \rho$, then $\Phi_t^0 = \Phi_t \, \forall \|x\| > \rho$. Hence $\forall (x, p) \in D_+$ there exists some T such that $\Phi_{-t} \circ \Phi_{T+t}^0(x, p)$ $= \Phi_T^0(x, p) \, \forall t > 0$. Consequently $\Phi_{-T-t} \circ \Phi_{T+t}^0(x, p) = \Phi_{-T} \circ \Phi_t^0(x, p)$ $\forall t > 0$, and so for all points of D_+ the limit is reached after a finite time, and is $\Omega_+ = \Phi_{-T} \circ \Phi_T^0$. To understand why the limit is also a diffeomorphism, note that for any compact subset K of D_+ there exists a T such that $\Omega_{+|K} = \Phi_{-T} \circ \Phi_{T|K}^0$, and that $\Phi_{-t} \circ \Phi_t^0$ is a diffeomorphism for all t (X_H and X_{H_0} are certainly complete). Similarly for Ω_- and $\Omega_\pm^{-1}$.

(b) $\forall x \in \mathcal{R}_+, \exists x_0 \in D_+ : x = \Phi_{-t} \circ \Phi_t^0 x_0 \, \forall t > T$, which $\Leftrightarrow \Phi_t x = \Phi_t^0 x_0 \, \forall t > T$. But for all n there exists a $t > T$ such that $\|\Phi_t^0 x_0\| > n$. Similarly for $\mathcal{R}_-$.

(c) Let $b_n^\pm = \{(x, p) \in T^*(\mathbb{R}^m) : \|\Phi_{\pm t} x\| < n \, \forall t > 0\}$. By Theorem (2.6.14), $\Omega(b_n^+ \cap b_n^-) = \Omega(b_n^+) = \Omega(b_n^-) \, \forall n \in \mathbb{Z}^+ :$ because $C\mathcal{R}_\pm = \bigcup_n b_n^\pm$, we conclude that $\Omega(C\mathcal{R}_+ \Delta C\mathcal{R}_-) = 0$, which is equivalent to $\Omega(\mathcal{R}_+ \Delta \mathcal{R}_-) = 0$. $\square$

If only the asymptotic parts of the trajectories are observed, one would like to know the relationships between them. For that purpose we make

Definition (3.4.9)

The local canonical transformation $S = \Omega_+ \circ \Omega_-^{-1}$ from $\mathcal{R}_-$ to $\mathcal{R}_+$ is called the **scattering transformation**.§

---

† Quantum Mechanics of Atoms and Molecules.

‡ We use the word "orbit" strictly for such trajectories.

§ In quantum mechanics this is the S-matrix of the Heisenberg picture, and is to be distinguished from $\Omega_+^{-1} \circ \Omega_-$, the S-matrix of the interaction picture.

Examples (3.4.10)

1. $M = \mathbb{R}^+$, $H = p^2/2 + \gamma/r^2$, and $H_0 = p^2/2 + \gamma_0/r^2$. According to (3.4.6; 2),
   $\mathscr{R}_+ = \mathscr{R}_- = T^*(\mathbb{R}^+)$, and $S = 1$.
2. $M = \mathbb{R}^2 \backslash \{0\}$, $H = |\mathbf{p}|^2/2 + \gamma/r^2$, and $H_0 = |\mathbf{p}|^2/2$. The domain of S is
   the standard domain $D_\pm$, and by (3.4.6; 3) it is

$$S: (r, \varphi, p_r, L) \to \left(r, \varphi - \pi \left(\frac{L}{\sqrt{2\alpha + L^2}} - 1 \right), p_r, L \right)$$

in polar coordinates.

Remarks (3.4.11)

1. $\Omega_\pm$ act like the asymptotic time-evolution in the sense that if $\tau_-^{-1} f \in \{H_0\}'$,
   then $S^*(f) = f \circ \Omega_- \circ \Omega_+^{-1} = \tau_+ \circ \tau_-^{-1} f$. But it is only then that S has
   the intuitive meaning of giving the changes of certain observables after
   long times. It thus happens that other quantities that are invariant
   under Φ_t are changed by S: only elements of $\{H_0\}' \cap \{H\}'$ are invariant
   under S. On the other hand, the radial momentum p_r remains unchanged
   under S, though it is not a constant of motion. This comes about because
   H and H_0 both asymptotically change p_r into $-p_r$.
2. Since $\mathbf{p} \in \{H_0\}'$, S gives the change in the momenta after long times. The
   intuitive meaning of S can be visualized if we represent points in phase
   space as points in configuration space with arrows (see Figure 28). The
   notation is such that $(\mathbf{x}, \mathbf{p})$ can be considered either as a point in phase
   space or as a function on phase space, if $\Omega^{-1}(p)$ is identified with $p \circ \Omega^{-1}$
   $= \Omega^* p$.
3. Points that are initially near each other in phase space can become
   separated arbitrarily far after an infinitely long time. That S is a diffeomorphism means only that the separation is small compared with the
   separation of points under the action of H_0.

One can thus calculate the angle between $\mathbf{p}_- \equiv \tau_- \mathbf{p}$ and $\mathbf{p}_+ \equiv \tau_+ \mathbf{p}$
from $S: S^*(\mathbf{p}_-) = \mathbf{p}_+$. Because in experiments the trajectories are only
statistically known, it is useful to make

Definition (3.4.12)

The angle Θ between $\mathbf{p}_-$ and $\mathbf{p}_+$ is called the **scattering angle**, and the **differential scattering cross-section** is defined as $d\sigma =$ (number of particles
scattered into the interval $[\Theta, \Theta + d\Theta]$ per unit time)/(number of incident
particles per time and per unit of surface area) $= 2\pi\sigma(\Theta)\sin \Theta \, d\Theta$.

Remarks (3.4.13)

1. If the initial (unnormalized) distribution of particles ρ (cf. (1.3.1)) specifies
   $\mathbf{p}_-$ precisely but leaves x completely unrestricted, i.e., $\rho(x, p) \cong \delta^3(\mathbf{p}_- - \mathbf{k})$,

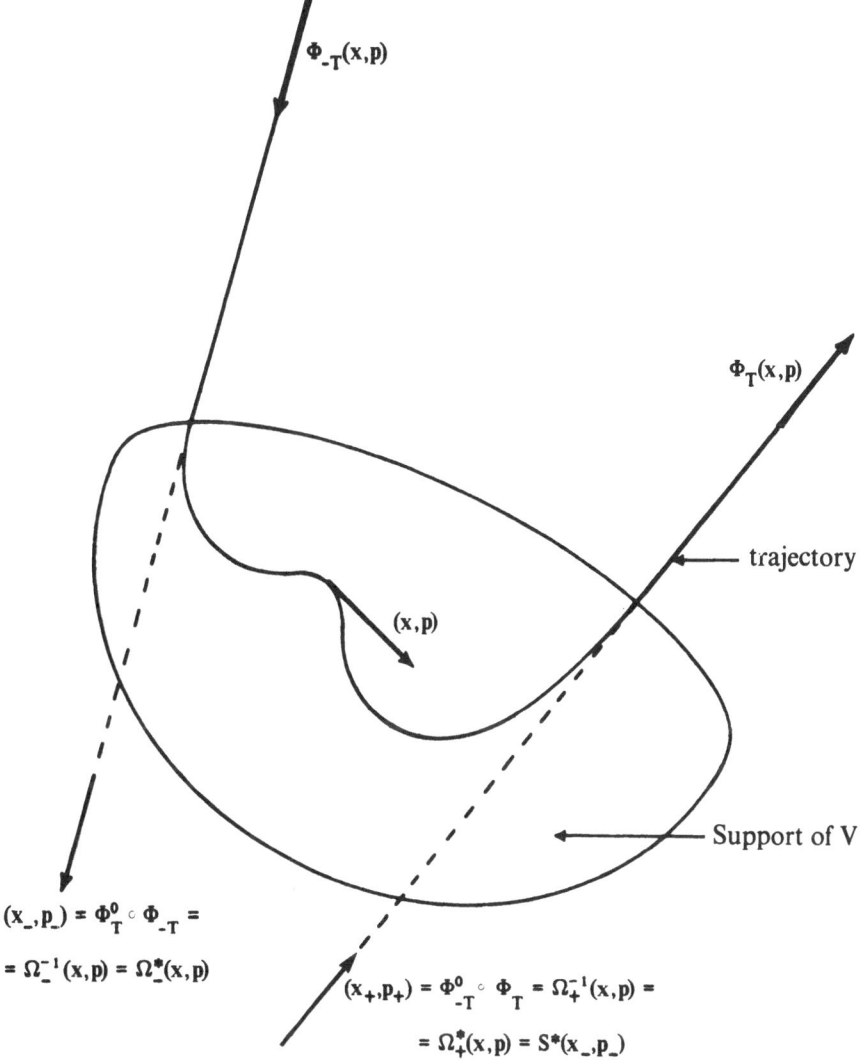

Figure 28   The meaning of the scattering transformation.

then the impact parameter $b = |\mathbf{L}|/|\mathbf{p}_-|$ has a probability distribution $2\pi b \, db \cdot f$, where f is the number of incident particles per time and per unit area. For a central potential in $\mathbb{R}^3$ the number of scattered particles per unit time is[†]

$$f 2\pi b \, db = f \frac{2\pi}{4E} d|\mathbf{L}|^2 = \frac{f\pi}{E} L \frac{dL}{d\Theta} \, d\Theta = 2\pi f \sigma(\Theta) \sin \Theta \, d\Theta,$$

where it is supposed that the relationship between Θ and L is known.

---

† If the mass m is not set to 1, E should be replaced with mE.

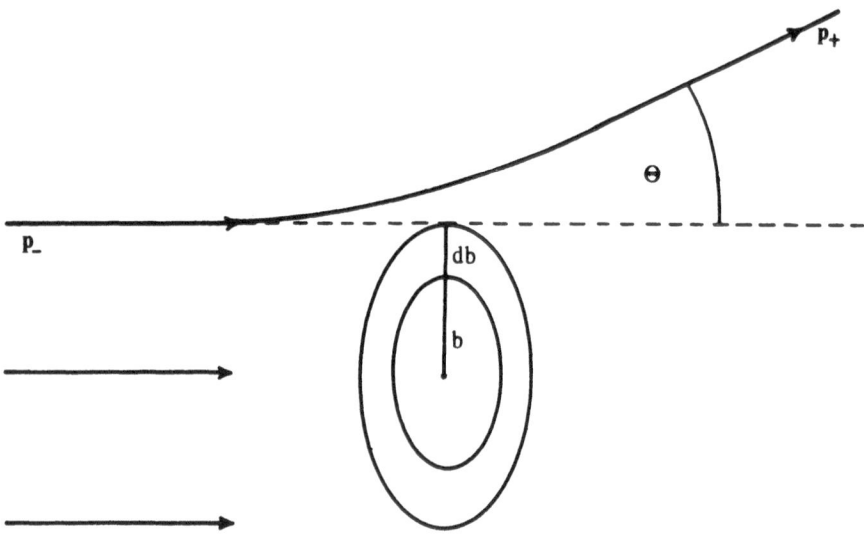

Figure 29   The scattering angle and scattering cross-section.

2. In $\mathbb{R}^2$ the unit of surface area becomes instead a unit of length, and
   correspondingly $2\pi b\, db \to db$ and $2\pi \sin \Theta\, d\Theta \to d\Theta$, and so

$$\sigma(\Theta) = \frac{1}{\sqrt{2E}} \frac{dL}{d\Theta}.$$

Examples (3.4.14)

1. The $1/r^2$ potential in $\mathbb{R}^2$. By $(3.4.10; 2)$, $\Theta = \pi((-L/\sqrt{2\alpha + L^2}) + 1)$,
   which implies

$$L^2 = 2\alpha \frac{(\Theta - \pi)^2}{\Theta(2\pi - \Theta)}$$

and hence

$$\sigma(\Theta) = \sqrt{\frac{\alpha}{E}} \frac{\partial}{\partial\Theta} \frac{\pi - \Theta}{\sqrt{\Theta(2\pi - \Theta)}}.$$

2. The $1/r^2$ potential in $\mathbb{R}^3$. By conservation of momentum this problem
   reduces to a planar one (see §5.3), so L becomes the same function of Θ
   as in Example 1, and therefore

$$\sigma(\Theta) = \frac{\pi\alpha}{E \sin \Theta} \frac{\partial}{\partial\Theta} \frac{(\pi - \Theta)^2}{\Theta(2\pi - \Theta)}.$$

Remarks (3.4.15)

1. The angular distribution is strongly bunched up in the forward direction
   and is not integrable as $\Theta \to 0$. This is because the particles with very
   large b can still be scattered, although not very much.

2. As $\alpha \to 0$, σ approaches 0, as it must. The cross-section in $\mathbb{R}^3$ is linear in α because only $\sqrt{\alpha/E}$ has the dimension of length.
3. Θ is independent of E because the canonical transformation $x \to \lambda x$, $p \to \lambda^{-1} p$ sends H to $\lambda^{-2} H$, but leaves Θ unchanged.

After having dealt with the trajectories that go off to infinity, we study those that always remain in compact neighborhoods of some equilibrium position. For such trajectories there is hope that the linearized theory (3.3.16; 3) will be useful as a basis of comparison. The intuitive notion of the stability of an equilibrium position is now made into a general

Definition (3.4.16)

Let S be a mapping of a topological space M into itself, and suppose that $S(p) = p$ for some $p \in M$ (a fixed point). We say that S is **stable** iff there exist arbitrarily small neighborhoods U of p such that $SU \subset U$. If the only subsets with this property are M and $\{p\}$, S is **unstable**.

Remarks (3.4.17)

1. It is clear that unstable $\Rightarrow$ not stable, but not vice versa. E.g., for $M = \mathbb{R}^2$ and $p = (0, 0)$, $S : (x, y) \to (x + y, y)$ is neither stable nor unstable. We refer to such cases as **mixed**.
2. If S is the time-evolution Φ_t for some t, then from the group property it follows that $\Phi_{nt} U \subset U \; \forall n \in \mathbb{Z}^+$. Whereas the general theorems about differential equations say only that a trajectory near the point p does not get very far away in a short time, stability requires this for all time. The phrase "stability under Φ_t" will be taken to mean stability under Φ_t for all t.

Convergence as $t \to \pm \infty$ can not be expected when the motion is periodic. In order to define an average time-dependence, let us assume that $T^*(M)$ equals $\mathbb{R}^{2m}$. Then it is possible to define the sum of two points of $T^*(M)$, and thereby to define the sum of diffeomorphisms. This enables us to make

Definition (3.4.18)

Given the canonical flows Φ_t and Φ_t^0 on $\mathbb{R}^{2m}$, let

$$C_\pm \equiv \lim_{T \to \pm\infty} \frac{1}{T} \int_0^T dt \; \Phi_{-t}^0 \circ \Phi_t,$$

assuming that this Cesàro average exists and is a local canonical transformation on a neighborhood $D_\pm \subset \mathbb{R}^{2m}$.

Remarks (3.4.19)

1. If $\Omega_\pm$ exists, then so does $C_\pm$, but not necessarily vice versa.
2. If $\bigcup_t \Phi^0_{-t} \circ \Phi_t$ is bounded†, then $C_\pm$ has the same effect as $\Omega_\pm$, as it maps both flows canonically onto each other; for

$$\Phi^0_\tau \circ C_\pm = \lim_{T \to \pm \infty} \frac{1}{T} \int_0^T dt\, \Phi^0_{-t+\tau} \circ \Phi_t = \lim_{T \to \pm \infty} \frac{1}{T} \int_{-\tau}^{T-\tau} dt'\Phi^0_{-t'} \circ \Phi_{t'} \Phi_\tau$$

$$= C_\pm \circ \Phi_\tau + \lim_{T \to \pm \infty} \frac{1}{T} \int_{-\tau}^0 dt'(\Phi^0_{-t'} \circ \Phi_{t'} - \Phi^0_{-T+t'} \circ \Phi_{T-t'}) \circ \Phi_\tau,$$

and the last term approaches zero.
3. Frequently the limit $\lim_{\alpha \to 0} \alpha \int_0^\infty \exp(-\alpha t)\Phi^0_{-t} \circ \Phi_t\, dt$ also equals C_+, and is easier to handle.

By using $C_\pm$ it is possible to reduce the problem of stability for complex analytic systems under Φ_t to the problem under Φ^0_t. Write $(x, p) \in \mathbb{C}^{2m}$ as the single variable z, and suppose that $z = 0$ is a fixed point of Hamilton's equations. Then for an analytic X_H we can write Hamilton's equations in the form

$$\dot{z} = f(z) = Az + \cdots, \tag{3.4.20}$$

where A is a constant matrix and ... denotes the Taylor series. The linearized theory is based on the equation $\dot{z} = Az$, which produces the comparison flow $\Phi^0_t: z \to \exp(tA)z$.

Theorem (3.4.21)

$\Phi_t: \mathbb{C}^{2m} \to \mathbb{C}^{2m}$ is stable at $z = 0$ iff

(a) A is diagonable and has purely imaginary eigenvalues, and
(b) Both $C_\pm$ exist on neighborhoods $\mathcal{R}_\pm$ of 0.

Remarks (3.4.22)

1. The canonical flow Φ^0_t comes from a series-expansion of H, and thus $\exp(tA)$ is a symplectic matrix. That does not imply that $\exp(tA)$ and A are diagonable (for instance $\begin{vmatrix} 1 & b \\ 0 & 1 \end{vmatrix}$ is symplectic but not diagonable); instead, it must be explicitly assumed. Distinguish between A diagonable $\Leftrightarrow A = T$ (diagonal matrix) T^{-1}, $T \in GL(m, \mathbb{C}) \Leftrightarrow$ any n-fold degenerate eigenvalue has n linearly independent eigenvectors; and A unitarily diagonable (T is unitary) $\Leftrightarrow AA^t = A^tA$ (i.e., A is normal) $\Leftrightarrow$ any n-fold degenerate eigenvalue has n orthogonal eigenvectors (see also (3.1.13; 4)).
2. Part (a) of the theorem is a necessary condition for stability, which goes

---

† I.e., $\bigcup_t \Phi^0_{-t} \circ \Phi_t(x)$ is a bounded subset of $\mathbb{R}^{2m}$ $\forall x \in \mathbb{R}^{2m}$.

back to Liapunov. It is not sufficient by itself, as shown by the flow $\dot{x} = -y + x(x^2 + y^2)$, $\dot{y} = x + y(x^2 + y^2)$ on $\mathbb{R}^2$. The fixed point $(x, y) = (0, 0)$ is unstable, because with polar coordinates the equations become $\dot{r} = r^3$, $\dot{\varphi} = 1$, which are solved by

$$r(t) = r(0)(1 - r(0)^2 2t)^{-1/2}, \qquad \varphi(t) = \varphi(0) + t.$$

If $r(0) \neq 0$, the particle spirals off to infinity in a finite time, whereas

$$A = \begin{vmatrix} 0 & -1 \\ 1 & 0 \end{vmatrix}$$

of the linearized equation is diagonable and has eigenvalues $\pm i$. It is obvious that condition (b) fails.

3. By "stability" we now mean stability for a complex neighborhood, which is a much stronger condition than stability for real values only. For instance, a pendulum is stable for real angles but not for complex angles. Hence the statement implied by the theorem, that stable systems are only linear systems written in awkward coordinates, is more of mathematical than physical interest.

4. Criterion (b) is not very useful for determining the stability of particular systems, since it is no easier to prove than the existence of C_+. There are a few other sufficient criteria for stability, but they are not applicable to the Hamiltonian systems that will interest us here. Hence our results about this question are somewhat deficient.

Proof

(i) (a) and (b) $\Rightarrow$ stable

As with Ω_t, we can suppose that $\mathscr{R}_\pm$ are invariant under Φ_t, and map this flow diffeomorphically by $C_\pm$ to $\exp(tA)$. Since stability is defined purely topologically, it is unaffected by diffeomorphisms, and we need only investigate the stability of $\exp(tA)$. This is guaranteed by (a) (cf. the following example and Problem 5).

(ii) (b) but not (a) $\Rightarrow$ not stable

If (a) fails, then Φ_t^0 is not stable at $(0, 0)$, and by the same argument as in (i), Φ_t is likewise not stable.

(iii) stable $\Rightarrow$ (b)

This part of the proof is somewhat involved, and will not be given here (see [14]). $\qquad \square$

Example (3.4.23)

Let $T^*(M)$ be $\mathbb{R}^2$. We investigate the form of $\exp(tA)$, which has to leave the canonical 2-form invariant. Taking the derivative of

$$\exp(tA^t)W \exp(tA) = W \quad (\text{see } (3.1.8; 2)),$$

results in the requirement that $A^t W + WA = 0$. This is plainly sufficient to make the first equation hold. Since

$$W = \frac{1}{2} \begin{vmatrix} 0 & 1 \\ -1 & 0 \end{vmatrix},$$

A has the form

$$A = \begin{vmatrix} a & b + c \\ b - c & -a \end{vmatrix}.$$

The eigenvalues are $\pm\sqrt{a^2 + b^2 - c^2}$. Thus we must distinguish among three cases:

(i) $a^2 + b^2 - c^2 > 0$. Φ_t^0 is like a dilatation:

$$A = T \begin{vmatrix} \lambda & 0 \\ 0 & -\lambda \end{vmatrix} T^{-1}, \qquad e^{At} = T \begin{vmatrix} e^{\lambda t} & 0 \\ 0 & e^{-\lambda t} \end{vmatrix} T^{-1};$$

(ii) $a^2 + b^2 - c^2 < 0$. Φ_t^0 is like a rotation:

$$A = T \begin{vmatrix} i\omega & 0 \\ 0 & -i\omega \end{vmatrix} T^{-1}, \qquad e^{At} = T \begin{vmatrix} e^{i\omega t} & 0 \\ 0 & e^{-i\omega t} \end{vmatrix} T^{-1}; \quad \text{and}$$

(iii) $a^2 + b^2 - c^2 = 0$. Φ_t^0 is like linear motion:

$$A = T \begin{vmatrix} 0 & \lambda \\ 0 & 0 \end{vmatrix} T^{-1}, \qquad e^{At} = T \begin{vmatrix} 1 & \lambda t \\ 0 & 1 \end{vmatrix} T^{-1}.$$

Here λ and $\omega \in \mathbb{R}$, and T is a similarity transformation. The fixed point $\mathbf{0}$ is unstable for (i) (the hyperbolic case—an oscillator with an imaginary frequency); stable for (ii) (the elliptic case—an oscillator with a real frequency); and mixed for (iii) (the linear case—an oscillator with frequency zero). The trajectories in phase space are shown in Figure 30.

Problems (3.4.24)

1. In (3.4.4) we assumed the convergence of $\Phi_{-t} \circ \Phi_t^0$ and $(\Phi_{-t} \circ \Phi_t^0)^{-1}$. Find homeomorphisms Ω_t of $D = \{(x, y) \in \mathbb{R}^2 : x^2 + y^2 \leq 1\}$ such that as $t \to \infty$, $\Omega_t \to 1$, but $\Omega_t^{-1} \not\to 1$.

2. Derive the formula

$$\Omega^*(f) = f \circ \Omega^{-1} = \left(1 + \int_0^\infty dt\, \tau_t \circ \tau_{-t}^0 L_{X_{\tau_t^0(H-H_0)}}\right) f.$$

3. Verify that the Møller transformation $(r, p) \to (\bar{r}, \bar{p})$ of (3.4.6; 2) is canonical.

4. Show that for a measure-preserving transformation, stability (3.4.16) is equivalent to

$$\forall W \,\exists V \subset W : SV = V$$

(where U and W are neighborhoods).

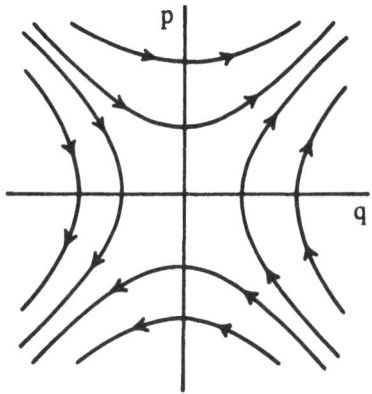

(i)   Hyperbolic fixed point.

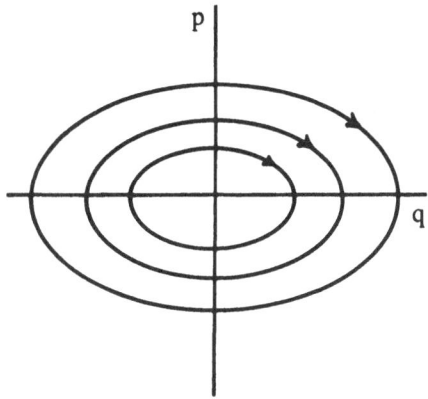

(ii)   Elliptic fixed point.

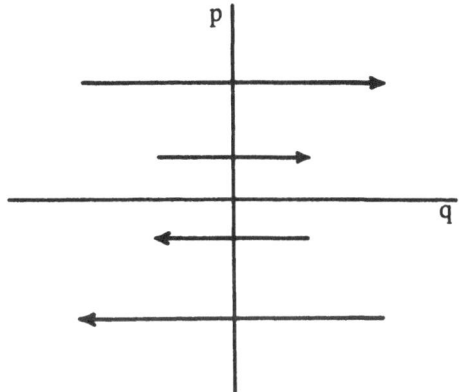

(iii)   Mixed fixed point.

Figure 30.

5. What is $\exp(tA)$ of (3.4.20) for free motion $\dot{q}^i = p_i, \dot{p}_i = 0$ on $\mathbb{R}^m$ like?

6. On $\mathbb{R}^2 \backslash \{0\} = \mathbb{R}^+ \times T^1$, let the Hamiltonian be

$$ H = \frac{p^2}{2} + \frac{2\alpha + L^2}{2r^2} - \frac{\beta}{r}, \qquad \beta > 0, $$

(the relativistic Kepler problem). If $E < 0$, every trajectory fills a 2-dimensional submanifold densely. Generalize the constants found in (3.4.3; 3) and find out why they are not globally definable for $E < 0$.

Solutions (3.4.25)

1.† Let Ω_t be $\mathbf{1}$ on $\Gamma_t \equiv \{(r, \varphi) : 1/t \le r \le 1 - 1/t, 0 \le \varphi \le 2\pi - 1/t\}$, and define it on $C\Gamma_t$ such that $\Omega_{t|\partial D} = \mathbf{1}, \Omega_t(0) = c_t \to 0, \Omega_t(1 - 1/t) = 0$. Since $\bigcup_t \Gamma_t = D \backslash \{0\} \backslash \partial D$, $\lim_{t \to \infty} \Omega_t(x) = x \, \forall x \in D$. But $\Omega_t^{-1} \nrightarrow \mathbf{1}$, because $\Omega_t^{-1}(0) \to 1$.

† For this example I am grateful to W. Schmidt, University of Colorado, Boulder.

2. Use $\tau_t = \exp(tL_H)$ and $\tau_t^0 = \exp(tL_{H_0})$. An expansion in a series shows

$$\frac{d}{dt} e^{tL_H} e^{-tL_{H_0}} = e^{tL_H}(L_H - L_{H_0})e^{-tL_{H_0}} = e^{tL_H}L_{H-H_0}e^{-tL_{H_0}} = e^{tL_H}e^{-tL_{H_0}} L_{\tau_t^0(H-H_0)}$$

(see (2.5.9; 2)), and integration by t gives the result.

3. It is elementary to calculate $dr \wedge dp = d\bar{r} \wedge d\bar{p}$.

4. By (3.4.16), S stable $\Leftrightarrow \forall W \exists U \subset W$ with $SU \subset U \Rightarrow V \equiv \bigcup_{n \geq 1} S^n U \subset U \subset W$ and $SV = V$. Conversely, if $\forall W \exists V \subset W$ with $SV = V$, then S is stable.

5. $A^2 = 0 \Rightarrow e^{At} = 1 + At$.

$$A = \begin{bmatrix} \begin{matrix} 0 & 1 \\ 0 & 0 \end{matrix} & & & \\ & \begin{matrix} 0 & 1 \\ 0 & 0 \end{matrix} & & \\ & & \ddots & \\ & & & \begin{matrix} 0 & 1 \\ 0 & 0 \end{matrix} \end{bmatrix} 2m$$

6. We find

$$\tau_+ p_x = \sqrt{2H} \cos\left(\varphi + \frac{L}{\sqrt{2\alpha + L^2}} \left(\frac{\pi}{2} - \arctan \frac{pr\sqrt{2\alpha + L^2}}{L^2 + 2\alpha - \beta r}\right)\right)$$

(by calculating $\varphi(r)$, for instance). At the equilibrium point $r = (L^2 + 2\alpha)/\beta$ the argument of the arctangent goes through ∞, and it is not possible to continue this function uniquely. For $\beta < 0$, $\tau_+ p_x$ is defined on $T^*(M)$.

3.5 Perturbation Theory: Preliminaries

Continuous changes in H influence the time-evolution for finite times continuously. However, quantities involving infinitely long times, such as constants of the motion, can exhibit behavior that is highly discontinuous.

In this section we look at Hamiltonians of the form $H = H_0 + \lambda H_1$, $\lambda \in \mathbb{R}$, and study how the dynamics depends on λ.

We first make some general observations about the time-evolution of an observable $f \to \exp(tL_H)f$, starting with the most tractable case, in which all functional dependence is analytic. If we differentiate the series for the exponential function by time term by term we get

$$\frac{d}{dt} e^{-tL_{H_0}} e^{tL_{H_0 + \lambda H_1}} f = e^{-tL_{H_0}}(L_{H_0 + \lambda H_1} - L_{H_0})e^{tL_{H_0 + \lambda H_1}} f$$

$$= \lambda e^{-tL_{H_0}}\{\exp(tL_{H_0 + \lambda H_1})f, H_1\}. \quad (3.5.1)$$

Integrating this by t,

$$e^{-tL_{H_0}}e^{tL_{H_0+\lambda H_1}}f = f + \lambda \int_0^t dt_1 \{\exp(-tL_{H_0})\exp(t_1 L_{H_0+\lambda H_1})f, H_1(-t_1)\},$$

$$(3.5.2)$$

where

$$g(t) \equiv e^{tL_{H_0}}g \qquad\qquad (3.5.3)$$

gives the time-evolution according to H_0. Operating on (3.5.2) with $\exp(tL_{H_0})$ yields

Estimate (3.5.4)

$$\forall f, H, H_0 \in C_0^\infty \ \exists c, k \in \mathbb{R}^+ : \|e^{tL_H}f - e^{tL_{H_0}}f\| \leq c\lambda t e^{k|t|} \ \forall t,$$

where

$$\|f\| = \sup_{x \in T^*(M)} |f(x)|.$$

Remarks (3.5.5)

1. This means that the effect of the perturbation can at first only grow linearly with t, but later it can grow exponentially.
2. Equation (3.5.2) still holds when H and H_0 are assumed in C_0^∞, which guarantees the existence of a flow.
3. Estimate (3.5.4) can be extended to $H = p^2 + V(x)$, $V \in C_0^\infty$. It does not, however, hold unrestrictedly; it is possible for particles to run off to infinity in finite times.

Proof

From (3.5.2),

$$\|e^{tL_H}f - e^{tL_{H_0}}f\| = \lambda \left\| \int_0^t dt_1 \{\exp(t_1 L_H)f, H_1\} \right\|$$

$$\leq \lambda t \sup_{0 \leq t_1 \leq t} \|\{\exp(t_1 L_H)f, H_1\}\|.$$

Now note that whereas $\exp(tL_H)f = f \circ \Phi_t$ has the same supremum as f, the derivative $d(f \circ \Phi_t) = df \circ T(\Phi_t)$ implied in the brackets $\{\ \}$ may grow with t. A general theorem about differential equations (see (2.3.6; 1)) says that for forces of finite range (compact support),

$$\|T(\Phi_t)\| = \left\| \frac{\partial u_i(t, x(0))}{\partial x_j(0)} \right\| \leq c e^{kt},$$

where $\sup_{x(0)}$ is taken over a compact set, and it is understood that the norm $\|\ \|$ includes taking the supremum over all elements of the matrix. What is left over is independent of t, and since all suprema are finite on the compact support of f, (3.5.4) follows.   □

Iterating (3.5.2) leads to

$$e^{-tL_{H_0}}e^{tL_{H_0+\lambda H_1}}f = f + \sum_{n=1}^{r-1} \lambda^n \int_0^t dt_1 \int_0^{t_1} dt_2 \cdots \int_0^{t_{n-1}} dt_n$$

$$\cdot \{\{\cdots\{f, H_1(-t_n)\}, \ldots, H_1(-t_2)\}, H_1(-t_1)\}$$

$$+ \lambda^r \int_0^t dt_1 \int_0^{t_1} dt_2 \cdots \int_0^{t_{r-1}} dt_r \{\{\cdots\{\exp(-t_r L_{H_0})$$

$$\exp(t_r L_H)f, H_1(-t_r)\}, \ldots, H_1(-t_2)\}, H_1(-t_1)\}.$$

$$(3.5.6)$$

If the latter term goes to 0 as $r \to \infty$, then we have shown the validity of the

Perturbation series (3.5.7)

$$e^{tL_{H_0+\lambda H_1}}f = f(t) + \sum_{n \geq 1} \lambda^n \int_0^t dt_1 \int_{t_1}^t dt_2 \cdots \int_{t_{n-1}}^t dt_n$$

$$\cdot \{\{\cdots\{f(t), H_1(t_n)\}, \ldots, H_1(t_2)\}, H_1(t_1)\}.$$

Remarks (3.5.8)

1. If only the first few terms of (3.5.7) are kept, it can give a completely false picture of the time-dependence. For an oscillator with a changed frequency, $|\sin(1 + \lambda)t - \sin t| \leq 2$ for all t, whereas the Taylor series in λ, $\sin(1 + \lambda)t = \sin t - \lambda t \cos t + 0(\lambda^2)$ appears to grow with t.
2. Even if the canonical flow can also be expanded in λ, it is not necessarily true that the constants of motion can be expanded. Consider free motion on T^2: $(\varphi_1, \varphi_2) \to (\varphi_1 + t\omega_1, \varphi_2 + t\omega_2)$. If the frequencies have a rational ratio, $\omega_1/\omega_2 = g_1/g_2$, then $\sin(g_2\varphi_1 - g_1\varphi_2)$ is a constant. The rationality of the ratio can be destroyed by an arbitrary small perturbation in the frequency, leaving no time-independent constants. Accordingly, a series expansion in λ with $g_1 \to g_1(1 + \lambda)$ leads to $\varphi_2 g_1 \cos(g_2\varphi_1 - g_1\varphi_2)$, which is not globally defined.

Examples (3.5.9)

1. $\{H_0, H_1\} = 0$, so $H_1(t) = H_1$. Then for the perturbation series (3.5.7),

$$e^{tL_{H_0+\lambda H_1}} = e^{t\lambda L_{H_1}}e^{tL_{H_0}},$$

which can also be shown by making a series expansion in t and using the formulas $L_{H_0+\lambda H_1} = L_{H_0} + \lambda L_{H_1}$ and $L_{H_0}L_{H_1} = L_{H_1}L_{H_0}$.

2. The driven oscillator,

$$H_0 = \tfrac{1}{2}(p^2 + q^2), \qquad H_1 = q : H_1(t) = q \cos t + p \sin t,$$

$$\left\{ H_1(t_1), \begin{vmatrix} q(t) \\ p(t) \end{vmatrix} \right\} = \begin{vmatrix} -\sin(t_1 - t) \\ \cos(t_1 - t) \end{vmatrix} \Rightarrow \left\{ H_1(t_1) \left\{ H_1(t_2), \begin{vmatrix} q \\ p \end{vmatrix} \right\} \right\} = 0 \Rightarrow$$

$$e^{tL_H} q = q \cos t + p \sin t + \lambda(\cos t - 1),$$
$$e^{tL_H} p = -q \sin t + p \cos t - \lambda \sin t,$$

which is the correct solution of $\dot{q} = p$, $\dot{p} = -q - \lambda$.

The most that can be said in complete generality about the influence of H_1 is (3.5.7). But often more precise estimates can be made for perturbations of integrable systems. Let $\bar{I}_j$ and $\bar{\varphi}_j$, $j = 1, \ldots, m$, be action and angle variables, and let us study

$$H(\bar{I}, \bar{\varphi}) = H_0(\bar{I}) + \lambda H_1(\bar{I}, \bar{\varphi}). \tag{3.5.10}$$

Since H_1 depends periodically on $\bar{\varphi}$, it can be expanded in a Fourier series:

$$H_1(\bar{I}, \bar{\varphi}) = \sum_k \tilde{H}_k(\bar{I}) e^{i(\bar{\varphi} \cdot k)},$$

$$\tilde{H}_k(\bar{I}) = (2\pi)^{-m} \int d\bar{\varphi}_1 \cdots d\bar{\varphi}_m e^{-i(\bar{\varphi} \cdot k)} H_1(\bar{I}, \bar{\varphi}), \tag{3.5.11}$$

$$k = (k_1, \ldots, k_m) \in \mathbb{Z}^m,$$
$$(\bar{\varphi} \cdot k) \equiv \bar{\varphi}_1 k_1 + \bar{\varphi}_2 k_2 + \cdots + \bar{\varphi}_m k_m.$$

Next we make a canonical transformation to the variables I, φ with the generator $S(I, \bar{\varphi})$:

$$\bar{I}_j = I_j + \lambda \frac{\partial S}{\partial \bar{\varphi}_j}, \qquad \varphi_j = \bar{\varphi}_j + \lambda \frac{\partial S}{\partial I_j}, \tag{3.5.12}$$

so that for some value of I, say $I = 0$, the system remains integrable to order λ. To do this, let

$$\omega_j \frac{\partial S}{\partial \bar{\varphi}_j} + H_1(I, \bar{\varphi}) = \tilde{H}_{k=0}(I), \qquad \omega_j = \left. \frac{\partial H_0(I)}{\partial I_j} \right|_{I=0}, \tag{3.5.13}$$

which is formally solved by

$$S(I, \bar{\varphi}) = -\sum_{k \neq 0} \frac{e^{i(k \cdot \bar{\varphi})}}{i(\omega \cdot k)} \tilde{H}_k(I). \tag{3.5.14}$$

Then

$$H = H_0(I) + \lambda \tilde{H}_{k=0}(I) + \lambda^2 H_2(I, \varphi), \tag{3.5.15}$$

where

$$H_2 = \lambda^{-2} \left[H_0 \left(I + \lambda \frac{\partial S}{\partial \bar{\varphi}} \right) - H_0(I) - \lambda \omega_j \frac{\partial S}{\partial \bar{\varphi}_j} \right]$$

$$+ \lambda^{-1} \left[H_1 \left(I + \lambda \frac{\partial S}{\partial \bar{\varphi}}, \bar{\varphi} \right) - H_1(I, \bar{\varphi}) \right].$$

Remarks (3.5.16)

1. If $|I| < \lambda$, then H_2 remains finite as $\lambda \to 0$.
2. The term with $k = 0$ is obviously to be dropped from the sum (3.5.14). Because we dropped it, we removed $\tilde{H}_{k=0}(I)$, and thereby caused a change in the frequency of order λ.
3. The denominator of (3.5.14) vanishes whenever the frequencies have rational ratios. When that happens we assume that $\mathrm{Det}(\partial^2 H/\partial I_i \partial I_j) \neq 0$, and instead of $I = 0$ we consider a nearby value of I for which the frequencies do not have rational ratios.
4. Even when the ratios of the frequencies are irrational, the denominators $(\omega \cdot k)$ may become arbitrarily small, and the convergence of the formal series (3.5.14) must be checked ("the problem of small denominators").

Lemma (3.5.17)

Let H_1 be analytic on the domain $|\mathrm{Im}\ \bar{\varphi}_j| < \rho, |\bar{I}_j| < r$. If there exist $\tau, c \in \mathbb{R}^+$ such that

$$|(\omega \cdot k)|^{-1} \leq c|k|^{\tau - m - 1}, \quad \forall k \neq 0,$$

where

$$|k| \equiv |k_1| + |k_2| + \cdots + |k_m|, \qquad k_i \in \mathbb{Z},$$

then the series (3.5.14) converges to an analytic function on the same domain, and the estimate

$$\|S\|_{\rho - \delta, r} \leq \bar{c} \delta^{-\tau + 1} \|H_1\|_{\rho, r}, \qquad 0 < \delta < \rho < 2,$$

where

$$\|F(I, \varphi)\|_{\rho, r} \equiv \sup_{|I_j| < r} \ \sup_{|\mathrm{Im}\ \varphi_j| < \rho} |F(I, \varphi)|,$$

$$\bar{c} = c 2^{3m} \left(\frac{2}{e} (\tau - m - 1) \right)^{\tau - m - 1},$$

holds.

Remarks (3.5.18)

1. The requirement of analyticity is not very severe.
2. By Cauchy's theorem, this estimate implies

$$\left\|\frac{\partial S}{\partial \bar{\varphi}_j}\right\|_{\rho-\delta,\,r} \leq \bar{c}\delta^{-\tau}\|H_1\|_{\rho,\,r}$$

$$\left\|\frac{\partial S}{\partial I_j}\right\|_{\rho-\delta,\,r-\varepsilon} \leq \bar{c}\delta^{-\tau+1}\varepsilon^{-1}\|H_1\|_{\rho,\,r}$$

$$\left\|\frac{\partial^2 S}{\partial I_i\,\partial\bar{\varphi}_j}\right\|_{\rho-\delta,\,r-\varepsilon} \leq \bar{c}\delta^{-\tau}\varepsilon^{-1}\|H_1\|_{\rho,\,r}.$$

Therefore the matrix

$$\frac{\partial \varphi_i}{\partial \bar{\varphi}_j} = \delta_{ij} + \lambda \frac{\partial^2 S}{\partial I_i\,\partial\varphi_j}$$

is invertible for small enough λ, as is required for (3.1.6).
3. The condition on the ω's means that they must be sufficiently rationally independent. When they have rational ratios, resonance behavior occurs, which can amplify the effect of the perturbation dramatically. Perturbation methods can founder even when the system is only in the vicinity of a resonance.
4. As $|k|$ increases, the condition of (3.5.17) becomes weaker and thus makes the complement of the set M of ω's that satisfy it smaller. To show that (3.5.14) converges, it suffices to have boundedness by any power of $|k|$. However, if $\tau > 2m$, then the measure of

$$CM \equiv \{\omega \in \mathbb{R}^m \colon \exists k \in \mathbb{Z}^m \backslash \{0\} \quad \text{such that} \quad (\omega \cdot k) < \frac{|k|^{m+1-\tau}}{C},$$

$$\text{and} \quad |\omega| < m \in \mathbb{R}^+\}$$

approaches zero as $1/C$ (Problem 7), although the set, which consists of points violating (3.5.17), contains all rational points. Thus CM is a strange example of an open, dense set of small measure. In this it resembles $\bigcup_{n=1}^{\infty}(v_n - \varepsilon 2^{-n}, v_n + \varepsilon 2^{-n})$, where $v_1, v_2, \ldots$ are the rational numbers and $\varepsilon > 0$. The measure of this set is less than $\sum_{n=1}^{\infty} \varepsilon 2^{-n+1} = \varepsilon/2$; that is, it is arbitrarily small, although the set is open and dense.

Proof

If the integration path for φ_j in (3.5.11) is displaced so that $\operatorname{Im} \varphi_j = \pm\rho$, with the same sign as k_j, there results

$$\|\tilde{H}_k(I)\|_r = \sup_{|I_j| < r} \left|\int \frac{d\phi_1 \cdots d\phi_m}{(2\pi)^m} e^{-i(k \cdot \varphi)} H_1(I, \varphi)\right| \leq e^{-|k|\rho}\|H_1\|_{\rho,\,r}.$$

This means that

$$\|S\|_{\rho-\delta,r} \le \sup_{|\mathrm{Im}\,\varphi_j|<\rho-\delta} \sum_{k\ne 0} e^{i(k\cdot\varphi)-|k|\rho} c|k|^{\tau-m-1}\|H_1\|_{\rho,r}$$

$$\le \sum_{k\ne 0} e^{-|k|\delta} c|k|^{\tau-m-1}\|H_1\|_{\rho,r},$$

In order to bound the sum $\sum_k$, use the inequality

$$|k|^\sigma \le \left(\frac{2\sigma}{e\delta}\right)^\sigma e^{|k|\delta/2} \quad \forall \sigma > 0, \delta > 0,$$

and the elementary calculation that

$$\sum_{k_1} e^{-|k_1|\delta/2} = \frac{2}{1-e^{-\delta/2}} - 1 < \frac{8}{\delta}, \quad \forall \delta < 2.$$

Together these produce the inequality of (3.5.17):

$$\|S\|_{\rho-\delta,r} \le c\|H_1\|_{\rho,r}\left(\frac{2(\tau-m-1)}{e\delta}\right)^{\tau-m-1}\sum_{k\ne 0} e^{-|k|\delta/2}$$

$$< c\|H_1\|_{\rho,r}8^m\delta^{-\tau+1}\left[\frac{2}{e}(\tau-m-1)\right]^{\tau-m-1}.$$

The same bound for the sum also shows the analyticity. □

Corollary (3.5.19)

With the assumptions of (3.5.17), the effect of the perturbation for trajectories with $|I_j| < \lambda$ is simply the change in the frequency

$$\omega_j \to \omega_j + \lambda \left.\frac{\partial \tilde{H}_{k=0}}{\partial I_j}\right|_{I=0},$$

to $O(\lambda^2)$. Consequently the condition $|I_j| < \lambda$ continues to hold for $|t| < c\lambda^{-1}$.

Remarks (3.5.20)

1. This corollary accords with intuition. For $\lambda = 0$, the I's are constant, and the φ's move with a finite angular velocity. Hence one would guess that for small λ, the average of H over φ is dominant.
2. Corollary (3.5.19) improves on (3.5.4), but with stronger assumptions, in that in (3.5.4) we only learned that $\forall |t| \le c\lambda^{-1}$, the difference between the flows Φ_t^0 and Φ_t is $O(1)$, whereas now we learn that for this range of times only the frequencies are changed, to accuracy $O(\lambda)$.
3. The independence of the frequencies is necessary in order to exclude some trivial counterexamples to this corollary. For $m = 1$, independence simply means $\omega \ne 0$. If $H_0 = 0$ and $H_1 = \cos\varphi$, and therefore $\tilde{H}_{k=0}(I) = 0$, then I and φ are constant in perturbation theory to $O(\lambda^2)$. On the other hand, this H generates the flow $\varphi \to \varphi$, $I \to I + \lambda t \sin\varphi$, which means that for $t \sim 1/\lambda$, the change is $O(1)$ rather than $O(\lambda)$.

Examples (3.5.21)

1. $H_0 = \omega_j I_j$ and $H_1(\bar{\varphi})$ is independent of the I's. In this case H_2 vanishes, and perturbation theory actually yields the exact solution:

$$\bar{\varphi}_j = \varphi_j, \qquad \bar{I}_j = I_j - \lambda \sum_{k \neq 0} \frac{e^{i(k \cdot \varphi)} k_j}{(\omega \cdot k)} H_k,$$

where $\dot{\varphi}_j = \omega_j$ and $\dot{I}_j = 0$, is the solution to

$$\dot{\bar{\varphi}}_j = \omega_j, \qquad \dot{\bar{I}}_j = -\lambda \sum_k i k_j e^{i(k \cdot \varphi)} \tilde{H}_k.$$

2. A driven oscillator. $H = (p^2 + q^2)/2 + \lambda q$. We must first transform to action and angle variables in such a way that the point $\bar{I} = 0$ can be conveniently chosen:

$$p = \sqrt{2(\bar{I} + \rho)} \cos \bar{\varphi}, \qquad q = \sqrt{2(\bar{I} + \rho)} \sin \bar{\varphi}, \qquad \rho \in \mathbb{R}^+,$$
$$\Rightarrow H = \bar{I} + \rho + \lambda \sqrt{2(\bar{I} + \rho)} \sin \bar{\varphi}, \qquad \tilde{H}_{k=0} = 0.$$

Thus the frequency remains 1, and we calculate

$$S = \sqrt{2(I + \rho)} \cos \bar{\varphi}, \qquad \bar{I} = I - \lambda \sqrt{2(I + \rho)} \sin \bar{\varphi},$$
$$\varphi = \bar{\varphi} + \lambda \cos \bar{\varphi} / \sqrt{2(I + \rho)}.$$

Note that the transformation $(q, p) \to (q - \lambda, p)$ changes H into $(p^2 + q^2)/2 - \lambda^2/4$. If we put

$$\sqrt{2(\bar{I} + \rho)}(\sin \bar{\varphi}, \cos \bar{\varphi}) = (\sqrt{2(I + \rho)} \sin \varphi - \lambda, \sqrt{2(I + \rho)} \cos \varphi),$$

i.e.,

$$\bar{I} = I - \lambda \sqrt{2(I + \rho)} \sin \varphi + \lambda^2/2,$$

$$\cos \bar{\varphi} = \left[1 - \lambda \sqrt{\frac{2}{I + \rho}} \sin \varphi + \frac{\lambda^2}{2(I + \rho)} \right]^{-1/2} \cos \varphi,$$

then H becomes $I + \rho - \lambda^2/4$. Perturbation theory produces the same transformation up to $0(\lambda^2)$. Since $[\cdots]^{-1/2}$ has singularities for nonreal λ, a power series in λ would only have a finite radius of convergence, although the action and angle variables exist $\forall \lambda \in \mathbb{R}$.

3. An oscillator with a changing frequency. $H = (p^2 + q^2)/2 + \lambda q^2 = \bar{I} + 2\bar{I} \sin^2 \bar{\varphi}$. Now $\tilde{H}_{k=0} = 1$, and the frequency changes to $1 + \lambda$, the first-order approximation to $\sqrt{1 + 2\lambda}$. Moreover,

$$S = \frac{I}{2} \sin 2\bar{\varphi} \Rightarrow \bar{I} = I(1 + \lambda \cos 2\bar{\varphi}), \qquad \varphi = \bar{\varphi} + \frac{\lambda}{2} \sin 2\bar{\varphi}.$$

Once again this constitutes the first two terms in a series in λ of a transformation, which turns H into $I\sqrt{1 + 2\lambda}$ (Problem 4). Perturbation theory converges for $|\lambda| < \frac{1}{2}$, after which point the behavior becomes exponential rather than periodic. The time-evolution is not singular at $\lambda = -\frac{1}{2}$ (Problem 2); but at that point action and angle variables no longer exist.

Problems (3.5.22)

1. For a linear differential equation, $\exp(tL_H)$ can be written as a matrix $\exp(tA)$. What is the formula analogous to (3.5.2) for $\exp(t(A_0 + \lambda A_1))$?

2. Apply (3.5.7) to $H_0 = p^2/2$, $H_1 = -q^2/2$.

3. Calculate H_2 from (3.5.21; 2), as a cautionary example.

4. For what I and φ does H from (3.5.21; 3) become $I\sqrt{1 + 2\lambda}$? Compare with perturbation theory.

5. For what frequencies ω is perturbation theory (3.5.19) applicable to an oscillator with a periodic external force, $H = (p^2 + q^2)/2 + \lambda q \cos \omega t$ (cf. (3.3.17; 3))?

6. For what values of λ would perturbation theory be expected to be useful for the pendulum $H = (\bar{I} + \omega)^2/2 + \lambda \cos \bar\varphi$?

7. For $m = 2$, show that the measure of CM (3.5.18; 4) is small for $\tau > 4$ and large C.

Solutions (3.5.23)

1.
$$\frac{d}{dt}e^{-tA_0}e^{t(A_0 + \lambda A_1)} = \lambda e^{-tA_0}A_1 e^{t(A_0 + \lambda A_1)} \to$$

$$e^{t(A_0 + \lambda A_1)} = e^{tA_0} + \lambda \int_0^t dt_1\, e^{(t-t_1)A_0}A_1 e^{t_1(A_0 + \lambda A_1)}.$$

2. $q(t) = q + tp$, $p(t) = p$. $H_1(t_n) = -(q + t_n p)^2/2$. $\{H_1(t_n), q(t)\} = (t_n - t)q(t_n) \Rightarrow$
   $\{\cdots \{q(t), H_1(t_n)\}, \ldots, H_1(t_2)\}, H_1(t_1)\} = (t_1 - t_2)(t_2 - t_3) \cdots (t_n - t)q(t_1)$.
   Using the formula $\int_0^1 dt\, t^p(1 - t)^q = p!q!/(p + q + 1)!$, integrating

$$\int_0^t dt_1 \int_{t_1}^t dt_2 \cdots \int_{t_{n-1}}^t dt_n$$

produces the series

$$q\frac{t^{2n}}{2n!} + p\frac{t^{2n+1}}{(2n + 1)!},$$

which are the Taylor coefficients of $q \cosh \lambda^{1/2}t + p\lambda^{-1/2} \sinh \lambda^{1/2}t$.

3. $H_2 = \lambda^{-1}[\sqrt{2(I + \rho - \lambda\sqrt{2(I + \rho)}\sin \bar\varphi)} - \sqrt{2(I + \rho)}]\sin \bar\varphi$.

4. $\sqrt{2\bar{I}}\sin \bar\varphi = \sqrt{2I}/\sqrt{1 + 2\lambda}\sin \varphi$, $\sqrt{2\bar{I}}\cos \bar\varphi = \sqrt{2I}\sqrt{1 + 2\lambda}\cos \varphi \to$

$$\bar{I} = I\sqrt{1 + 2\lambda}\{\cos^2 \varphi + \sin^2 \varphi/(1 + 2\lambda)\}, \cos \varphi = \sqrt{\bar{I}/I}\sqrt{1 + 2\lambda}\cos \bar\varphi.$$

Expanding in a power series in λ gives (3.5.21; 3).

5. Introduce action and angle variables for (q, p), and make the substitution $(t, E) \to (t/\omega, \omega E)$. Then $\mathscr{H} = I - \omega E + \lambda\sqrt{2I} \sin \varphi \cos t$. The irrationality of the ratios of the frequencies is only assured when $\omega \notin \mathbb{Q}$.

6. By (3.3.16), the action variable for H equals

$$I(E) = \frac{1}{2\pi} \oint d\varphi \sqrt{E + \lambda \cos \varphi},$$

which is analytic in λ for $|\lambda| < E$.

7. We may suppose $0 \le \omega_1/\omega_2 \equiv \alpha \le 1$. Consider the measure of

$$B_\varepsilon \equiv \{\alpha \in [0, 1] : \exists (k_1, k_2) \in \mathbb{Z}^2 \setminus \{(0, 0)\} : |\alpha k_1 - k_2| < \varepsilon |\mathbf{k}|^{-n}\},$$

where $\varepsilon = 1/c\omega_2 < 1$ and $n = \tau - 3 > 1$. This is contained in the set of α's with $|\alpha - k_2/k_1| < \varepsilon |k_1|^{-n-1}$. The k's must have the same sign, and $|k_2| < |k_1| + 1$. Thus

$$\mu(B_\varepsilon) \le 2 \sum_{\substack{k_1 > 0 \\ k_1 + 1 > k_2 > 0}} \varepsilon k_1^{-n-1} \le 2\varepsilon \sum_{k_1 > 0} \frac{k_1 + 1}{k_1^{n+1}} < \infty, \text{ for } n > 1.$$

3.6 Perturbation Theory: The Iteration

If an integrable system is perturbed, many of the invariant tori are completely destroyed, while others are only deformed. If the perturbation is sufficiently small, the ones that are only deformed fill up most of phase space.

In §3.5 a perturbed integrable system was transformed into another integrable system, up to $O(\lambda^2)$. The question arises of whether this procedure can be repeated to completely eliminate the perturbation. There has long been a wide-spread opinion that with an arbitrarily small perturbation—"just the least little bit"—all the constants other than H are destroyed, and the trajectory winds around densely through the energy surface (is ergodic). Thanks to the work of Kolmogoroff, Arnold, and Moser, it is now known that it is not so. Even if there exist no constants other than H, for small λ, enough m-dimensional submanifolds exist so that in most cases the system acts virtually like an integrable system.

Here we shall discuss the simplest nontrivial case

$$
\begin{aligned}
H_0(\bar{I}, \bar{\varphi}) &= \bar{I}_j \omega_j + \tfrac{1}{2} C_{ij}(\bar{\varphi}) \bar{I}_i \bar{I}_j \\
H_1(\bar{I}, \bar{\varphi}) &= A(\bar{\varphi}) + B_j(\bar{\varphi}) \bar{I}_j, \qquad j = 1, \dots, m,
\end{aligned}
\tag{3.6.1}
$$

and try to reduce H_1 to zero by a series of transformations. If only H_0 survives, then the torus $I_j = 0$ is time-invariant, and is filled up by the trajectory $\varphi_j \to \varphi_j + \omega_j t$.

Remarks (3.6.2)

1. We may allow C to depend on $\bar{\varphi}$, because the equations of motion for $H = H_0(I^{(\infty)}, \varphi^{(\infty)})$,

$$\dot{\varphi}_j^{(\infty)} = \omega_j + C_{j\ell}^{(\infty)} I_\ell^{(\infty)}, \qquad \dot{I}_j^{(\infty)} = -\frac{1}{2} \frac{\partial C_{i\ell}^{(\infty)}}{\partial \varphi_j} I_i^{(\infty)} I_\ell^{(\infty)},$$

are satisfied by $I_j^{(\infty)}(t) = 0$, $\varphi_j^{(\infty)}(t) = \varphi_j^{(\infty)}(0) + \omega_j t$.

2. The analysis made below works just as well when H_0 contains terms of higher orders in I and when H_1 is more complicated.

3. To a first approximation the frequencies are changed by $\tilde{B}_j(k = 0)$, which could affect the rational independence required in (3.5.17). If $\mathrm{Det}\, C \neq 0$, however, we can recover the old frequencies ω_j for $\partial H_0^{(1)}/\partial I_{j|I=0}$ by shifting the I coordinate.

4. One might attempt to continue to expand in powers of H_1; this series does not converge, however, due to resonance denominators. Therefore one has to use a more rapidly converging procedure, which takes the larger steps: $H_1, H_1^2, H_1^4, H_1^8, \ldots$.

The generator S from (3.5.12) is now chosen so as to produce a point transformation combined with a φ-dependent shift in I. This leaves the form of $H_0 + H_1$ invariant. Next we try to make the new A and B quadratic in the old ones, which we consider as small to first order.

Step 1 (3.6.3)

If the X and Y of the generators (for $\xi \in \mathbb{R}^m$)

$$S(I, \bar{\varphi}) \equiv \xi_j \bar{\varphi}_j + X(\bar{\varphi}) + I_j Y_j(\bar{\varphi}),$$

$$\varphi_j = \bar{\varphi}_j + Y_j(\bar{\varphi}), \qquad \bar{I}_j = I_j + \frac{\partial X}{\partial \bar{\varphi}_j} + I_\ell \frac{\partial Y_\ell}{\partial \bar{\varphi}_j} + \xi_j$$

satisfy the equations

$$A(\bar{\varphi}) + \omega_j \frac{\partial X}{\partial \bar{\varphi}_j} = \tilde{A}(k = 0),$$

$$B_j(\bar{\varphi}) + \omega_i \frac{\partial Y_j}{\partial \bar{\varphi}_i} + C_{ji}\left(\frac{\partial X}{\partial \bar{\varphi}_i} + \xi_i\right) = 0,$$

then†

$$H = H_0 + H_1 = \omega_j I_j + \tfrac{1}{2} C_{ij}^{(1)} I_i I_j + A^{(1)} + B_j^{(1)} I_j + \tilde{A}(k = 0) + \omega_j \xi_j,$$

$$A^{(1)} = \tfrac{1}{2} C_{ij}\left(\frac{\partial X}{\partial \bar{\varphi}_i} + \xi_i\right)\left(\frac{\partial X}{\partial \bar{\varphi}_j} + \xi_j\right) + B_j\left(\xi_j + \frac{\partial X}{\partial \bar{\varphi}_j}\right),$$

$$B_\ell^{(1)} = C_{ij}\left(\frac{\partial X}{\partial \bar{\varphi}_i} + \xi_i\right)\frac{\partial Y_\ell}{\partial \bar{\varphi}_j} + B_j \frac{\partial Y_\ell}{\partial \bar{\varphi}_j},$$

$$C_{\ell m}^{(1)} = C_{ij}\left(\delta_{i\ell} + \frac{\partial Y_\ell}{\partial \bar{\varphi}_i}\right)\left(\delta_{jm} + \frac{\partial Y_m}{\partial \bar{\varphi}_j}\right).$$

† $\tilde{A}(0)$ and $\omega_j \xi_j$ are independent of φ and I, and therefore inessential.

Remarks (3.6.4)

1. This follows the notation of (3.5.11) for the Fourier coefficients. It is easy to obtain an expression for H by substitution.
2. If we want to express the coefficients of H in terms of φ, we must first convince ourselves that the mapping $\bar{\varphi} \rightarrow \varphi$ is bijective.
3. Although S is not periodic in φ, dS is, and hence it can be defined globally.

Estimate of S (3.6.5)

In order to solve the equations for X, ξ, and Y, let us suppose as we did in (3.5.17) that A, B, and C are analytic for $|\operatorname{Im} \bar{\varphi}_j| < \rho$. The first equation of (3.6.3) is satisfied if we write

$$X(\bar{\varphi}) = -\sum_{k \neq 0} \frac{e^{i(k \cdot \bar{\varphi})} \tilde{A}(k)}{i(\omega \cdot k)}, \tag{3.6.6}$$

from which the quantity

$$E_i(\bar{\varphi}) \equiv C_{ij}(\bar{\varphi}) \frac{\partial X(\bar{\varphi})}{\partial \bar{\varphi}_j} \tag{3.6.7}$$

which appears in the second equation can be calculated. The $k = 0$ and $k \neq 0$ parts of this equation become

$$\tilde{C}_{ji}(0)\xi_i = -\tilde{B}_j(0) - \tilde{E}_j(0),$$
$$Y_j(\bar{\varphi}) = -\sum_{k \neq 0} \frac{e^{i(k \cdot \bar{\varphi})}}{i(\omega \cdot k)} [\tilde{B}_j(k) + \tilde{C}_{ji}(k)\xi_i + \tilde{E}_j(k)]. \tag{3.6.8}$$

Lemma (3.5.17) gives bounds for X and Y. The I-dependence is now explicit, and we can forget about the r in the norm $\| \quad \|_{\rho,r}$: (assuming that τ is always ≥ 1)

$$\|X\|_{\rho-h} \leq ch^{-\tau+1}\|\tilde{A}\|_\rho,$$
$$\left\|\frac{\partial X}{\partial \bar{\varphi}_j}\right\|_{\rho-h} \leq ch^{-\tau}\|A\|_\rho, \qquad j = 1, \ldots, m. \tag{3.6.9}$$

With the resulting inequality,

$$\|E_j\|_{\rho-h} \leq cmh^{-\tau}\|C\|_\rho\|A\|_\rho,$$
$$\|C\|_\rho \equiv \max_{i,j} \|C_{ij}\|_\rho, \tag{3.6.10}$$

the remaining quantities can be bounded:

$$|\xi_i| \leq m\|\tilde{C}(0)^{-1}\|\{\|B\|_\rho + cmh^{-\tau}\|C\|_\rho\|A\|_\rho\},$$
$$\left\|\frac{\partial Y_j}{\partial \bar{\varphi}_i}\right\|_{\rho-2h} \leq \{\|A\|_\rho\|C\|_\rho mc^2 h^{-2\tau} + c\|B\|_\rho h^{-\tau}\} \tag{3.6.11}$$
$$\cdot \{1 + m\|C\|_\rho\|\tilde{C}(0)^{-1}\|\}.$$

Remarks (3.6.12)

1. Define the norm $\|v\| = \max_i \|v_i\|$ for a vector (v_i) and $\|M\| = \max_{i,j} |M_{ij}|$ for a matrix (M_{ij}). Then for a product of a matrix and a vector, $\|Mv\| \leq m\|M\| \cdot \|v\|$, and for the product of two matrices, $\|M_1 \cdot M_2\| \leq m\|M\|_1 \cdot \|M_2\|$.

2. The bound for $\partial Y_i/\partial \bar\varphi_j$ shows that the matrix $\partial \varphi_i/\partial \bar\varphi_j$ is invertible for small enough A and B, and $\bar\varphi \to \varphi$ is a diffeomorphism.

These bounds will now be used to determine by how much $H_2 \equiv A^{(1)} + B_j^{(1)} I_j$ is reduced with respect to H_1.†

Estimate for H_2 (3.6.13)

Let

$$\|H_1\|_\rho = \max\{\|A\|_\rho, \|B\|_\rho\}, \quad \|H_2\|_{\rho-3h} = \sup_{|\operatorname{Im}\varphi_j| < \rho-3h} \max\{\|A^{(1)}\|, \|B^{(1)}\|\},$$

and $\Gamma = \max\{1, m\|C\|_\rho, m\|\tilde{C}(0)^{-1}\|\}$. If $\|H_1\|_\rho c^2 \Gamma^3 h^{-2\tau} \leq \frac{1}{4}ch^{-\tau} \geq 1$, then $\|H_2\|_{\rho-3h} \leq \|H_1\|_\rho^2 16c^3 h^{-3\tau}\Gamma^6$.

Proof

If we substitute from (3.6.9) and (3.6.11) into (3.6.3), we must bear in mind that the $\sup_{|\operatorname{Im}\bar\varphi_j| < \rho-2h}$ is taken in (3.6.11) and that it must be rewritten in terms of φ. If the condition on $\|H_1\|_\rho$ holds, then $|\bar\varphi - \varphi(\bar\varphi)| < h$ $\forall |\operatorname{Im}\bar\varphi_j| < \rho - 2h$, and the strip $\{\varphi : |\operatorname{Im}\varphi_j| < \rho - 3h\}$ is contained in the image of the strip $\{\bar\varphi : |\operatorname{Im}\bar\varphi_j| < \rho - 2h\}$; and the estimate then follows. □

The simple recursive form of (3.6.13) invites us to repeat the procedure n times. However, note that in the norm ρ is reduced by $3h$. In the n-th step h_n must be chosen small enough so that $\sum_{n=1}^\infty h_n < \rho$. Moreover, C is changed by the transformation, and we have to check whether Γ continues to be bounded. But at any rate, the above analysis can be repeated, producing an

Estimate of H_n (3.6.14)

Let $h_n \equiv h3^{-n+1}$ and $\rho_n \equiv \rho - 3\sum_{j=1}^{n-1} 3^{-j+1}h > \rho - 9h/2$. Then, in the notation of (3.6.13) (and writing $\Gamma_{n-1} \equiv \max\{1, m\|C^{(n-1)}\|_{\rho_{n-1}}, m\|\tilde{C}^{(n-1)}(0)^{-1}\|\}$),

$$\|H_n\|_{\rho_n} \leq \|H_{n-1}\|_{\rho_{n-1}}^2 16c^3 h^{-3\tau}3^{3\tau n}\Gamma_{n-1}^6 3^{-6\tau}.$$

This enables us to let $n \to \infty$:

Convergence of the Iteration (3.6.15)

The recursion formula

$$x_n = x_{n-1}^2 \gamma \delta^n$$

---

† Recalling that $\|C\|_\rho \geq \|\tilde{C}(0)\| \geq (m\|\tilde{C}(0)^{-1}\|)^{-1}$.

is solved by

$$x_n = \frac{(\gamma \delta^3 x_1)^{2^{n-1}}}{\gamma \delta^{n+2}}.$$

Thus

$$\|H_n\|_{\rho_n} \le (16c^3 h^{-3\tau} 3^{3\tau} \Gamma^6 \|H_1\|_\rho)^{2^{n-1}} (16c^3 h^{-3\tau} \Gamma^6 3^{3\tau n})^{-1},$$

and if $\Gamma > \Gamma_n \; \forall n$ and

$$\|H_1\|_\rho < \frac{h^{3\tau}}{16c^3 3^\tau \Gamma^6},$$

then

$$\|H_n\|_{\rho_n} \ge \|H_n\|_{\rho - 9h/2}$$

converges to zero.

Remarks (3.6.16)

1. The generalization of the condition in (3.6.13), i.e., that at the n-th step φ is changed by less than h_n, now reads

$$\|H_n\|_{\rho_n} c^2 \Gamma^3 h^{-2\tau} 3^{2\tau(n-1)} \le \tfrac{1}{4}$$

and is guaranteed when the iteration converges.
2. This procedure converges uniformly in the strip $|\operatorname{Im} \varphi_j| < \rho - 9h/2$, and thus the limiting function is analytic in that region.

We must next convince ourselves that we can bound Γ at each step, in which case the small denominators are under control.

Estimate for $C^{(n)}$ (3.6.17)

Inequality (3.6.11) implies that

$$\|C^{(n)}\|_{\rho_n} \le \|C^{(n-1)}\|_{\rho_{n-1}} (1 + 4\|H_n\|_{\rho_n} 4c^2 h^{-2\tau} 3^{2\tau(n-1)} \Gamma^3)^2.$$

Estimating $C^{(n)}$ is thus a matter of checking the convergence of

$$\prod_{n=1}^{\infty} (1 + x_n)^2 = \exp\left[2 \sum_{n=1}^{\infty} \ln(1 + x_n)\right] \le \exp\left[2 \sum_{n=1}^{\infty} x_n\right],$$

where

$$x_n \le \frac{(16c^3 3^{3\tau} h^{-3\tau} \Gamma^6 \|H_1\|_\rho)^{2^{n-1}}}{4\Gamma^3 h^{-\tau} 3^{\tau(n+2)}}.$$

But this clearly converges, because for $16c^3 h^{-3\tau} 3^{3\tau} \Gamma^6 \|H_1\|_\rho < \tfrac{3}{4}$, it is easy to see that

$$\|C^{(n)}\|_{\rho_n} < 2\|C\|_\rho, \quad \forall n.$$

It is somewhat more troublesome to deal with $\tilde{C}^{(n)}(0)^{-1}$:

Estimate of $\tilde{C}^{(n)}(0)^{-1}$ (3.6.18)

We begin with

$$\tilde{C}_{ij}^{(1)}(0) = (2\pi)^{-m} \int d\varphi_1 \cdots d\varphi_m (\delta_{i\ell} + Y_{i,\ell}) C_{\ell k}(\delta_{kj} + Y_{j,k}).$$

This is averaged over the φ's, but Y and C are given as functions of $\bar{\varphi}$. Transforming to barred variables and using matrix notation ($Y' = \partial Y_i/\partial\bar{\varphi}_j$, etc.),

$$\|\tilde{C}(0) - \tilde{C}^{(1)}(0)\| = \|(2\pi)^{-m} \int d\bar{\varphi}_1 \cdots d\bar{\varphi}_m$$

$$\cdot \{C - (1 + Y')C(1 + Y'^\ell)\mathrm{Det}(1 + Y')\}\|$$

$$= \|(2\pi)^{-m} \int d\bar{\varphi}_1 \cdots d\bar{\varphi}_m \{C(1 - \mathrm{Det}(1 + Y'))$$

$$-(CY'^\ell + Y'C + Y'CY'^\ell)\mathrm{Det}(1 + Y')\}\|$$

$$\leq \|C\|_{\bar{\rho}} \frac{3m\|Y'\|_\rho + m^2\|Y'\|_\rho^2}{1 - m\|Y'\|_{\bar{\rho}}}, \qquad \forall\bar{\rho} > 0,$$

where we have used $|\mathrm{Det}(1 + Y')| \leq (1 - m\|Y'\|)^{-1}$ (Problem 1). From this we get a bound for $\tilde{C}^{(1)}(0)^{-1}$, for

$$\|\tilde{C}^{(1)}(0)^{-1}\| \leq \|\tilde{C}(0)^{-1}\|(1 - m^2\|\tilde{C}(0)^{-1}\| \cdot \|\tilde{C}(0) - \tilde{C}^{(1)}(0)\|)^{-1}$$

(Problem 2). By (3.6.11), $\|Y'\|_{\rho - 2h} \leq 4\Gamma^3 c^2 h^{-2\tau}\|H_1\|_\rho/m$, and generalizing this to the n-th step yields

$$\frac{\|\tilde{C}^{(n)}(0)^{-1}\|}{\|\tilde{C}^{(n-1)}(0)^{-1}\|} \leq \left(1 - 2m^2\|\tilde{C}^{(n-1)}(0)^{-1}\|\|C\|_{\rho_n} \cdot 2\right.$$

$$\left.\cdot \frac{12\Gamma^3 c^2 h^{-2\tau}3^{2\tau(n-1)}\|H_n\|_{\rho_n} + 16\Gamma^6 c^4 h^{-4\tau}3^{4\tau(n-1)}\|H_n\|_{\rho_n}^2}{1 - 4\Gamma^3 c^2 h^{-2\tau}3^{2\tau(n-1)}\|H_n\|_{\rho_n}}\right)^{-1}.$$

Under the same circumstances as in (3.6.17), $\|\tilde{C}^{(n)}(0)^{-1}\|$ is bounded in n (cf. Problem 3).

Now that we have convinced ourselves of the boundedness of Γ, we collect the results of Kolmogorov, Arnold, and Moser in the

K–A–M Theorem (3.6.19)

For H as in (3.6.1), suppose that

(a) $|(\omega \cdot k)|^{-1} < c|k|^{\tau - m - 1} \forall k \in \mathbb{Z}^m \backslash \{0\}$ *for some $c > 0$ and $\tau \geq 1$.*
(b) *A, B, and C are analytic in $|\mathrm{Im}\,\bar{\varphi}_j| < \rho$.*

(c)
$$\sup_{|\operatorname{Im}\bar{\varphi}_j|<\rho} \max(|A|,|B_j|) \le h^{3\tau}\Bigg(16c^3 3^\tau m^6$$

$$\times \Bigg(2 \sup_{|\operatorname{Im}\bar{\varphi}_j|<\rho} \max\{\|C\|, \|\tilde{C}(0)^{-1}\|\}\Bigg)^6\Bigg)^{-1},$$

and h is less than both $2\rho/9$ and $c^{-1/\tau}$.

Then there exists a canonical transformation to $\varphi^{(\infty)}$, $I^{(\infty)}$, which is analytic in φ for $|\operatorname{Im}\varphi_j| < \rho - 9h/2$ and affine† in I, such that

$$H = \omega_j I_j^{(\infty)} + \tfrac{1}{2}C_{ij}^{(\infty)}(\varphi^{(\infty)})I_i^{(\infty)}I_j^{(\infty)},$$

where $C^{(\infty)}$ is analytic for $|\operatorname{Im}\varphi_j^{(\infty)}| < \rho - 9h/2$.

Remarks (3.6.20)

1. As already mentioned, this theorem can be extended to a wider class of H's, and analyticity can be weakened to sufficiently-often differentiability.
2. Condition (c) is by no means necessary, but as yet no one has been able to improve it in such a way that the strong fall-off with m is essentially any better. For systems with many degrees of freedom, the perturbation has to be so ridiculously small for this theorem to apply that it is questionable whether there is any physical relevance in such cases.
3. This result shows that the torus $I = 0$, which is invariant under Φ_t^0, gets deformed only moderately when the perturbation stays small, which happens when there is a bound on the independence of the frequencies. But tori that are invariant under Φ_t^0 completely fill phase space. With small perturbations, those tori that are far away from resonance continue to be invariant submanifolds.
4. Although the open regions on which the conditions of (3.6.19) are violated are dense, their total measure on compact sets goes to zero for vanishingly small perturbations. The motion on these sets is chaotic. They are divided into disconnected components by the invariant surfaces only when $m = 2$, in which case the trajectory can not leave the part of phase space bounded by the invariant tori.
5. Measure-theoretical ideas are usually more relevant than topological ones, when one wants to determine whether the invariant submanifolds are mathematically pathological. Thinking topologically, one would call the invariant submanifolds exceptional, as complements of open, dense sets. However, since they have large Liouville measures, they are more the rule than the exception.

We conclude by summarizing the general facts learned about Hamiltonian systems and seeing what impression they give us about the global structure of a canonical flow.

---

† Affine = inhomogeneous linear, that is, of the form $I \to aI + b$.

Figure 31   The invariant tori.

In the standard case for physics, where forces decrease with distance, particles with enough energy can escape. Their momenta will approach constants as $t \to \pm\infty$, with which the system will be integrable in the parts of phase space where enough particles escape. All trajectories will be diffeomorphic to $\mathbb{R}$ in these regions, and the canonical flow can be transformed into a linear flow.

Orbits that always return on themselves fill up higher-dimensional manifolds, and it is in this situation that the most complicated things can happen. In the special case of an integrable system, these manifolds are m-dimensional tori, which collapse to lower-dimensional figures in those parts of phase space where the frequencies have rational ratios. Perturbations often destroy many of these tori, making the trajectory cover a $2m - 1$-dimensional region, while many other tori remain, for which the system acts like an integrable system.

Problems (3.6.21)

1. For a Hermitian matrix M such that $\|M\| < 1/m$, show that $|\mathrm{Det}(1 + M)| < 1/(1 - m\|M\|)$.

2. For M_i as in Problem 1, show that $\|(M_0 + M_1)^{-1}\| \le \|M_0^{-1}\|(1 - m^2\|M_1\| \cdot \|M_0\|^{-1})^{-1}$.

3. Complete Estimate (3.6.18).

4. Investigate the convergence of the perturbation series for

$$H = \frac{\omega}{2}(p^2 + q^2) + \frac{c}{8}(p^2 + q^2)^2 + \frac{b}{2}q^2$$

on $T^*(\mathbb{R})$. (With dimensionless action and angle variables, ω, c, and b have the same dimensions as H.)

Solutions (3.6.22)

1. Putting M into diagonal form, we see

$$\text{Det}(1 + M) = \exp[\text{tr} \ln(1 + M)] = \exp\left[\sum_{j=1}^{\infty} \frac{(-1)^{j+1}}{-j} \text{tr } M^j\right]$$

$$\leq \exp\left[\sum_{j=1}^{\infty} \frac{1}{j} (m\|M\|)^j\right] = \frac{1}{1 - m\|M\|}.$$

2. $(M_0 + M_1)^{-1} = M_0^{-1} - M_0^{-1}M_1(M_0 + M_1)^{-1} \Rightarrow \|(M_0 + M_1)^{-1}\|$

$$\leq \|M_0^{-1}\| + \|(M_0 + M_1)^{-1}\|m^2\|M_0\|^{-1}\|M_1\|.$$

3. Substituting from (3.6.15) into the last formula of (3.6.18), recalling that Γ, τ, and $ch^{-\tau} \geq 1$, and generously conceding factors, the recursion formula simplifies to

$$x_n \leq \frac{x_{n-1}}{1 - 3^{-\tau(n+2)}},$$

where

$$x_n \equiv \|\tilde{C}^{(n)}(0)^{-1}\|.$$

Since $-\ln(1 - x) \leq x2 \ln 2 \ \forall 0 \leq x \leq 1/2$,

$$\prod_{n=1}^{\infty} (1 - 3^{-\tau(n+2)})^{-1} \leq \exp\left(2 \ln 2\, 3^{-2\tau} \sum_{n=1}^{\infty} 3^{-\tau n}\right) = \exp\left(2 \ln 2\, \frac{3^{-2\tau}}{3^\tau - 1}\right) < 2,$$

and so $\|\tilde{C}^{(n)}(0)^{-1}\| < 2\|\tilde{C}(0)^{-1}\| \ \forall n$.

4. In (3.6.19), (a) holds for $\tau = 1$ and $c = 1/\omega$ and (b) holds for all h. If $m = 1$, we can replace $\|C\|\|\tilde{C}(0)^{-1}\|$ with 1, so that the term $(\ldots)^6$ of (c) simplifies to $(2c)^2$. The condition is then satisfied if

$$\sup_{|\text{Im } \varphi| < h} |b \sin^2 \varphi| = |b|\frac{e^{2h}}{4} \leq \frac{h^3 c^{-2}\omega^3}{3 \cdot 16 \cdot 2^2},$$

i.e., for $h = \frac{3}{2}$,

$$b \leq \frac{\omega^3}{c^2} \frac{9}{e^3 2^7} = \frac{\omega^3}{c^2} \cdot 0.003.$$

The numerical factors of this calculation could be easily improved.

4 Nonrelativistic Motion

4.1 Free Particles

The study of free particles is the foundation of kinematics, and can be used as a basis of comparison for realistic systems. The canonical flow for free particles is linear.

In this section we apply the mathematical methods that we have developed to the problems posed in (1.1.1) and (1.1.2). We begin with the trivial case of a free particle, in order to illustrate the various concepts. In other words, let $M = \mathbb{R}^3$ and $T^*(M) = \mathbb{R}^6$, and choose a chart $(\mathbb{R}^6, \mathbf{1})$, calling the coordinates x_i and p_i. By (2.3.26) H has the simple form

$$H = \frac{|\mathbf{p}|^2}{2m}. \tag{4.1.1}$$

Theorem (4.1.2)

The point transformations that leave H invariant are the elements of the Euclidean group E_3:

$$x_i \to M_{ij}x_j + \lambda_i, \qquad MM^t = \mathbf{1}, \qquad \lambda_i \in \mathbb{R}^3.$$

On $T^*(\mathbb{R}^3)$ they induce the mapping $p_i \to M_{ij}p_j$.

Proof:

See Problem 4. □

Corollary (4.1.3)

To each one-parameter subgroup of E_3 there corresponds a canonical flow that leaves H invariant, and hence there is a constant of the motion. As discussed in (3.2.6; 2 and 3), the six generators are

$$p_i, \qquad L_i = \varepsilon_{ijk} x_j p_k.$$

Remarks (4.1.4)

1. Since $T^*(\mathbb{R}^3)$ has only six dimensions, and the trajectory takes up at least one dimension, the six constants can not be algebraically independent. The relation connecting them is $(\mathbf{p} \cdot \mathbf{L}) = 0$.
2. The important question is now whether there exists a 5-parameter subgroup of E_3 that furnishes five independent constants. One might think of the following sort of construction: for fixed (x, p), the mapping $E_3 \colon (x, p) \to (x', p')$ gives a mapping $E_3 \to$ the energy shell, which is surjective but not injective. The stabilizer of $(\mathbf{0}, \mathbf{p})$ is $\{M, \lambda \colon \lambda_i = 0, M\mathbf{p} = \mathbf{p}\}$, which is a one-parameter subgroup, and so the factor group should be 5-dimensional and mapped bijectively onto the energy surface. The flaw in this argument is that E_3 is not Abelian, and thus not every subgroup is an invariant subgroup, and this 5-parameter factor group does not exist.

For extended phase space, the transition to a uniformly moving frame of reference,

$$\begin{aligned} \mathbf{x} &\to \mathbf{x} + \mathbf{v}t \\ t &\to t \end{aligned} \tag{4.1.5}$$

is a good candidate for a 3-parameter, Abelian invariance group. Taking (4.1.5) as a point transformation, the momenta transform according to (2.4.35; 3) as

$$\begin{aligned} \mathbf{p} &\to \mathbf{p} \\ E &\to E + (\mathbf{p} \cdot \mathbf{v}), \end{aligned} \tag{4.1.6}$$

but this fails to leave $\mathscr{H} = |\mathbf{p}|^2/2m - E$ invariant. However, suppose that the momenta are transformed as

$$\begin{aligned} \mathbf{p} &\to \mathbf{p} + m\mathbf{v} \\ E &\to E + (\mathbf{p} \cdot \mathbf{v}) + \frac{m|\mathbf{v}|^2}{2}. \end{aligned} \tag{4.1.7}$$

With (4.1.5), this is a 3-parameter group of canonical transformations Φ_v that leaves $\mathscr{H}$ invariant:

$$\Phi_v^* \mathscr{H} = \mathscr{H} \circ \Phi_v^{-1} = \mathscr{H}, \qquad \Phi_{v_1} \circ \Phi_{v_2} = \Phi_{v_1 + v_2}, \tag{4.1.8}$$

although it is not a point transformation.

We combine these transformations with the previous ones in

Definition (4.1.9)

The transformations of extended phase space,

$$x_i \to M_{ij}x_j + v_i t + \lambda_i, \qquad v_j, \lambda_j \in \mathbb{R}^3, \qquad MM^t = 1,$$
$$t \to t + c, \qquad c \in \mathbb{R},$$
$$p_i \to M_{ij}p_j + mv_i,$$
$$E \to E + p_i v_i + \frac{m}{2} v_i^2,$$

form the **Galilean group**. The ten generators, p_i, L_i, H, and $K_j \equiv p_j t - x_j m$, correspond to the ten parameters, λ_i, M_{ij}, c, and v_j.

Remarks (4.1.10)

1. The condition $\mathscr{H} = 0$ defines a 7-dimensional manifold in extended phase space. Hence six constants suffice to determine a trajectory, and there must be four relationships connecting the ten generators. Specifically, we have $H = |\mathbf{p}|^2/2m$, and $[\mathbf{p} \times {}^{\mathbf{K}}] = m\mathbf{L}$. The relationships $(\mathbf{p} \cdot \mathbf{L}) = \mathbf{K} \cdot \mathbf{L}) = 0$ follow from these.
2. The Poisson brackets of the generators can not be expressed in terms of the constants alone. The mass appears on the right side of $\{p_i, K_i\} = \delta_{ij}m$, but as a numerical constant it is not the generator of a transformation; $dm = 0 = X_m$, and the Poisson bracket of the mass with any quantity is zero. Together with m, the 10 constants of motion generate an 11-parameter group, the factor group of which by the center† is the Galilean group.
3. There is a subgroup that furnishes six independent constants, and from which all ten generators of the Galilean group can be constructed. It is generated by $\mathbf{p}$ and $\mathbf{K}$ (see Remark 1). The mass m again appears in the Poisson brackets

$$\{p_i, K_j\} = \delta_{ij}m, \qquad \{p_i, p_j\} = \{K_i, K_j\} = 0,$$

and so the group of transformations,

$$\mathbf{x} \to \mathbf{x} + \lambda + \mathbf{v}t, \qquad t \to t,$$

$$\mathbf{p} \to \mathbf{p} + m\mathbf{v}, \qquad E \to E + (\mathbf{p} \cdot \mathbf{v}) + \frac{m|\mathbf{v}|^2}{2}, \qquad (4.1.11)$$

on phase space is isomorphic to the factor group by the center. The subgroup is minimal in the sense that with fewer parameters it is impossible to produce six constants of motion. The Galilean group (4.1.9) is not distinguished by any special property: it is neither the largest invariance group nor the smallest one that produces all the constants of the motion.

---

† I.e., the one-parameter group generated by m.

N Free Particles (4.1.12)

In order to generalize the above discussion to cover

$$H = \sum_{i=1}^{N} \frac{|\mathbf{p}_i|^2}{2m_i},$$

use the canonical transformation $\mathbf{x}_i \rightarrow \mathbf{x}_i/\sqrt{m_i}$, $\mathbf{p}_i \rightarrow \mathbf{p}_i\sqrt{m_i}$ to put it into the form

$$H = \frac{1}{2} \sum_{i=1}^{N} |\mathbf{p}_i|^2.$$

Then on $T^*(M) = T^*(\mathbb{R}^{3N}) = \mathbb{R}^{6N}$ there are the $3N + 3N(3N - 1)/2$ constants of motion p_μ and $p_\mu x_\nu - x_\mu p_\nu$, $\mu, \nu = 1, \ldots, 3N$. The system is integrable with the p_μ, although the submanifold $p_\mu = \alpha_\mu = $ constant is not compact, but instead diffeomorphic to $\mathbb{R}^{3N}$. All trajectories are submanifolds of the form $x_\mu \rightarrow x_\mu + tp_\mu$, and for $p \neq 0$ they are diffeomorphic to $\mathbb{R}$.

Problems (4.1.13)

1. Show that $\bar{x} = x + vt$, $\bar{p} = p + mv$, $\bar{E} = E + pv + mv^2/2$ is a canonical transformation both with and without the term $mv^2/2$. What is the reason that $mv^2/2$ appears in (4.1.7)?

2. Calculate the Poisson brackets of the generators of the Galilean group.

3. Discuss the group of canonical transformations generated by $D = (\mathbf{x} \cdot \mathbf{p})$.

4. Show that the Euclidean group is the largest group of point transformations that leaves H invariant.

Solutions (4.1.14)

1. We verify that $\{\bar{x}_i, \bar{E}\} = v_i - v_i = 0$, which is the only nontrivial Poisson bracket. Without the $mv^2/2$, the canonical transformations do not form a group:

$$\frac{d}{d\lambda} \exp[\lambda L_{(\mathbf{K}\cdot\mathbf{v})}]E_{|\lambda=0} = \{E, \mathbf{K} \cdot \mathbf{v}\} = (\mathbf{p} \cdot \mathbf{v}),$$

$$\frac{d^2}{d\lambda^2} \exp[\lambda L_{(\mathbf{K}\cdot,\mathbf{v})}]E_{|\lambda=0} = \{(\mathbf{p}\cdot\mathbf{v}), (\mathbf{K} \cdot \mathbf{v})\} = m|\mathbf{v}|^2.$$

2.
$$\{p_i, p_j\} = \{p_i, E\} = \{L_i, E\} = \{K_i, K_j\} = 0,$$

$$\{L_i, L_j\} = \varepsilon_{ijk} L_k, \{p_i, L_j\} = \varepsilon_{ijk} p_k, \{K_i, L_j\} = \varepsilon_{ijk} K_k,$$

$$\{E, K_j\} = p_j, \{p_i, K_j\} = m\delta_{ij}.$$

3. D generates the group of dilatations $(\mathbf{x}, \mathbf{p}) \rightarrow (e^\lambda \mathbf{x}, e^{-\lambda}\mathbf{p})$. This leaves $\mathbf{L}$ invariant, so the dilatations together with the Euclidean group generate a 7-parameter group of canonical transformations. A dilatation changes H by $e^{-2\lambda}$, and hence the equation $\ddot{x} = 0$ remains unchanged.

4. The new coordinates $\bar{x}$ have to satisfy $\bar{x}_{m,i}(x)\bar{x}_{m,k}(x) = \delta_{ik}$ for all x, which implies

$$\bar{x}_{m,ij}\bar{x}_{m,k} + \bar{x}_{m,i}\bar{x}_{m,kj} = 0 \quad \updownarrow$$

$$\bar{x}_{m,jk}\bar{x}_{m,i} + \bar{x}_{m,j}\bar{x}_{m,ik} = 0 \quad \updownarrow$$

$$-\bar{x}_{m,ki}\bar{x}_{m,j} - \bar{x}_{m,k}\bar{x}_{m,ji} = 0$$

$$\overline{\qquad\qquad\qquad\qquad\qquad}$$

$$2\bar{x}_{m,i}\bar{x}_{m,kj} = 0 \Rightarrow \bar{x}_{m,kj} = 0.$$

Consequently the transformation is inhomogeneous-linear, and thus of the form (4.1.2). Substitution yields $MM' = \mathbf{1}$.

4.2 The Two-Body Problem

Mathematical physics was born when Newton solved the Kepler Problem. This problem has lost none of its attraction over the centuries, especially as it has remained soluble while the theory has become more and more refined (with relativistic electrodynamics and gravitation, and both nonrelativistic and relativistic quantum mechanics).

For equations (2.3.21) through (2.3.26) in the case $N = 2$, we have

$$M = \mathbb{R}^3 \times \mathbb{R}^3 \setminus \{(\mathbf{x}_1, \mathbf{x}_2): \mathbf{x}_1 = \mathbf{x}_2\},$$

$$H = \frac{|\mathbf{p}_1|^2}{2m_1} + \frac{|\mathbf{p}_2|^2}{2m_2} + \frac{\alpha}{|\mathbf{x}_1 - \mathbf{x}_2|}, \qquad \alpha = e_1 e_2 - \kappa m_1 m_2. \qquad (4.2.1)$$

We give the solution in several steps:

Separation into center-of-mass and relative coordinates (4.2.2)

The point transformation $\mathbb{R}^6 \to \mathbb{R}^6$:

$$\mathbf{x}_{cm} = \frac{m_1 \mathbf{x}_1 + m_2 \mathbf{x}_2}{m_1 + m_2}, \qquad \mathbf{x} = \mathbf{x}_1 - \mathbf{x}_2,$$

induces the transformation

$$\mathbf{p}_{cm} = \mathbf{p}_1 + \mathbf{p}_2, \qquad \mathbf{p} = \mathbf{p}_1 \frac{m_2}{m_1 + m_2} - \mathbf{p}_2 \frac{m_1}{m_1 + m_2}$$

on $T^*(\mathbb{R}^6)$. This makes

$$H = \frac{|\mathbf{p}_{cm}|^2}{2(m_1 + m_2)} + \frac{|\mathbf{p}|^2(m_1 + m_2)}{2m_1 m_2} + \frac{\alpha}{|\mathbf{x}|} \equiv H_s + H_r.$$

Remarks (4.2.3)

1. In these coordinates M has been restricted to $\mathbb{R}^3 \times (\mathbb{R}^3 \setminus \{0\})$. $H \in C^\infty(T^*(M))$, in order to remove the singularity of the potential.
2. Since H_{cm} depends only on $\mathbf{p}_{cm}$, and H_r only on $\mathbf{x}$ and $\mathbf{p}$, the time-evolution is the Cartesian product of the flows determined by H_{cm} and H_r, and $\{H_{cm}, H_r\} = 0$.
3. H_{cm} has the form of the H of §4.1 (with mass $m_1 + m_2$), so we consider this part of the problem solved and only work with the second part.
4. H_r is a limiting case of (4.2.1), in which one particle has infinite mass, and the other has the reduced mass $m \equiv m_1 m_2/(m_1 + m_2)$.

Constants of the Motion (4.2.4)

$$\{H_r, \mathbf{L}\} = \{H_r, \mathbf{F}\} = 0,$$

$$\mathbf{L} = [\mathbf{x} \times \mathbf{p}], \qquad \mathbf{F} = [\mathbf{p} \times \mathbf{L}] + m\alpha \frac{\mathbf{x}}{r},$$

where $m = m_1 m_2/(m_1 + m_2)$ and $r = |\mathbf{x}|$. $\mathbf{F}$ is known as Lenz's vector.

Proof

The angular momentum $\mathbf{L}$ is constant due to the invariance of H_r under $x_i \to M_{ij} x_j$, $p_i \to M_{ij} p_j$. The constancy of $\mathbf{F}$ can be directly verified by calculating

$$\dot{\mathbf{x}} = \{\mathbf{x}, H_r\} = \frac{\mathbf{p}}{m}, \qquad \dot{\mathbf{p}} = \frac{\alpha \mathbf{x}}{r^3}:$$

$$\dot{\mathbf{F}} = \frac{\alpha}{r^3}[\mathbf{x} \times [\mathbf{x} \times \mathbf{p}]] + \alpha \frac{\mathbf{p}}{r} - \alpha \frac{\mathbf{x}}{r^3}(\mathbf{x} \cdot \mathbf{p}) = 0. \qquad \square$$

Remarks (4.2.5)

1. The only elements of E_3 that remain as invariances are the rotations, because H_r is not left unchanged by displacements. If $\alpha \neq 0$, the constants $[\mathbf{p} \times \mathbf{L}]$ generalize to $\mathbf{F}$. See Problem 1 for the transformations that are generated by $\mathbf{F}$ and leave H_r invariant.
2. $\mathbf{L}$ should be thought of as the internal angular momentum (spin), and should be distinguished from the total angular momentum $[\mathbf{x}_{cm} \times \mathbf{p}_{cm}]$. Both of these angular momenta are conserved, but the angular momenta of the individual particles, $[\mathbf{x}_1 \times \mathbf{p}_1]$ and $[\mathbf{x}_2 \times \mathbf{p}_2]$, are not.
3. Only five of the seven constants of motion in the second factor, $H_r, \mathbf{L},$ and $\mathbf{F}$, can be independent. Two relationships between them are

$$(\mathbf{L} \cdot \mathbf{F}) = 0, \quad \text{and} \quad |\mathbf{F}|^2 = 2m|\mathbf{L}|^2 H_r + m^2 \alpha^2. \tag{4.2.6}$$

Thus $\mathbf{F}$ lies in the plane perpendicular to $\mathbf{L}$ (the plane of motion), and its length is fixed by L and H_r.

The Invariance Group (4.2.7)

The invariance group is determined by the Poisson brackets of the constants of motion, which can be calculated as

$$\{L_i, L_j\} = \varepsilon_{ijk} L_k, \qquad \{L_i, F_j\} = \varepsilon_{ijk} F_k,$$
$$\{F_i, F_j\} = -2mH_r \varepsilon_{ijk} L_k, \qquad \{H_r, L_i\} = \{H_r, F_i\} = 0, \qquad (4.2.8)$$

(Problem 2). Since they may be expressed in terms of $\mathbf{L}$, $\mathbf{F}$, and H, the flows generated by these quantities form a group, the center of which is the time-evolution. The factor group generated by $\mathbf{L}$ and $\mathbf{F}$ is isomorphic respectively to $SO(4)$, E_3, or $SO(3, 1)$ on the submanifold $H_r = E$, when $E < 0$, $= 0$, or > 0. In order to see this, suppose that $E < 0$, and define

$$A_i = \frac{1}{2}\left(L_i + \frac{F_i}{\sqrt{-2mE}}\right), \qquad B_i = \frac{1}{2}\left(L_i - \frac{F_i}{\sqrt{-2mE}}\right).$$

Since

$$\{A_i, A_j\} = \varepsilon_{ijk} A_k, \qquad \{B_i, B_j\} = \varepsilon_{ijk} B_k, \qquad \{A_i, B_j\} = 0,$$

it is apparent that the invariance group is isomorphic to $SO(3) \times SO(3) = SO(4)$. For $E = 0$ the claim follows from (4.2.8), and if $E > 0$, A and B can be defined as above, with the appropriate signs.

Remarks (4.2.9)

1. It is not possible to factorize $SO(4)$ into the form (rotations generated by $\mathbf{L}$) × (some other rotations); the situation is more complicated than that.
2. When $E = 0$, $\mathbf{F}$ plays the same role as $\mathbf{p}$ in E_3.
3. Since the flows provide continuous mappings of the group into phase space, the noncompactness of E_3 and $SO(3,1)$ is equivalent to the existence of unbounded trajectories for $H_r \geq 0$.

The Shapes of the Trajectories (4.2.10)

The most convenient way to calculate the projection of a trajectory onto M is with the aid of (4.2.4):

$$\mathbf{F} \cdot \mathbf{x} = |\mathbf{L}|^2 + m\alpha r \Rightarrow r = \frac{|\mathbf{L}|^2}{|\mathbf{F}| \cos \varphi - m\alpha}, \qquad \varphi = \sphericalangle(\mathbf{F}, \mathbf{x}).$$

These are conic sections, which lie in the region $\{(\mathbf{L} \cdot \mathbf{x}) = 0$ (the plane of motion$\} \cap \{(\mathbf{F} \cdot \mathbf{x}) > m\alpha r\}$. There are three cases to be distinguished (cf. Figure 32):

(a) $H_r > 0$. According to (4.2.6), $|\mathbf{F}| > |m\alpha|$, and r becomes infinite when $\varphi = \arccos(m\alpha/|\mathbf{F}|)$. The trajectory is hyperbolic (or linear, if $\alpha = 0$).
(b) $H_r = 0$. $|\mathbf{F}| = |m\alpha|$, and r becomes infinite when $\varphi = \pi$, if $\alpha = 0$. The trajectory is parabolic.
(c) $H_r < 0$. $|\mathbf{F}| < |m\alpha|$, and r is always finite. The trajectory is elliptic if $\alpha < 0$, or a point if $\alpha = 0$.

Remarks (4.2.11)

1. Cases (b) and (c) only occur when $\alpha \leq 0$.
2. Trajectories that pass through the origin have $\mathbf{L} = 0$. The canonical flow exists on the invariant submanifold $T^*(\mathbb{R}^3 \setminus \{0\}) \setminus \{(\mathbf{x}, \mathbf{p}): [\mathbf{x} \times \mathbf{p}] = \mathbf{0}\})$, where H_r generates a complete vector field.
3. The trajectory of $\mathbf{p}(t)$ always lies on some circle (Problem 6).

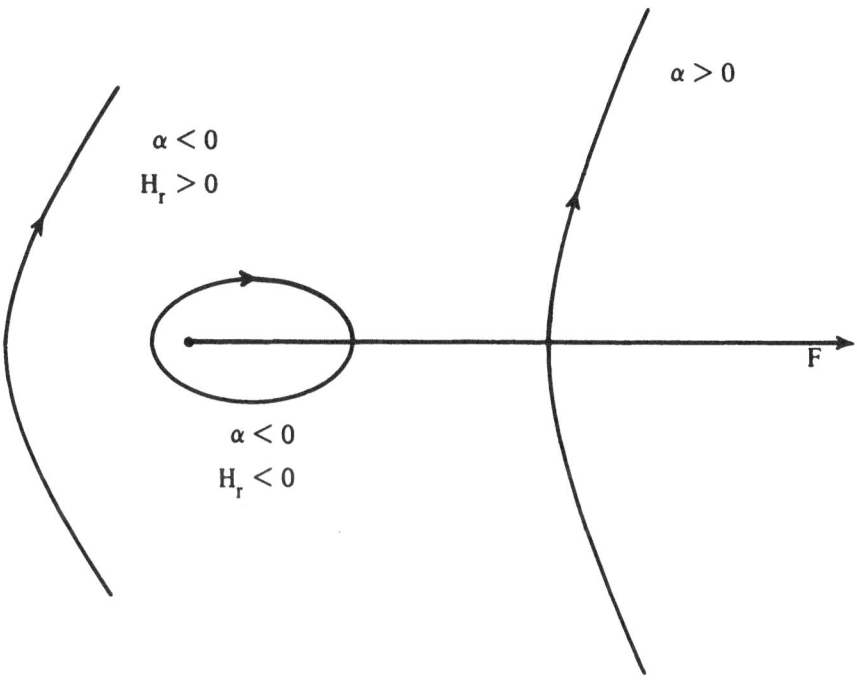

Figure 32   The Kepler trajectories.

The Elapsed Time (4.2.12)

The momentum canonically conjugate to r is $p_r \equiv (\mathbf{x} \cdot \mathbf{p})/r$, in terms of which

$$H_r = \frac{p_r^2}{2m} + \frac{|\mathbf{L}|^2}{2mr^2} + \frac{\alpha}{r}.$$

Thus the radial motion on the invariant submanifold $\mathbf{L} = $ constant is like a one dimensional motion where the original potential gets an additional term from the centrifugal force, $|\mathbf{L}|^2/2mr^2$. Integrating

$$\dot{r} = \frac{p_r}{m} = \sqrt{\frac{2}{m}} \sqrt{E - \frac{\alpha}{r} - \frac{|\mathbf{L}|^2}{2mr^2}}$$

yields (see Problem 7)

$$t - t_0 = \int_{r_0}^{r} \frac{dr' r' \sqrt{m/2}}{\sqrt{r'^2 E - \alpha r' - |\mathbf{L}|^2/2m}} = \left| r \sqrt{\frac{m}{2E}} \sqrt{1 - \frac{\alpha}{rE} - \frac{|\mathbf{L}|^2}{2mr^2 E}} \right.$$

$$+ \frac{\alpha\sqrt{m/2}}{2|E|^{3/2}} \left| \begin{array}{l} \ln\left(Er - \frac{\alpha}{2} + \sqrt{E\left(r^2 E - \alpha r - \frac{|\mathbf{L}|^2}{2m}\right)} \right) \\[2mm] \text{arc}\sin \dfrac{2Er - \alpha}{\sqrt{\alpha^2 + 2E|\mathbf{L}|^2/m}} \end{array} \right| \begin{array}{l} \left.\vphantom{\begin{array}{c}a\\b\end{array}}\right|^t \\ \\ \left.\vphantom{\begin{array}{c}a\\b\end{array}}\right|_{t_0} \end{array} \quad \begin{array}{l} E > 0 \\ \text{for} \\ E < 0 \end{array}$$

$$(4.2.13)$$

Corollary (Kepler's third law) (4.2.14)

If $E < 0$, *then* $r = r_0$ *when* $t = t_0 + \tau$, *where* $\tau = 2\pi a^{3/2}\sqrt{m/|\alpha|}$, *and* $a = |\alpha|/2|E|$ *is the major semiaxis of the ellipse.*

Remarks (4.2.15)

1. Conservation of angular momentum reduces any problem with a central force to a one-dimensional problem. That the orbits are in fact closed, the radial and angular frequencies being degenerate, is a consequence of the richer invariance group of the $1/r$ potential.
2. Since τ is proportional to $a^{3/2}$, and thus grows faster than the circumference of the orbit, the speed is less for larger orbits. This is in accordance with the virial theorem, which we shall discuss later, which says that kinetic energy $\sim$ potential energy, and thus speed $\sim a^{-1/2}$.
3. Corollary (4.2.14) is particularly well illustrated in the solar system by the extremes, Mercury and Pluto. The radii of their orbits and their periods are roughly 1:100 and 1:1000, respectively.
4. The time-evolution $r(t)$ is the inverse function of the elapsed time (4.2.13), which is not expressible in terms of elementary functions.

When $H_r \geq 0$, the particles escape to infinity, and a number of quantities approach constants. Hence we utilize the concepts introduced in (3.4.1), making the restriction to the invariant submanifold $H_r > 0$.

Asymptotic Constants of Motion (4.2.16)

$$\mathbf{p}, \frac{\mathbf{x}}{r}, \frac{1}{r} \in \mathscr{A}.$$

Proof

To see that the momenta are asymptotically constant when $E > 0$, we use the equation $|\dot{\mathbf{p}}| = |\alpha|/r^2$, and conclude from (4.2.13) that as $t \to \infty, r$ becomes proportional to t:

$$|\mathbf{p}(2T) - \mathbf{p}(T)| = \left| \int_T^{2T} dt\, \dot{\mathbf{p}}(t) \right| \leq |\alpha| \int_T^{2T} \frac{dt}{r(t)^2} \leq \frac{\text{const.}}{T} \to 0 \quad \text{as } T \to \infty,$$

and likewise as $t \to -\infty$. Hence $\mathbf{p}_\pm \equiv \tau_\pm \mathbf{p}$ exist for $E > 0$. This implies that the particles escape at definite angles:

$$\tau_\pm \frac{\mathbf{x}}{r} = \lim_{T \to \pm\infty} \frac{(1/T)\mathbf{x}(0) + (1/T)\int_0^T dt\ \mathbf{p}(t)/m}{|(1/T)\mathbf{x}(0) + (1/T)\int_0^T dt\ \mathbf{p}(t)/m|} = \pm\frac{\mathbf{p}_\pm}{|\mathbf{p}_\pm|}.$$

Lastly, $\tau_\pm(1/r) = 0$ when $E > 0$.   □

Constants of the form $K(\mathbf{x}/|\mathbf{x}|, \mathbf{p})$ converge trivially as $t \to \pm\infty$, and have the limits $K = K(\pm\mathbf{p}_\pm/|\mathbf{p}_\pm|, \mathbf{p}_\pm)$.

Corollaries (4.2.17)

1. $E = |\mathbf{p}_\pm|^2/2m$.
2. $\mathbf{F} = [\mathbf{p}_\pm \times \mathbf{L}] \pm \eta\mathbf{p}_\pm$, where $\eta \equiv \alpha\sqrt{m/2E}$.

Remarks (4.2.18)

1. The latter equation is easily solved for the limiting momenta:

$$\mathbf{p}_\pm = \frac{[\mathbf{L} \times \mathbf{F}]}{L^2 + \eta^2} \pm \frac{\eta}{L^2 + \eta^2}\mathbf{F}.$$

This implies what is intuitively obvious, that $\mathbf{p}_+$ and $-\mathbf{p}_-$ are related by reflection about $\mathbf{F}$.
2. The Møller transformation (3.4.4) using the free motion for the comparison flow Φ_t^0 simply does not exist in this case. By (4.2.13), for large times

$$r - t\sqrt{\frac{2E}{m}} \sim \frac{|\alpha|\sqrt{m/2}}{2E^{3/2}}\ln t,$$

whereas for Φ_t^0, $r - t\sqrt{2E/m} \sim$ constant. If the potential fell off as $r^{-1-\varepsilon}$, there would be no logarithmic term in t, and $\Phi_{-t} \circ \Phi_t^0$ would converge.
3. There exist other simple kinds of time-evolution that the flow approaches asymptotically;

$$H_0 = \frac{|\mathbf{p}|^2}{2m} + \frac{m\alpha}{|\mathbf{p}|t}$$

generates such a flow. However, it depends on t explicitly, and so Φ_t^0 is not a one-parameter group, which causes the consequences of definition (3.4.4) to lose some of their elegance.

The Scattering Transformation (4.2.19)

Since Ω_+ do not exist, we define S by using an algebra $\mathscr{A}_s \subset \mathscr{A}$ on which τ_- is bijective.

$$\tau_+ \circ \tau_-^{-1} f = f \circ S^{-1}, \quad \forall f \in \tau_- \mathscr{A}_s = \tau_+ \mathscr{A}_s = \mathscr{A}_\infty = \{H_r\}'/\!\!\!\!\!\!\!\! \wedge\ H_r > 0.$$

Since we are interested in the momenta, a reasonable choice for $\mathscr{A}_s$ is the algebra generated by $\mathbf{p}_-$ and $\mathbf{L}$. Then from (4.2.18; 1) we can calculate

$$\mathbf{p}_+ = \mathbf{p}_- \circ S^{-1} = \frac{L^2 - \eta^2}{L^2 + \eta^2}\,\mathbf{p}_- + \frac{2\eta}{L^2 + \eta^2}\,[\mathbf{p}_- \times \mathbf{L}]. \qquad (4.2.20)$$

Remarks (4.2.21)

1. Schematically, the situation looks like this:

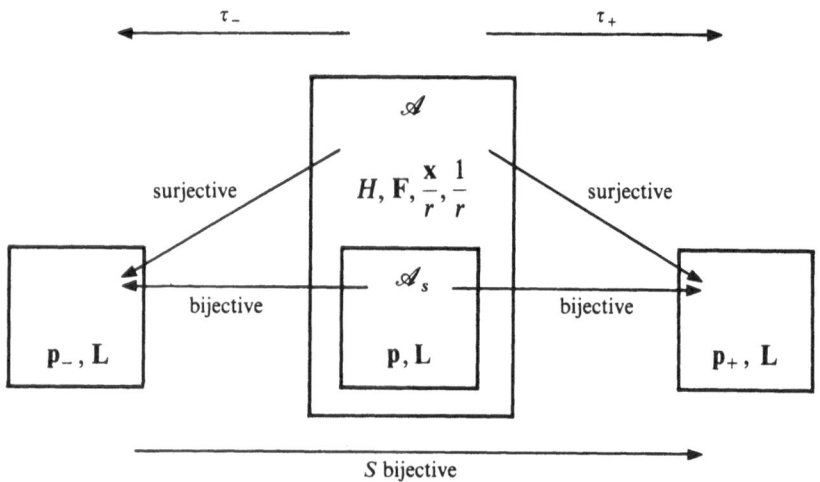

2. S depends on the choice of $\mathscr{A}_s$. If $\mathscr{A}_s$ were chosen as the set of constants of motion, S could be set to **1**.
3. The choice made in (4.2.20) does not fix S uniquely. A more precise specification of S, however, would be arbitrary.
4. As stressed in (3.4.11; 1), S does not leave all the constants of motion invariant. For example,

$$\mathbf{F} \circ S^{-1} = \mathbf{F} + 2\eta\mathbf{p}_+ .$$

Proposition (4.2.22)

By (4.2.20), the scattering angle is

$$\Theta = \text{arc cos}\,\frac{\mathbf{p}_+ \cdot \mathbf{p}_-}{2mE} = \text{arc cos}\,\frac{L^2 - \eta^2}{L^2 + \eta^2},$$

so

$$b^2 = \frac{|\mathbf{L}|^2}{|\mathbf{p}_-|^2} = \frac{\eta^2}{|\mathbf{p}_-|^2}\,\frac{1 + \cos\Theta}{1 - \cos\Theta}$$

(cf. (3.4.13; 1)). *Then the differential scattering cross-section (3.4.12) can be calculated as*

$$\sigma(\Theta) = \frac{\alpha^2}{16E^2} \sin^{-4} \frac{\Theta}{2}.\dagger$$

Remarks (4.2.23)

1. σ is independent of the sign of α, although if $\alpha > 0$ the particle turns around on the near side of the scattering center, while if $\alpha < 0$ it turns around on the far side.
2. It is because of the way that the potential changes under dilatations that α and E occur in Θ only in the combination $\alpha/\sqrt{E}$. A dilatation can be used to put H into the form

$$E\left(\frac{p^2}{2m} + \frac{\alpha}{\sqrt{E}}\frac{1}{r}\right).$$

To summarize, we have learned that the canonical flow exists globally on phase space as restricted in (4.2.11; 2), and the trajectories are submanifolds diffeomorphic to $\mathbb{R}$ when $H_r \geq 0$, and to T^1 when $H_r < 0$.

Problems (4.2.24)

1. Calculate the canonical transformation generated by $\mathbf{F}_i$. Is it a point transformation?

2. Verify equations (4.2.8).

3. Show that the trajectories (4.2.10) are conic sections with foci at the origin.

4. Use (4.2.14) to calculate how long a body with r_0 equal to the radius of Earth's orbit and $v_0 = 0$ takes to fall into the sun.

5. Calculate the scattering angle for (4.2.10; (a)).

6. Calculate the projection of the trajectory onto the second factor of $T^*(\mathbb{R}^3) = \mathbb{R}^3 \times \mathbb{R}^3$ (i.e., $\mathbf{p}(t)$).

7. With the variables $u: r = a(1 + \varepsilon \cos u)$, $a = |\alpha/2E| = $ the major semiaxis, and $\varepsilon = |\mathbf{F}|/m\alpha = $ the eccentricity, equation (4.2.13) is written as

$$\frac{2|E|^{3/2}}{\alpha\sqrt{m/2}} (t - t_0) = u - \varepsilon \sin u$$

(Kepler's equation). Interpret this geometrically.

Solutions (4.2.25)

1.
$$\{F_i, p_k\} = \delta_{ik}|\mathbf{p}|^2 - p_i p_k + m\alpha\left(\frac{\delta_{ik}}{r} - \frac{x_i x_k}{r^3}\right),$$

$$\{F_i, x_k\} = \delta_{ik}(\mathbf{p} \cdot \mathbf{x}) - 2x_i p_k + x_k p_i.$$

It is not a point transformation, because $\mathbf{F}$ is not linear in $\mathbf{p}$.

$\dagger$ α/E is the turning radius, i.e., the minimum distance from the particle to the scattering center.

2. Since L_i generates rotations and $\mathbf{L}$ and $\mathbf{F}$ are vectors,
$$\{L_i, L_j\} = \varepsilon_{ijk} L_k, \qquad \{L_i, F_j\} = \varepsilon_{ijk} F_k.$$
For the calculation of $\{F_i, F_j\}$, use: $F_i = x_i |\mathbf{p}|^2 - p_i(\mathbf{x} \cdot \mathbf{p}) + m\alpha x_i/r$.

3. $r \pm \sqrt{(x - A)^2 + y^2} = C \Leftrightarrow (C - r)^2 = A^2 - 2Ax + r^2 \Leftrightarrow$
$r = (A^2 - C^2)/(2A \cos \varphi - 2C)$, where $x = r \cos \varphi$.

4. The major semiaxis of the trajectory is half the radius of Earth's orbit, and it takes half an orbital period to fall into the sun, so the answer is $2^{-5/2}$ years.

5. $\Theta = \pi - 2\varphi$, where φ is the angle at $r = \infty$. Hence, according to (4.2.6),
$$\cos \Theta = -\cos 2\varphi = 1 - 2\cos^2 \varphi = 1 - 2\frac{m^2\alpha^2}{|\mathbf{F}|^2} = 1 - 2\frac{\eta^2}{L^2 + \eta^2}.$$

6.
$$[\mathbf{L} \times \mathbf{F}] = \mathbf{p}L^2 + m\alpha\left[\mathbf{L} \times \frac{\mathbf{x}}{r}\right] \Rightarrow (\mathbf{p} \cdot [\mathbf{L} \times \mathbf{F}])$$
$$= L^2\left(p^2 + m\frac{\alpha}{r}\right) \Rightarrow \left|\mathbf{p} - \frac{[\mathbf{L} \times \mathbf{F}]}{L^2}\right|^2 = p^2 - 2\left(\mathbf{p} \cdot \frac{[\mathbf{L} \times \mathbf{F}]}{L^2}\right) + \frac{|\mathbf{F}|^2}{L^2}$$
$$= -2mE + \frac{|\mathbf{F}|^2}{L^2} = \frac{m^2\alpha^2}{L^2}.$$

Thus $\mathbf{p}(t)$ lies on a circle.

7. $r \cos \varphi = a(\cos u + \varepsilon) \Rightarrow r = a(1 - \varepsilon^2)/(1 - \varepsilon \cos \varphi)$.

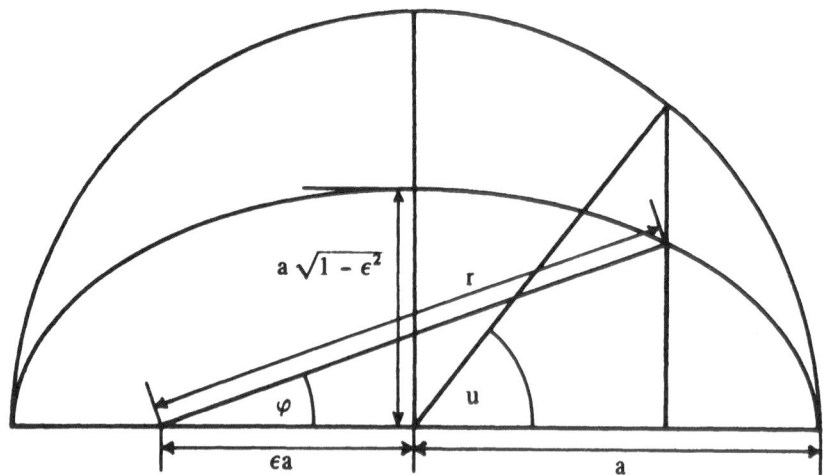

The variables used in Kepler's equation.

4.3 The Problem of Two Centers of Force

This is the connecting link between one-body problems and the restricted three-body problem. There are no longer five independent constants, but only three. Even so, the system is integrable.

In the last section we saw that the two-body problem can be reduced to the problem of a single particle in the force-field of an infinitely heavy, immovable object. This suggests that the first three-body problem to study is one where one particle is so light that it does not influence the motion of the other two. At this point the nature of the problem depends on whether it involves the electrical or the gravitational force. Let M be the mass of the heavy particles and m the mass of the light one. Then from (4.2.13) we can easily estimate the order of magnitude of the orbital frequencies ω_H of the heavy particles and ω_L of the light one. This is just because the centrifugal and centripetal forces are balanced; thus if R is the orbital radius,

$$MR\omega_H^2 \cong \frac{\kappa M^2 + e^2}{R^2}$$

$$mR\omega_L^2 \cong \frac{\kappa mM + e^2}{R^2}.$$

(4.3.1)

If gravitation predominates, that is, $\kappa mM \gg e^2$, then $\omega_H^2 = \omega_L^2 = \kappa M/R^3$, and the motion of the heavy particles can not be neglected when one studies the motion of the light one. This is a direct consequence of the fact discovered by Galileo, that all masses are accelerated equally strongly in a gravitational field. The case of dominant gravitational forces is known as the restricted three-body problem. It is of obvious interest for space travel, but is rather difficult to attack analytically; we shall study it in the next section. It is somewhat simpler when the electrical force predominates, $\kappa M^2 \ll e^2$, as happens with elementary particles. In that case, $\omega_L^2/\omega_H^2 = M/m$, and the heavy particles move slowly compared with the light one when M/m is large. This would be appropriate for the simplest kinds of molecules, with two nuclei and one electron, except that the important physical properties lie outside the domain of classical physics. We shall return to this problem when we treat the quantum theory.

The Hamiltonian (4.3.2)

For mathematical convenience we can set the two centers of force at $(1, 0, 0)$ and $(-1, 0, 0)$ without loss of generality, and start off with the manifold

$$M_0 = \mathbb{R}^3 \backslash \{(1, 0, 0), (-1, 0, 0)\}.$$

(4.3.3)

It will be necessary to restrict M_0 and $T^*(M_0)$ further in order to avoid some complications. Let r_1 and r_2 be the distances of the light particle from the two centers, of strengths α_1 and α_2 (see Figure 33), and suppose that $m = 1$. Then the Hamiltonian of the problem becomes

$$H = \frac{|\mathbf{p}|^2}{2} - \frac{\alpha_1}{r_1} - \frac{\alpha_2}{r_2}.$$

(4.3.4)

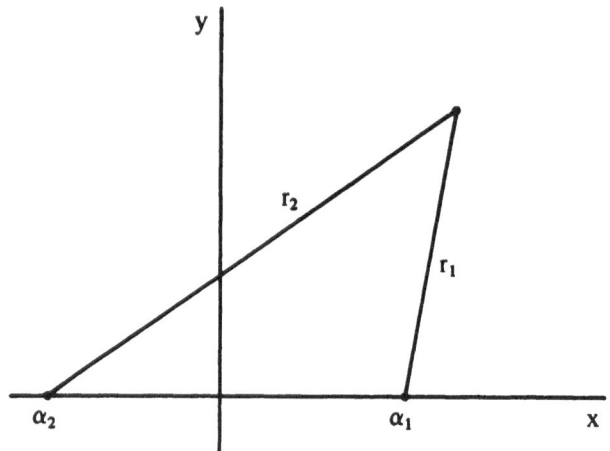

Figure 33   The centers of force.

Since r_1 and r_2 have the rather unwieldy form $\sqrt{(x \pm 1)^2 + y^2 + z^2}$ in Cartesian coordinates, it is convenient to introduce new coordinates that take advantage of the rotational symmetry.

Elliptic Coordinates (4.3.5)

Elliptic coordinates use the chart (M_1, Ψ):

$$M_1 = \mathbb{R}^3 \backslash (\mathbb{R}, 0,0) \subset M_0,$$
$$\Psi(M_1) = \mathbb{R}^+ \times (0, \pi) \times T^1 \ni (\xi, \eta, \varphi),$$

$$\Psi: (x, y, z) \to (\cosh \xi \cos \eta, \sinh \xi \sin \eta \cos \varphi, \sinh \xi \sin \eta \sin \varphi),$$

with which H is written as

$$H = \tfrac{1}{2}(\cosh^2 \xi - \cos^2 \eta)^{-1}$$

$$\times \left[p_\xi^2 + p_\eta^2 + p_\varphi^2 \left(\frac{1}{\sinh^2 \xi} + \frac{1}{\sin^2 \eta} \right) - \alpha \cosh \xi - \beta \cos \eta \right],$$

$$\tfrac{1}{2}\alpha = \alpha_1 + \alpha_2, \qquad \tfrac{1}{2}\beta = -\alpha_1 + \alpha_2 \qquad (4.3.6)$$

(Problem 1). Since H would be integrable without the factor $(\cosh^2 \xi - \cos^2 \eta)^{-1}$, we recall (3.2.14; 6), according to which such a factor can be eliminated by a transformation to a new time variable. Thus if we let

$$\mathscr{H} = 2(\cosh^2 \xi - \cos^2 \eta)(H - E) = H_\xi + H_\eta, \text{ where}$$

$$H_\xi = p_\xi^2 + \frac{p_\varphi^2}{\sinh^2 \xi} - \alpha \cosh \xi - E \cosh^2 \xi, \text{ and} \qquad (4.3.7)$$

$$H_\eta = p_\eta^2 + \frac{p_\varphi^2}{\sin^2 \eta} - \beta \cos \eta + E \cos^2 \eta,$$

on extended phase space; then on the submanifold $\mathscr{H} = 0$, $\mathscr{H}$ describes the time-evolution with a parameter s such that $dt/ds = 2(\cosh^2 \xi - \cos^2 \eta) > 0$.

Constants of the Motion (4.3.8)

On extended phase space, $\mathscr{H}$, E, H_ξ, and p_φ are all constant and mutually independent.

Remarks (4.3.9)

1. Because $dt/ds > 0$, anything that is constant in s is also constant in t. Restricting ourselves again to $T^*(M_1)$, we can use H, p_φ, and $H_\xi = -H_\eta$ as three independent constants of motion, replacing E with H.
2. The conservation of p_φ comes from the cylindrical symmetry of the problem. The canonical flows coming from H_ξ and H_η are rather complicated.
3. Since the Poisson brackets between any two of the four conserved quantities (or respectively H, p_φ, and H_ξ) vanish, the system is integrable on $T^*(M_1)$.
4. No additional constants can be found, so the invariance group of $\mathscr{H}$ (respectively H) is a 4-parameter (3-parameter) Abelian group.

Effective Potentials (4.3.10)

Integration of the equations of motion with $\mathscr{H}$ from (4.3.7) leads to two one-dimensional problems with the potentials:

$$V_\xi = \frac{p_\varphi^2}{\sinh^2 \xi} - \alpha \cosh \xi - E \cosh^2 \xi, \xi \in \mathbb{R}^+,$$

$$V_\eta = \frac{p_\varphi^2}{\sin^2 \eta} - \beta \cos \eta + E \cos^2 \eta, \eta \in (0, \pi) \tag{4.3.11}$$

(see Figures 34–37).

Remarks (4.3.12)

1. If $p_\varphi \neq 0$, the effective potential V_ξ becomes infinite as $\xi \to 0$ (as does V_η as $\eta \to 0$ or $\eta \to \pi$), and the trajectory can never leave M_1. On $T^*(M_1)\backslash\{(\mathbf{x}, \mathbf{p}): yp_z - zp_y = 0\}$, H generates a complete vector field, and the canonical flow exists (but see (4.3.17)).
2. If $E < 0$, then V_ξ goes to $+\infty$ as $\xi \to \infty$, and the trajectory remains in a compact set. In this case the conditions of the recurrence theorem hold, and almost all orbits are almost periodic.
3. There are equilibrium positions in ξ and η, so there exist some strictly periodic orbits.

When $E < 0$, the invariant submanifolds $N: H = E$, $p_\varphi = L$, $H_\xi = K$, are compact, and therefore diffeomorphic to T^3. We can determine the frequencies as we did following (3.3.14):

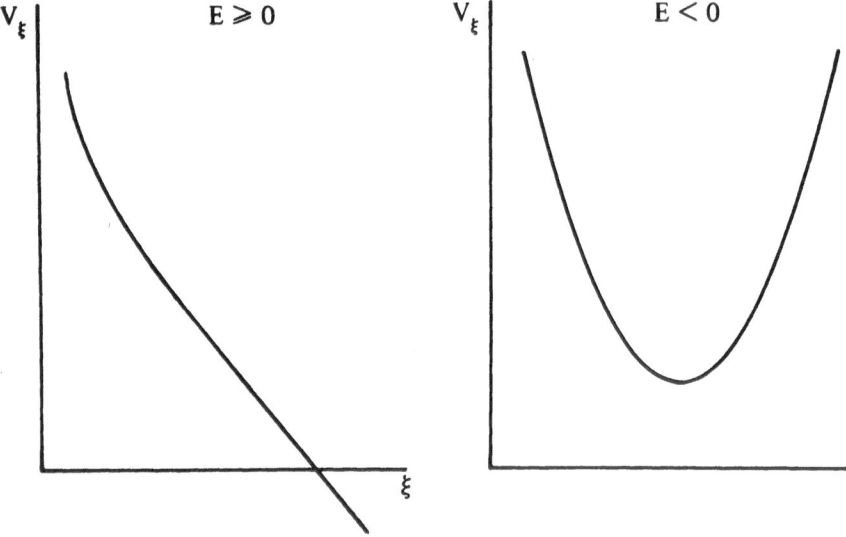

Figure 34   Effective potential.   Figure 35   Effective potential.

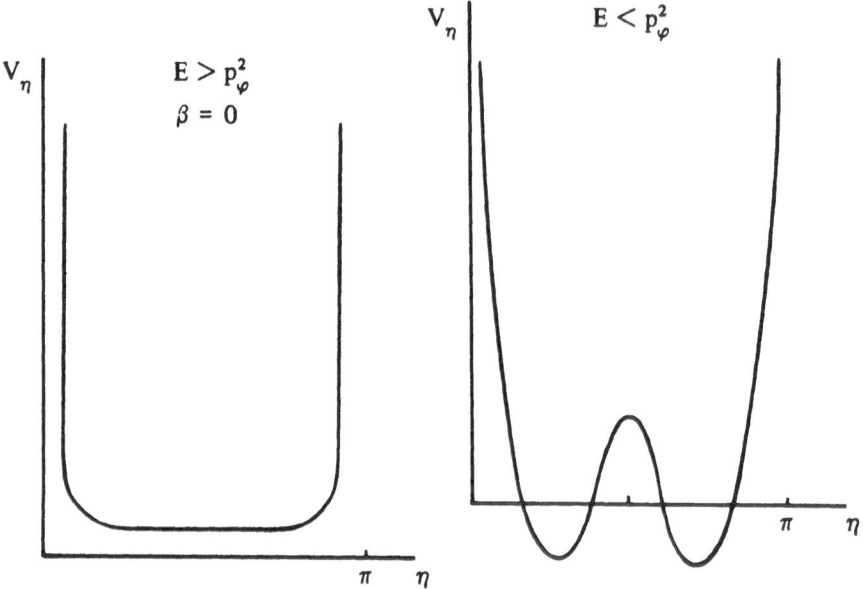

Figure 36   Effective potential.   Figure 37   Effective potential.

Action and Angle Variables (4.3.13)

The constants of motion

$$I_{\varphi} = \frac{1}{2\pi} \oint d\varphi \, p_{\varphi} = L,$$

$$I_{\xi} = \frac{1}{2\pi} \oint d\xi \, p_{\xi} = \frac{1}{2\pi} \oint d\xi \, \sqrt{K + E \cosh^2 \xi + \alpha \cosh \xi - \frac{L^2}{\sinh^2 \xi}},$$

$$I_{\eta} = \frac{1}{2\pi} \oint d\eta \, p_{\eta} = \frac{1}{2\pi} \oint d\eta \, \sqrt{-K - E \cos^2 \eta + \beta \cos \eta - \frac{L^2}{\sin^2 \eta}}$$

$$\tag{4.3.14}$$

are mutually independent, since

$$J \equiv \frac{\partial(I_{\varphi}, I_{\xi}, I_{\eta})}{\partial(L, E, K)} = I_{\xi,E} I_{\eta,K} - I_{\xi,K} I_{\eta,E}$$

$$= \oint \frac{d\xi \, d\eta (\cos^2 \eta - \cosh^2 \xi)}{16\pi^2 p_{\xi} p_{\eta}} < 0.$$

$$\tag{4.3.15}$$

Accordingly, $E(I_{\varphi}, I_{\xi}, I_{\eta})$ exists, and, as in (3.3.15; 4), its derivatives by the action variables are the three frequencies of the motion.

Remarks (4.3.16)

1. N is the Cartesian product of three tori, on each of which two of the three variables (φ, ξ, η) are held fixed while the third runs through its allowed domain, i.e., $0 < \varphi \leq 2\pi$, $\xi_1 \leq \xi \leq \xi_2$, and $\eta_1 \leq \eta \leq \eta_2$, where the boundaries for ξ and η are the values for which $p_{\xi} = 0$ and $p_{\eta} = 0$.† The frequencies correspond to the orbital periods of these variables.
2. The frequencies cannot be written explicitly as functions of the I's or of $E, K,$ and L, because the integrals in (4.3.13) cannot be expressed in terms of ordinary functions. The frequencies can be written as

$$\omega_{\varphi} = \frac{I_{\xi,K} I_{\eta,L} - I_{\xi,L} I_{\eta,K}}{J},$$

$$\omega_{\xi} = \frac{I_{\eta,K}}{J}, \qquad \omega_{\eta} = \frac{I_{\xi,K}}{J}.$$

Since these functions of $E, K,$ and L vary continuously from one torus to the next one, they are generally not in rational ratios to one another, and the trajectory fills some three-dimensional region densely. Hence

† In Figures 34 and 36 these are given by the intersections with horizontal lines at heights K and $E - K$.

there are no additional constants of the motion in that part of phase space.

3. Since the curves ξ = constant and η = constant are respectively ellipses and hyperbolas in Cartesian coordinates, the projection of N to the plane $\varphi = 0$ is the region with the points shown in Figure 38, which should be pictured three-dimensionally as if it were rotated about the x-axis.

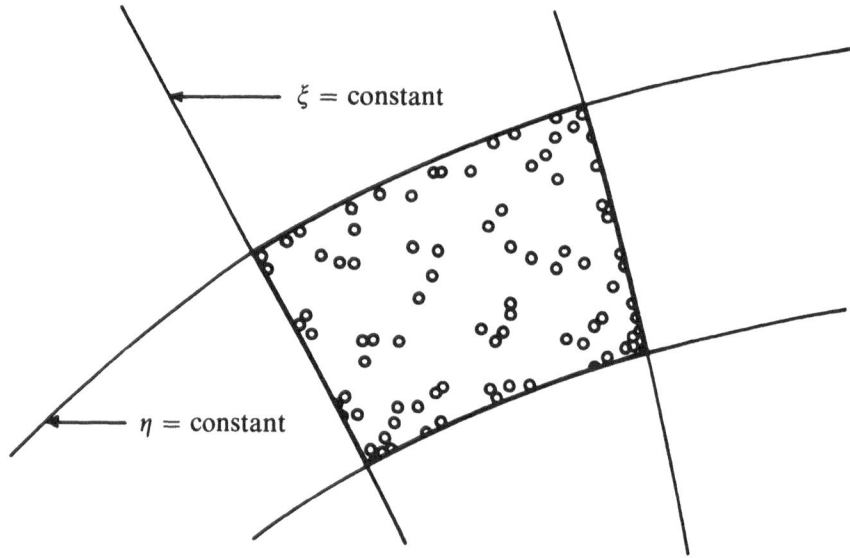

Figure 38   A computer calculation of the points at which the trajectory intersects the plane $\varphi = 0$. These points fill the projection of an invariant torus to the plane densely.

$E > 0$: **Unbound Trajectories** (4.3.17)

When $E > 0$, the vector field generated by H_ξ is not complete; instead, the particles reach infinity at a finite value of s,

$$ s_\infty = \int_{\xi_0}^{\infty} \frac{d\xi}{\sqrt{K - (p_\varphi^2/\sinh^2 \xi) + \alpha \cosh \xi + E \cosh^2 \xi}} < \infty $$

(cf. Figure 34 and (4.3.7) with $H_\xi = K$).

Remarks (4.3.18)

1. Depending on whether $p_\xi > 0$ or $p_\xi < 0$ when $s = 0$, the above integral either runs directly from ξ_0 to infinity, or else first passes through the turning point. Other than that, on the surface $H = E$, $H_\xi = K$, p_φ = constant, s_∞ depends only on the initial value ξ_0.

2. This does not mean that the particles reach infinity after a finite time t. Because $dt/ds = 2(\cosh^2 \xi - \cos^2 \eta)$, t goes to infinity at s_∞, as $(s_\infty - s)^{-1}$ (Problem 5).

Even in the Kepler problem, free time-evolution was not good enough as an asymptotic basis of comparison for the unbound trajectories, so we cannot expect much of it in this case either. However, there is a good expectation that at large distances the separation of the centers hardly matters, and the time-evolution of the problem with one force-center can be used for Φ_t^0.

Theorem (4.3.19)

The Møller transformations

$$\Omega_{\pm} = \lim_{t \to \pm \infty} \Phi_{-t} \circ \Phi_t^0$$

exist, where Φ_t^0 is the flow with $\alpha_1' = 0$ and $\alpha_2' = \alpha_1 + \alpha_2$ (that is, (α, β) becomes (α, α)). The domains are

$$D_{\pm} = \left\{ (\mathbf{x}, \mathbf{p}) : \frac{|\mathbf{p}|^2}{2} - \frac{\alpha}{r_2} > 0, \quad yp_z - zp_y \neq 0 \right\}.$$

Remarks (4.3.20)

1. Φ_t^0 is arbitrary in that the single force-center can be put anywhere at all. The Møller transformations exist as long as the strength of the force center is α.
2. Because $H_0 \circ \Omega_{\pm}^{-1} = H_{|\mathcal{R}_{\pm}}$ (3.4.5; 1), we see

$$\mathcal{R}_{\pm} = \left\{ (\mathbf{x}, \mathbf{p}) : \frac{|\mathbf{p}|^2}{2} - \frac{\alpha_1}{r_1} - \frac{\alpha_2}{r_2} > 0, (y, z) \neq 0, yp_z - zp_y \neq 0 \right\}.$$

   The remaining free constant in H is determined by the condition that if $(\mathbf{x} \cdot \mathbf{p}) > 0$ and $|\mathbf{x}| \to \infty$, then $\Omega_{\pm}(x)$ approaches x.

Proof

Although the proof is not difficult in principle, it requires some involved calculations, and will not be done here.   □

Corollaries (4.3.21)

1. $\Omega_{\pm}$ map the flows $\Phi_{t|D_{\pm}}^0$ to $\Phi_{t|\mathcal{R}_{\pm}}$. Since $\Omega_{\pm}$ are diffeomorphisms, Φ_t must have five independent constants on $\mathcal{R}_{\pm}$, just like Φ_t^0 on $D_{\pm}$. From (3.4.5; 3) and (4.2.4), these constants are $\tau_{\pm}(\mathbf{L}) = \mathbf{L} \circ \Omega_{\pm}^{-1} = \lim_{t \to \pm \infty} \mathbf{L}(t)$ and $\tau_{\pm}(\mathbf{F}) = \mathbf{F} \circ \Omega_{\pm}^{-1} = \lim_{t \to \pm \infty} \mathbf{F}(t)$, where $(\tau_{\pm}(\mathbf{L}) \cdot \tau_{\pm}(\mathbf{F})) = 0$.
2. $\Omega_{\pm}(\beta)$ depend on β continuously, and $\Omega_{\pm}(\alpha) = 1$. The trajectories of Φ_t with $E > 0$ are mapped by $\Omega_{\pm}^{-1}$ homotopically (see (2.6.17; 5)) onto those of Φ_t^0. Since the one set of trajectories is continuously deformed into the other, no knots can form. With the Møller transformations it is easy to make such global statements, which are otherwise hard to find.

To summarize what we have learned about how the separation into two centers of force affects the flow: The unbound trajectories $(E > 0)$ are only moderately deformed, and the flow can be transformed diffeomorphically to a linear one. The periodic orbits get wound up like balls of string, most of them filling three-dimensional regions densely.

Problems (4.3.22)

1. Derive (4.3.6) by calculating $T(\Psi)$.

2. Derive (4.3.6) by introducing elliptic coordinates in $L = |\mathbf{x}|^2/2 - V$.

3. Use the equations of motion to verify that $dK_\xi/dt = 0$.

4. Use the Hamilton–Jacobi equation (3.2.16) to separate this problem.

5. Show that if $E > 0$, then t goes to infinity as $1/2E(s_\infty - s)$.

Solutions (4.3.23)

1. Note that

$$\cosh^2 \xi \begin{vmatrix} \cos^2 \eta \\ \sin^2 \eta \end{vmatrix} + \sinh^2 \xi \begin{vmatrix} \sin^2 \eta \\ \cos^2 \eta \end{vmatrix} = \cosh^2 \xi - \begin{vmatrix} \sin^2 \eta \\ \cos^2 \eta \end{vmatrix} : \Rightarrow$$

$$\tfrac{1}{2}(r_1 + r_2) = \cosh \xi, \ \tfrac{1}{2}(r_1 - r_2) = \cos \eta \Rightarrow r_1 r_2 = \cosh^2 \xi - \cos^2 \eta.$$

That takes care of the potential energy. For the kinetic energy, calculate

$$T(\Psi)^t \cdot T(\Psi) = \begin{vmatrix} \cosh^2 \xi - \cos^2 \eta & 0 & 0 \\ 0 & \cosh^2 \xi - \cos^2 \eta & 0 \\ 0 & 0 & \sinh^2 \xi \sin^2 \eta \end{vmatrix} \Rightarrow$$

$$|\mathbf{p}|^2 = (p_\xi, p_\eta, p_\varphi)(T(\Psi)^{-1} \cdot T(\Psi)^{-1t}) \begin{vmatrix} p_\xi \\ p_\eta \\ p_\varphi \end{vmatrix}$$

$$= \frac{p_\xi^2 + p_\eta^2}{\cosh^2 \xi - \cos^2 \eta} + \frac{p_\varphi^2}{\sinh^2 \xi \sin^2 \eta}.$$

2. $\qquad |\dot{\mathbf{x}}|^2 = (\dot{\xi}^2 + \dot{\eta}^2)(\cosh^2 \xi - \cos^2 \eta) + \dot{\varphi}^2 \sinh^2 \xi \sin^2 \eta \Rightarrow$

$$(p_\xi, p_\eta, p_\varphi) = (\dot{\xi}(\cosh^2 \xi - \cos^2 \eta), \dot{\eta}(\cosh^2 \xi - \cos^2 \eta), \dot{\varphi} \sinh^2 \xi \sin^2 \eta).$$

3. $\dfrac{dH_\xi}{dt} = -2p_\xi H_{,\xi} - \dfrac{p_\xi}{\cosh^2 \xi - \cos^2 \eta} \left[\dfrac{p_\varphi^2 2\cosh \xi}{\sinh^3 \xi} + \alpha \sinh \xi + 2H \sinh \xi \cosh \xi \right] = 0.$

4. Let $f = S - Et$. The equation

$$\left(\frac{\partial S}{\partial \xi}\right)^2 + \left(\frac{\partial S}{\partial \eta}\right)^2 + \left(\frac{\partial S}{\partial \varphi}\right)^2 (\sinh^{-2} \xi + \sin^{-2} \eta)$$

$$- \alpha \cosh \xi - \beta \cos \eta - E(\cosh^2 \xi - \cos^2 \eta) = 0$$

can be solved by supposing that S separates as

$$S = S_1(\xi) + S_2(\eta) + \varphi p_\varphi.$$

5. Asymptotically,

$$\frac{d\xi}{ds} = e^\xi \sqrt{E} \Rightarrow s = s_\infty - \frac{e^{-\xi}}{\sqrt{E}}, \qquad t = 2 \int ds \frac{e^{2\xi}}{4} = \frac{1}{2E(s_\infty - s)}.$$

4.4 The Restricted Three-Body Problem

The motion of a particle in the field of two rotating masses is already so complicated—even when all the motion is in a plane—that only rather fragmentary statements can be made about it.

As mentioned above, the restricted three-body problem has to do with the motion of a particle that is so light that its influence on the motion of the other two particles is negligible. Such an assumption is justified for, say, the flight of a spacecraft to the moon. We need a way to take the motion of the centers of force (with masses m and μ and coordinates $\mathbf{x}_m(t)$ and $\mathbf{x}_\mu(t)$) into account. So let us define a time-dependent

Hamiltonian (4.4.1)

$$H = \tfrac{1}{2}|\mathbf{p}|^2 - \kappa \left(\frac{m}{|\mathbf{x}_m(t) - \mathbf{x}|} + \frac{\mu}{|\mathbf{x}_\mu(t) - \mathbf{x}|} \right).$$

Remarks (4.4.2)

1. We shall only consider the case in which the heavy particles move in circular orbits, and therefore $\mathbf{x}_m$ and $\mathbf{x}_\mu$ describe circles about the center of mass, which we may take as the origin.
2. If both $\mathbf{p}$ and $\mathbf{x}$ are in the orbital plane at any time, then they are in it for all times. This will be the main variant of the problem treated here.
3. In (4.4.1) the mass of the light particle has been set to 1, since it factors out of the equations of motion. For simplicity we shall henceforth use units where $R = \kappa = m + \mu = 1$. In these units the frequency of the heavy particles $\omega = \sqrt{(m + \mu)}/R^{3/2}$ equals 1, and the only essential parameter that remains is $\mu/m \leq 1$. The masses m and μ are respectively at distances μ and m from the origin.

Rotating Coordinates (4.4.3)

In a coordinate system that rotates with the heavy particles, the centers of force are fixed. As in Example (3.2.15; 2), H gains a term from the centrifugal force, and if it is written out in components it is

$$H = \tfrac{1}{2}(p_x^2 + p_y^2) - xp_y + yp_x - \frac{m}{[(x + \mu)^2 + y^2]^{1/2}} - \frac{\mu}{[(x - m)^2 + y^2]^{1/2}}$$

(4.4.4)

$$\dot{x} = p_x + y, \qquad \dot{y} = p_y - x,$$

$$\dot{p}_x = p_y - V_{,x}, \qquad \dot{p}_y = -p_x - V_{,y},$$

$$V_{,x} = \frac{m(x + \mu)}{[(x + \mu)^2 + y^2]^{3/2}} + \frac{\mu(x - m)}{[(x - m)^2 + y^2]^{3/2}},$$

$$V_{,y} = \frac{ym}{[(x + \mu)^2 + y^2]^{3/2}} + \frac{y\mu}{[(x - m)^2 + y^2]^{3/2}}.$$

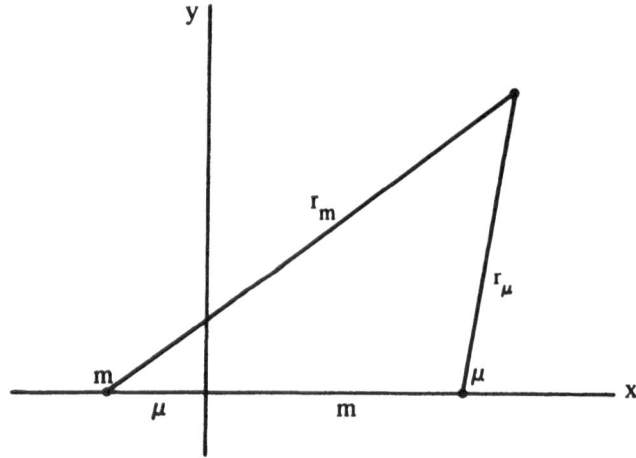

Figure 39   The coordinates used in the restricted three-body problem.

Remarks (4.4.5)

1. Changing to a rotating coordinate system is a point transformation in extended phase space, but we can just as well use H on the phase space $T^*(\mathbb{R}^2 \setminus \{(m, 0)\} \setminus \{(-\mu, 0)\})$.
2. Since H does not depend explicitly on the time in the rotating system, it is a constant, known as **Jacobi's constant**. However, no other constant, which would make the system integrable, is to be found.
3. In order for H to generate a complete vector field, i.e., for collisions to be avoided, we shall have to restrict phase space more than this, but it is not yet clear exactly how this is to be done.

Equilibrium Configurations (4.4.6)

Although the gravitational force is nowhere equal to zero, it is possible for the centrifugal force to balance it in a rotating system. Therefore there exist critical points in phase space where $dH = 0$, at which, by (4.4.4),

(i) $p_x = -y$, (iv) $x = (x + \mu)mr_m^{-3} + \mu(x - m)r_\mu^{-3}$,

(ii) $p_y = x$, (v) $y = ymr_m^{-3} + y\mu r_\mu^{-3}$,

(iii) $r_m^2 = (x + \mu)^2 + y^2$, (vi) $r_\mu^2 = (x - m)^2 + y^2$.

These equations have

(a) *Two equilateral solutions*

If $y \neq 0$, then equation (v) implies that $1 = mr_m^{-3} + \mu r_\mu^{-3}$, and then (iv) implies that $m\mu(r_m^{-3} - r_\mu^{-3}) = 0$, and thus $r_\mu = r_m$. Because $m + \mu = 1$, it follows that $r_m = r_\mu = 1$, independently of m/μ. Consequently both configurations for which the three particles are at the corners of an equilateral triangle are in equilibrium.

(b) *Three collinear solutions*

If $y = 0$, then there are clearly three solutions, since the curves $f = x$ and

$$f = m\frac{x + \mu}{|x + \mu|^3} + \mu\frac{x - m}{|x - m|^3}$$

have three points of intersection:

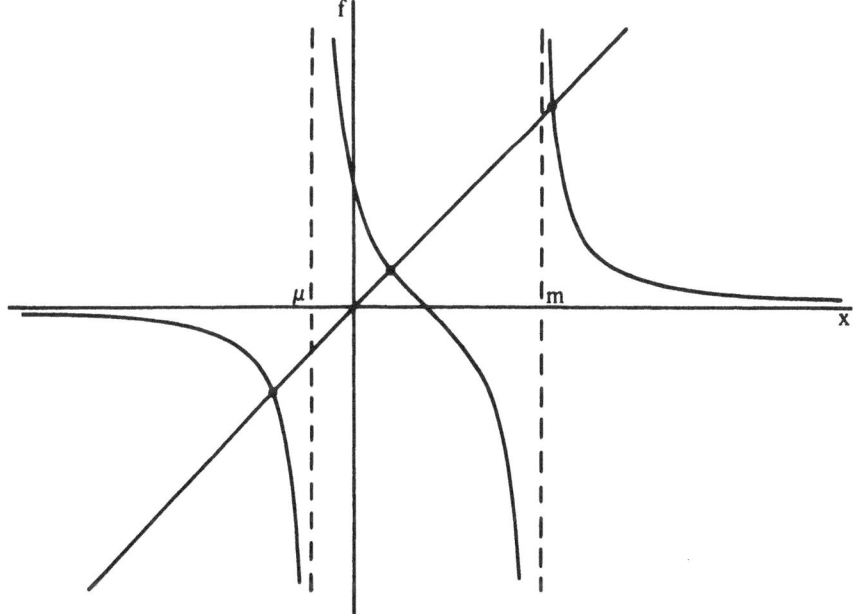

Figure 40   Determination of the collinear solutions.

Remarks (4.4.7)

1. The equilibrium configurations are zero-dimensional trajectories. In the original system, they correspond to circular orbits with frequency 1.
2. The equilateral solutions were known to Lagrange, though he drew no conclusions about their meaning for astronomy. It turns out, however, that there are real bodies in the solar system approximately in such a configuration. A group of asteroids, the Trojans, nearly make an equilateral triangle with the sun and Jupiter. Since their masses are vanishingly small on this scale, and since all motion in the solar system is roughly in a plane, the system of equations (4.4.4) is applicable.
3. The collinear solutions do not ever appear to be realized, probably because of their instability. Other large planets have an appreciable influence on the asteroids, and it is important to study whether they deform the asteroids' orbits only moderately, or destroy them altogether. We shall see below that the collinear solutions are always unstable, whereas the equilateral ones are stable so long as the mass of Jupiter is less than 4% that of the sun, which happens to be the case.

Periodic Orbits (4.4.8)

At this point it is of interest to ask whether there are other periodic solutions in addition to these five. In the special case $\mu = 0$ we already know that a great many trajectories are periodic—in the rotating system these are the Kepler orbits of a single force-center and a mass m with a rational frequency. According to the following argument, which goes back to Poincaré, under the right circumstances, for small enough μ, there exist periodic orbits with the same frequencies. Let $u(t, u_0; \mu)$, where u stands for x, y, p_x, or p_y, be the solution of the equation of motion with initial condition $u(0, u_0; \mu) = u_0$, and suppose an orbital period τ is specified. We ask for what initial values u_0 the orbit has the period τ. For these values the equation

$$u(\tau, u_0, \mu) = u_0 \qquad (4.4.9)$$

should hold, and we can regard it as the equation for $u_0(\mu)$. From our study of the Kepler problem we know that $\forall \tau \in \mathbb{R}^+$ equation (4.4.9) has a solution for $\mu = 0$. From the theory of differential equations [(1), 10.7], we learn that u is differentiable in μ and u_0. As long as $u(\tau, u_0, 0)$ satisfies

$$\mathrm{Det}\left(\frac{\partial(u - u_0)}{\partial u_0}\right) \neq 0, \qquad (4.4.10)$$

the implicit function theorem guarantees solutions $u_0(\mu)$ of (4.4.9) in a neighborhood of $\mu = 0$. Therefore, at each rational frequency there exists a periodic orbit for sufficiently small μ.

Remarks (4.4.11)

1. This is not a trivial statement, because periodicity can be destroyed by arbitrarily small perturbations, and yet here for every positive μ the force is changed by an arbitrarily large amount. (See Problem 1.)

2. We do not discover how large the neighborhood of $\mu = 0$ that allows periodic orbits is.
3. Whenever there exists a constant of the motion K, as is always the case for Hamiltonian systems, condition (4.4.10) is violated:

$$K(u(\tau, u_0)) = K(u_0) \ \forall u_0 \Rightarrow \frac{\partial K}{\partial u}(u(\tau, u_0)) \frac{\partial u(\tau, u_0)}{\partial u_0} = \frac{\partial K(u_0)}{\partial u_0}.$$

This means that the vector

$$\frac{\partial K(u_0)}{\partial u_0} = \frac{\partial K(u(\tau, u_0))}{\partial u}$$

is an eigenvector of the transposed matrix $\partial(u(\tau, u_0) - u_0)/\partial u_0$ with the eigenvalue 0, and so the determinant of the matrix vanishes. The problem can be surmounted, however, because if there exists a constant of the motion, then any one of the equations (4.4.9) automatically holds when the other three do.

Example (4.4.12)

We investigate the orbits of the unperturbed problem, $\mu = 0$, and $m = 1$. In plane polar coordinates,

$$H = \frac{1}{2}\left(p_r^2 + \frac{p_\varphi^2}{r^2}\right) - p_\varphi - \frac{1}{r}$$

and the equations of motion become

$$(\dot{r}, \dot{\varphi}, \dot{p}_r, \dot{p}_\varphi) = \left(p_r, \frac{p_\varphi}{r^2} - 1, -\frac{1}{r^2} + \frac{p_\varphi^2}{r^3}, 0\right).$$

The solution

$$u \equiv (r, \varphi, p_r, p_\varphi) = ((\omega + 1)^{-2/3}, \omega t, 0, (\omega + 1)^{-1/3})$$

is a circular orbit. Defining $\gamma = (\omega + 1)^{1/3}$ and $\tau = 2\pi/\omega$, we next calculate (Problem 2) that

$$\frac{\partial u}{\partial u_0}\bigg|_{t = \tau, u_0 = (\gamma^{-2}, 0, 0, \gamma^{-1})}$$

$$= \begin{bmatrix} \cos \tau\gamma^3 & 0 & \frac{1}{\gamma^3}\sin \tau\gamma^3 & \frac{2}{\gamma}(1 - \cos \tau\gamma^3) \\ -2\gamma^2 \sin \tau\gamma^3 & 1 & \frac{2}{\gamma}(1 - \cos \tau\gamma^3) & -3\gamma^4\tau + 4\gamma \sin \tau\gamma^3 \\ -\gamma^3 \sin \tau\gamma^3 & 0 & \cos \tau\gamma^3 & 2\gamma^2 \sin \tau\gamma^3 \\ 0 & 0 & 0 & 1 \end{bmatrix}.$$

The matrix $(\partial u / \partial u_0) - 1$ has determinant 0, because

$$\left.\frac{\partial H}{\partial u}\right|_{u_0} = \left(\frac{1}{r^2} - \frac{p_\varphi^2}{r^3}, 0, p_r, \frac{p_\varphi}{r^2} - 1\right) = (0, 0, 0, \omega),$$

where $(r, \varphi, p_r, p_\varphi) = u_0 = (\gamma^{-2}, 0, 0, \gamma^{-1})$, and $(0, 0, 0, 1)$ is an eigenvector of $(\partial u / \partial u_0)^t$ with eigenvalue 1. Let us look only at the r, φ, and p_r components of u and consider

$$\begin{aligned}
u_r(\tau; r, 0, p_r, p_\varphi; \mu) - r &= 0 \\
u_\varphi(\tau; r, 0, p_r, p_\varphi; \mu) - 2\pi &= 0 \qquad\qquad (4.4.13) \\
u_{p_r}(\tau; r, 0, p_r, p_\varphi; \mu) - p_r &= 0
\end{aligned}$$

as equations for the initial values (r, p_r, p_φ). The Jacobian in this case can be calculated as

$$24\pi \frac{(\omega + 1)}{\omega} \sin^2\left(\frac{\pi}{\omega}\right)$$

(Problem 3), which is nonzero for $1/\omega \notin \mathbb{Z}$. It is then easy to verify (Problem 4) that the three equations (4.4.13) suffice to prove that $u_{p_\varphi}(\tau)$ returns to its initial value even if $\mu \neq 0$. Thus, unless ω is in resonance with the rotating force-centers, there are periodic orbits with frequency ω for nonzero μ.

Remarks (4.4.14)

1. When $\mu = 0$, p_φ is also a constant of the motion. The vector $\partial p_\varphi / \partial u = (0, 0, 0, 1)$, which has the same direction as $\partial H / \partial u|_{u_0}$. There is no other eigenvector of $(\partial u / \partial u_0)^t$ with eigenvalue 1, and thus $(\partial u / \partial u_0) - 1$ has a submatrix of rank 3.
2. The mass ratios in the solar system are so extreme that it is not sheer madness to believe in the relevance of this result for astronomy without knowing exactly how large μ is allowed to be.

Stability of the Periodic Orbits (4.4.15)

The equilibrium configurations are fixed points of the canonical flow Φ_t, and in §3.4 it was explained precisely what stability means in this context. Periodic orbits are invariant under some Φ_τ, and so (3.4.16) defines stability. Then Theorem (3.4.21) gives a necessary criterion for the stability of the orbits: A must be diagonable and have purely imaginary eigenvalues. In this case the variable z is (x, y, p_x, p_y), and the matrix is

$$A = \begin{bmatrix} 0 & 1 & 1 & 0 \\ -1 & 0 & 0 & 1 \\ -V_{,xx} & -V_{,xy} & 0 & 1 \\ -V_{,xy} & -V_{,yy} & -1 & 0 \end{bmatrix}. \qquad (4.4.16)$$

Looking at the equilateral equilibrium configurations,

$$x = \tfrac{1}{2} - \mu, \qquad y = \frac{\sqrt{3}}{2},$$

we see that

$$V_{,xx} = \tfrac{1}{4}, \qquad V_{,xy} = -\frac{3\sqrt{3}}{4}(1 - 2\mu), \qquad V_{,yy} = -\tfrac{5}{4},$$

and the eigenvalues λ of A satisfy the equation

$$\lambda^2 = -\tfrac{1}{2} \pm \tfrac{1}{2}\sqrt{1 - 27\mu(1 - \mu)} \qquad (4.4.17)$$

(Problem 5). If $\mu(1 - \mu) < 1/27$, i.e., $\mu/m < 0.040$, then all the eigenvalues are imaginary and nondegenerate, and the orbits are possibly stable. Otherwise, they are certainly not stable.

Remarks (4.4.18)

1. A similar calculation for the collinear solutions reveals that the eigenvalues of A always have nonzero real parts. Thus those orbits fail to be stable for any value of μ/m.
2. The necessary stability condition of (3.4.21) is too strong for our purposes, since we only need real, and not complex, stability. To decide whether periodic orbits are stable, one can apply the K-A-M theorem (3.6.19),† which gives invariant two-dimensional tori for the perturbed system. On the three-dimensional surface $H = $ constant, the orbits that are within the tori can never lead outside them, and so stability can be proved as soon as some frequency condition is fulfilled [(6), §34]. The appropriate variant of the theorem would require that: (frequency of the small masses)/(frequency of the large masses) $\neq p/q$, for $|p - q| \leq 4$ and p and q relatively prime, i.e. $\neq \ldots, 4, 3, \tfrac{5}{2}, \tfrac{7}{3}, 2, \ldots$ The fact is that the asteroids move in the sun–Jupiter system with frequencies between 2 and 4 times the frequency of Jupiter's orbit around the sun, and there are gaps at the forbidden frequencies 4, 3, $\tfrac{5}{2}$, 2.

Unbound Trajectories (4.4.19)

In the foregoing examples, particles with large energies escape to infinity, and intuitively the same thing can be expected to happen in this case, since the potential approaches $1/r$ at large distances. This feeling is supported if we write the Hamiltonian of the rotating system in the form

$$H = \tfrac{1}{2}[(p_x + y)^2 + (p_y - x)^2] + \Omega,$$

$$\Omega = -\tfrac{1}{2}(x^2 + y^2) - \frac{m}{[(x + \mu)^2 + y^2]^{1/2}} - \frac{\mu}{[(x - m)^2 + y^2]^{1/2}}. \qquad (4.4.20)$$

---

† μ is the perturbation parameter, as the system is integrable for $\mu = 0$.

Remarks (4.4.21)

1. The motion is not the same as if there were a potential Ω, although if $H = E$ it is restricted to the region in phase space where $\Omega \leq E$.
2. $\Delta\Omega = -2$ on $\mathbb{R}^2 \setminus (\{m, 0\} \cup \{-\mu, 0\})$. Consequently Ω has only maxima (no minima), which are at the five equilibrium points (4.4.6).

For sufficiently negative E, the region where $\Omega \leq E$ consists of three disconnected components, a neighborhood of each force-center and a neighborhood of infinity (Figure 41). A particle that starts off sufficiently far away will certainly not approach the centers, and ought to run off to infinity. Recall that: (H in the rotating system) = (H in the fixed system) − (angular momentum). Thus E could be very negative either because the particle is near a

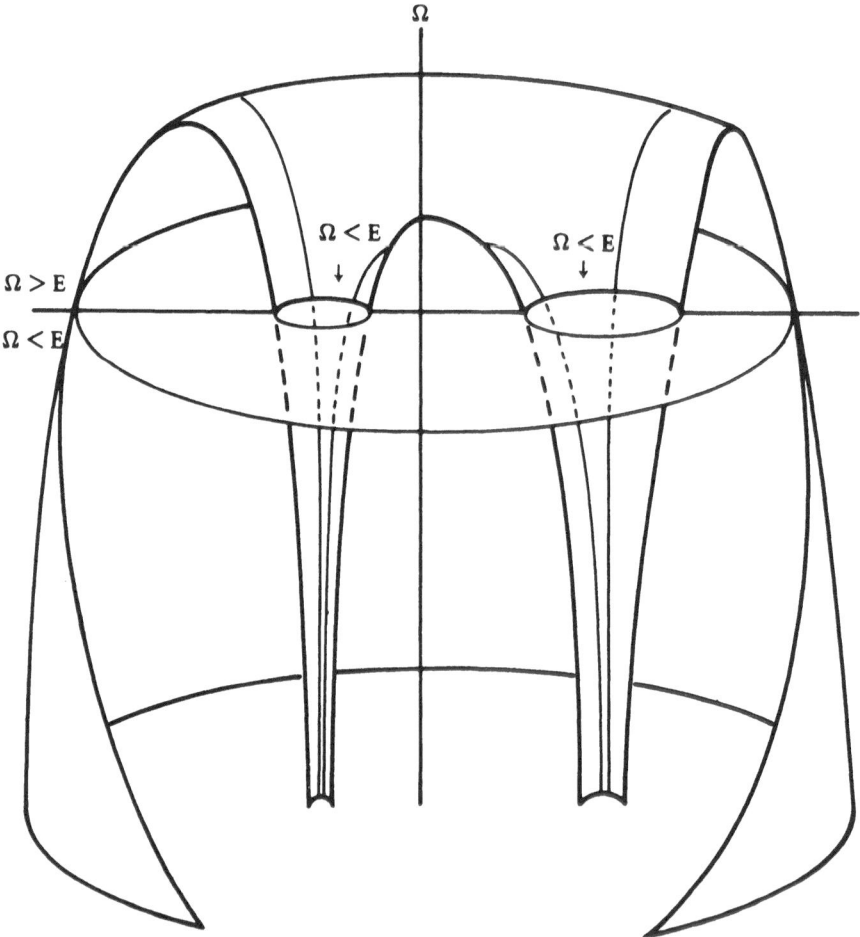

Figure 41   A cross-section of parts of the surface $\Omega(x, y)$ (somewhat like a volcano with two craters).

force-center, or because it has a large angular momentum. It is the latter possibility that produces the neighborhood of infinity, where the unbound trajectories are to be found.

We need some estimates to demonstrate this fact by calculation:

(a) Bounding the external region (say for $r \geq 2$).

$$\Omega \geq -\frac{r^2}{2} - \frac{m}{r+1-m} - \frac{1-m}{r-m}$$

$$= -\frac{r^2}{2} - \frac{1}{r}\left(1 + \frac{m(1-m)}{(r-m)(r+1-m)}\right) \quad \text{for } m > \mu.$$

If $r \geq 2$, then $m(1-m)/(r-m)(r+1-m) < \frac{1}{8}$ $\forall m$ such that $\frac{1}{2} \leq m \leq 1$. The curve $r^3 - 2|\Omega|r + \frac{9}{4} = 0$ has its minimum at $r = \sqrt{2|\Omega|/3}$. The minimum is negative if $|\Omega|^{3/2} \frac{4}{3}\sqrt{\frac{2}{3}} \geq \frac{9}{4}$. In that case r is restricted to the region

$$r > \sqrt{\tfrac{2}{3}|\Omega|} \geq \sqrt{\tfrac{2}{3}|E|}.$$

(b) Bounding the angular momentum.
Since the angular momentum L is conserved for the $1/r$ potential, it ought not to vary much in this case for trajectories at large distances. It follows from the equations of motion (Problem 6) that if $r(t') > 2$ $\forall t'$ such that $0 \leq t' \leq t$, and $L \equiv xp_y - yp_x$, then

$$|L(t) - L(0)| \leq \int_0^t \frac{dt'}{4I(t')}, \qquad \text{where } I \equiv \frac{r^2}{2}.$$

(c) Convexity of the moment of inertia.
For free particles, $I(t)$ is a quadratic function, and in certain other situations it is possible to show that it is at least convex. It is easy to discover (Problem 7) that

$$\ddot{I}(t) \geq H + L(0) - \int_{-\infty}^t dt' \frac{1}{4I(t')} - \frac{1}{4I(t)}.$$

This information can be used to specify in which regions of phase space $\lim_{t \to \pm\infty} I(t) = \infty$. If $\ddot{I}(t) \geq 2B^2 > 0$ $\forall t \in (0, \tau)$ then clearly

$$I(t) \geq \left(Bt - \frac{\dot{I}(0)}{2B}\right)^2 + I_{\min} \quad \forall t \in (0, \tau),$$

and by (a), $I_{\min}$ must be greater than $|E|/3$. It then follows that

$$\frac{1}{4}\int_{-\infty}^\tau \frac{dt}{I(t)} < \frac{\pi}{4B\sqrt{I_{\min}}} \leq \frac{\pi\sqrt{3}}{4B\sqrt{|E|}},$$

and (c) implies

$$\ddot{I}(t) \geq E + L(0) - \frac{\pi\sqrt{3}}{4B\sqrt{|E|}} - \frac{\sqrt{3}}{4|E|}, \quad \forall t \in (0, \tau).$$

Defining $B^2 \equiv (E + L(0) - (3/4|E|))/6 \geq \pi^{2/3}3^{1/3}|E|^{-1/3}2^{-8/3}$, we get

$$\ddot{I} \geq 6B^2 - \frac{\pi\sqrt{3}}{4B\sqrt{|E|}} \geq 2B^2, \quad \forall t \in (0, \tau),$$

because $2B^2 + \alpha/B \leq 6B^2 \; \forall B^2 \geq (\alpha/4)^{2/3}$. Since this bound is independent of τ, it yields a

Criterion for escaping to infinity (4.4.22)

If for some trajectory

$$E < -6, \qquad L(0) \geq |E| + \frac{3}{4|E|} + \frac{\pi^{2/3}3^{4/3}}{2^{5/3}|E|^{1/3}},$$

then $\lim_{t \to \pm\infty} r(t) = \infty$.

Remarks (4.4.23)

1. As with the problem with two fixed centers of force, the Møller transformation using the flow generated by a $1/r$ potential as Φ_t^0 exists. There are three constants of the motion (five in the three-dimensional case) in this part of phase space, and the trajectories are homotopic to Kepler hyperbolas.
2. Trajectories that get near the force-centers can become quite complicated. For instance, the following rather surprising statement can be made about trajectories perpendicular to the plane of motion of the heavy particles, if they travel in ellipses [(14), III.5]: $\exists m > 0$ such that for every sequence $s_k > m$, there exists a trajectory for which the time between the k-th and $(k + 1)$-th intersection with the plane of the ellipse is exactly s_k.
3. In the situation depicted in Figure 4, any trajectory once in the vicinity of one of the force-centers always remains nearby. One might be tempted to apply the recurrence theorem (2.6.13) in this case, but it does not work, because collisions can not be avoided, and no time-invariant region in phase space that is compact in the momentum coordinates as well as the spatial ones can be found.

The flow Φ_t of the restricted three-body problem is, as we see, not known in full detail, and our analysis only gives us the impression that trajectories at a respectable distance from the force-centers evolve smoothly. But if a particle happens to approach too near, it can dance around in a completely crazy way.

Problems (4.4.24)

1. With the two-dimensional harmonic oscillator, it is easy to see that even for arbitrarily small $\mu \neq 0$, the orbits of

$$H(\mu) = \tfrac{1}{2}(p_x^2 + p_y^2 + x^2 + y^2 + \mu(x^2 - y^2))$$

   that are periodic when $\mu = 0$ can be destroyed, so that no periodic orbits at all remain. Why doesn't Poincaré's argument work in this case?

2. Calculate $\partial u / \partial u_0$ from (4.4.12).

3. Same problem for (4.4.13). What is the determinant of this matrix?

4. Prove that $u_{p_\varphi}(\tau)$ is in fact equal to p_φ in (4.4.12).

5. Calculate the eigenvalues of A from (4.4.16) for the equilateral equilibrium con-figurations.

6. Let $p \equiv xp_y - yp_x$ and $I \equiv r^2/2$. Use (4.4.4) to show the following bound for the angular momentum in terms of the moment of inertia:

$$|p_\varphi(0) - p_\varphi(t)| \leq \int_0^t \frac{dt'}{4I(t')},$$

   when $I(t') > 2 \ \forall t'$ such that $0 \leq t' \leq t$.

7. Use Problem 6 to show that

$$\ddot{I}(t) \geq H + L(0) - \int_{-\infty}^{\infty} dt' \frac{1}{4I(t')} - \frac{1}{4I(t)}$$

   if $I(t') > 2 \ \forall t'$ such that $0 \leq t' \leq t$.

Solutions (4.4.25)

1. In polar coordinates,

$$H(0) = \tfrac{1}{2}\left(p_r^2 + \frac{p_\varphi^2}{r^2} + r^2\right).$$

   Letting $u \equiv (r, \varphi, p_r, p_\varphi)$ and the solution $u_0 \equiv u(\tau; r, 0, 0, r^2) = (r, \tau, 0, r^2)$, we cal-culate from the equation for $H(0)$ that

$$A = \frac{\partial X_H}{\partial u}\bigg|_{u = u_0} =
\begin{array}{|c|c|c|c|}
\hline
0 & 0 & 1 & 0 \\
\hline
-2/r & 0 & 0 & 1/r^2 \\
\hline
-4 & 0 & 0 & 2/r \\
\hline
\end{array}$$

   Then $A \cdot (a, b, c, d) = (c, -2a/r + d/r^2, -4a + 2d/r, 0)$, and thus $\mathrm{rank}(A) = \dim(A \cdot \mathbb{R}^4) = 2$. Consequently, $\mathrm{rank}(e^{tA} - 1) = \sum_{n=1}^{\infty} (tA)^n/n!$ is also equal to 2, and there is no nonsingular 3×3 submatrix of $\partial(u - u_0)/\partial u_0$.

2. Because $\dot{u}(t, u_0) = X_H(u(t, u_0))$, the matrix of derivatives satisfies the homogeneous differential equation

$$\frac{d}{dt}\frac{\partial u}{\partial u_0} = \frac{\partial X_H}{\partial u}\frac{\partial u}{\partial u_0},$$

in which $\partial X_H/\partial u$ depends on the solution $u(t, u_0)$, which is assumed known. For circular orbits this matrix is independent of t, and since $u(0, u_0) = u_0$ implies the initial condition $\partial u/\partial u_0|_{t=0} = 1$, the solution of the differential equation is simply

$$\frac{\partial u}{\partial u_0} = \exp\left(t\,\frac{\partial X_H}{\partial u}\right).$$

Hence we have to calculate the matrix

$$\partial X_H(u(t, u_0))/\partial u \equiv A$$

(u being given by the circular orbit), and then exponentiate it. From (4.4.12) it follows that

$$X_H = \left(p_r, \frac{p_\varphi}{r^2} - 1, \ -\frac{1}{r^2} + \frac{p_\varphi}{r^3}, 0\right),$$

and

$$\frac{\partial X_H}{\partial u} = \begin{bmatrix} 0 & 0 & 1 & 0 \\ \dfrac{-2p_\varphi}{r^3} & 0 & 0 & \dfrac{1}{r^2} \\ \dfrac{2}{r^3} - \dfrac{3p_\varphi}{r^4} & 0 & 0 & \dfrac{2p_\varphi}{r^3} \\ 0 & 0 & 0 & 0 \end{bmatrix},$$

and in particular, for the circular orbit

$$A = \begin{bmatrix} 0 & 0 & 1 & 0 \\ -2\gamma^5 & 0 & 0 & \gamma^4 \\ -\gamma^6 & 0 & 0 & 2\gamma^5 \\ 0 & 0 & 0 & 0 \end{bmatrix}.$$

In order to calculate $\exp(\tau A)$, we put A into Jordan normal form with a nonsingular matrix C (which is not necessarily unitary, since A is not Hermitian):

$$A = CNC^{-1},$$

where N is a matrix the diagonal elements of which are the eigenvalues of A, and which may have nonvanishing elements immediately above the diagonal, but all other elements are zero. Such a matrix can be easily exponentiated, and $\exp(\tau A) = C \exp(\tau N)C^{-1}$. The eigenvalues of A can be calculated from $\det|A - \lambda I| = \lambda^2(\lambda^2 + \gamma^3) = 0$ to be $\lambda = 0, 0$, and $\pm i(\omega + 1)$.

Explicit calculations of the various matrices are:

$$
C = \begin{bmatrix}
0 & \dfrac{2}{3} & 1 & 1 \\[2mm]
1 & 0 & 2i\gamma^2 & -2i\gamma^2 \\[2mm]
0 & 0 & i\gamma^3 & -i\gamma^3 \\[2mm]
0 & \dfrac{\gamma}{3} & 0 & 0
\end{bmatrix},
\qquad
C^{-1} = \begin{bmatrix}
0 & 1 & \dfrac{-2}{\gamma} & 0 \\[2mm]
0 & 0 & 0 & \dfrac{3}{\gamma} \\[2mm]
\dfrac{1}{2} & 0 & \dfrac{1}{2i\gamma^3} & \dfrac{-1}{\gamma} \\[2mm]
\dfrac{1}{2} & 0 & \dfrac{-1}{2i\gamma^3} & \dfrac{-1}{\gamma}
\end{bmatrix},
$$

$$
N = \begin{bmatrix}
0 & -\gamma^5 & & \\[2mm]
 & 0 & 0 & \\[2mm]
 & & i\gamma^3 & 0 \\[2mm]
 & & & -i\gamma^3
\end{bmatrix},
\qquad
e^{\tau N} = \begin{bmatrix}
1 & -\gamma^5\tau & & \\[2mm]
 & 1 & 0 & \\[2mm]
 & & e^{i\tau\gamma^3} & 0 \\[2mm]
 & & & e^{-i\tau\gamma^3}
\end{bmatrix}.
$$

Multiplication of the matrices yields the result given in (4.4.12).

3. We need to calculate the determinant of

$$
\begin{bmatrix}
u_{r,r} - 1 & u_{r,p_r} & u_{r,p_\varphi} \\[2mm]
u_{\varphi,r} & u_{\varphi,p_r} & u_{\varphi,p_\varphi} \\[2mm]
u_{p_r,r} & u_{p_r,p_r} - 1 & u_{p_r,p_\varphi}
\end{bmatrix}.
$$

If $\mu = 0$, and

$$
S \equiv \sin \tau\gamma^3, \quad \text{and} \quad C \equiv \cos \tau\gamma^3,
$$

then this matrix equals

$$
\begin{bmatrix}
C - 1 & \dfrac{S}{\gamma^3} & \dfrac{2}{\gamma}(1 - C) \\[3mm]
-2\gamma^2 S & -\dfrac{2}{\gamma}(1 - C) & -3\gamma^4\tau + 4\gamma S \\[3mm]
-\gamma^3 S & C - 1 & 2\gamma^2 S
\end{bmatrix}.
$$

Writing $\tau = 2\pi/\omega$ and $\gamma^3 = \omega + 1$ and taking the determinant yields

$$
24\pi \frac{(\omega + 1)^{4/3}}{\omega} \sin^2 \pi/\omega.
$$

4. We know that $H(r, 2\pi, p_r, u_{p_\varphi}(\tau)) = H(r, 0, p_r, p_\varphi)$, and that if $\mu = 0$,

$$
\frac{\partial H}{\partial p_\varphi} = \frac{p_\varphi}{r^3} - 1 = \omega \neq 0.
$$

Hence in some neighborhood of $\mu = 0$, $\partial H/\partial p_\varphi$ has a definite sign between p_φ and $u_{p_\varphi}(\tau)$; but then $u_{p_\varphi}(\tau)$ must equal p_φ for the first equality to hold.

5. For a block matrix

$$\begin{bmatrix} a & b \\ c & d \end{bmatrix},$$

if $db = bd$, then

$$\det \begin{vmatrix} a & b \\ c & d \end{vmatrix} = \det|ad - bc|.$$

Therefore

$$\det|A - \lambda| = \lambda^4 + \lambda^2 + \tfrac{27}{16}(1 - (1 - 2\mu)^2),$$

and the eigenvalues of A are

$$\lambda^2_{1,2} = -\tfrac{1}{2} \pm \tfrac{1}{2}\sqrt{1 - 27\mu(1 - \mu)}.$$

6. The amount of nonconservation of the angular momentum is

$$\dot{L}_z = -xV_{,y} + yV_{,x} = m\mu y\{[(x + \mu)^2 + y^2]^{-3/2} - [(x - m)^2 + y^2]^{-3/2}\} \Rightarrow$$

$$|\dot{L}_z| \leq \frac{m\mu}{(r - m)^2} \leq \frac{1}{2r^2} \forall r \geq 2 \Rightarrow |L_z| \leq \frac{1}{4I}, \quad \text{if } I \geq 2.$$

7.
$$\dot{I} = \frac{d}{dt} \mathbf{x} \cdot \mathbf{p} = p_x^2 + p_y^2 - xV_{,x} - yV_{,y}$$

$$= H + L + \tfrac{1}{2}(p_x^2 + p_y^2) + m\mu\left\{\frac{x + \mu}{[(x + \mu)^2 + y^2]^{3/2}} - \frac{x - m}{[(x - m)^2 + y^2]^{3/2}}\right\}.$$

As in Problem 6,

$$|\{\ \}| \leq \frac{1}{(r - m)^2} \leq \frac{1}{2r^2} \quad \text{and} \quad L(t) \geq L(0) - \int_{-\infty}^{\infty} \frac{dt}{4I(t)}, p_x^2 + p_y^2 \geq 0.$$

4.5 The N-Body Problem

Although the system of equations appears hopelessly complicated, it is possible not only to find exact solutions, but even to make some general propositions.

Since time immemorial many of the top minds have applied their mathematical skills to equations (1.1.1) and (1.1.2) with $n \geq 3$, but without great success. We will pick up some of the more amusing pieces from their efforts, and by doing so we hope to illustrate how one typically approaches the problem. We restrict ourselves to the case of pure gravitation; the inclusion of a Coulomb force requires only trivial changes and brings hardly any new insight. As usual we begin with the

Constants of Motion (4.5.1)

As in the two-body problem, the flow factorizes into the motion of the center of mass and relative coordinates. The center-of-mass part has the maximal number of constants, because of the Galilean invariance, while the relative part has only the conserved angular momentum, from invariance under rotations.

The Case $N = 3$ (4.5.2)

The overall phase space is 18-dimensional, while the phase space of the relative motion is only 12-dimensional. In the latter there are only four constants, the angular momentum and the energy, which are not enough for the equations to be integrable.

Remarks (4.5.3)

1. Whereas it can be proved in the restricted three-body problem [14, VI.8] that there are no additional constants of motion other than Jacobi's constant, in this case a classic theorem of Bruns implies that there are no other integrals that are algebraic functions of the Cartesian coordinates **x** and **p**. But since we attribute no special status to any coordinate system, it is not clear that this statement is of much value.
2. The known integrals do not separate off any compact part of phase space on which the recurrence theorem might be applied. It is even possible that collision trajectories are dense in regions of positive measure.
3. Additional constants certainly exist in the parts of phase space where scattering theory operates[19].
4. Computer studies of the restricted three-body problem have found parts of phase space where the trajectories—even trajectories that remain finite—form manifolds of a lower dimension, as if there existed more constants of the motion[16]. This phenomenon could also show up in the n-body problem.

Exact Solutions (4.5.4)

If all N particles move in a plane, it can easily happen that gravity and the centrifugal force balance each other. Let us consider the Cartesian coordinates in the plane as complex numbers, and set

$$x_j(t) + iy_j(t) = z(t)z_j,$$

$$z(t): \mathbb{R} \to \mathbb{C}, \quad \text{and} \quad z_j \in \mathbb{C}.$$

This assumption means that the configuration of the particles in the plane has the same shape at all times. Letting $\kappa = 1$, the equations of motion,

$$\ddot{z}z_i = \sum_{j \neq i} \frac{(z_j - z_i)m_j}{|z_j - z_i|^3} \frac{z}{|z|^3}, \tag{4.5.5}$$

can be decomposed into the Kepler problem in the plane,

$$\ddot{z} = -\omega^2 \frac{z}{|z|^3},$$ (4.5.6)

and the algebraic equation,

$$-\omega^2 z_i = \sum_{j \neq i} \frac{z_j - z_i}{|z_j - z_i|^3} m_j.$$ (4.5.7)

Hence, each particle moves in a Kepler trajectory about the collective center of mass (since (4.5.7) $\Rightarrow \sum_i z_i m_i = 0$).

Remarks (4.5.8)

1. Since we know that there are solutions to (4.5.6) for all $\omega \in \mathbb{R}^+$, only (4.5.7) needs to be discussed.
2. The total energy of the motion is

$$E = |\dot{z}|^2 \sum_i |z_i|^2 \frac{m_i}{2} - \frac{1}{2|z|} \sum_{i \neq j} \frac{m_i m_j}{|z_i - z_j|}$$

$$= \left\{ \frac{|\dot{z}|^2}{2} - \frac{\omega^2}{|z|} \right\} \sum_i |z_i|^2 m_i,$$

   i.e., the energy of the Kepler trajectory times the moment of inertia. The particles remain in a bounded region iff $E < 0$.

The Specialization to $N = 3$ (4.5.9)

It is necessary to distinguish two cases:
(a) All $|z_i - z_j| = R$, making an equilateral triangle. Adding the three equations (4.5.7) gives

$$(-\omega^2 R^3 + m_1 + m_2 + m_3) \sum_i z_i = 3 \sum_i z_i m_i = 0,$$

which is solved by

$$\omega^2 R^3 = m_1 + m_2 + m_3.$$

But then all three equations are satisfied.
(b) $|z_1 - z_3| \neq |z_2 - z_3|$. If the coordinate system is chosen so that $z_3 = 0$, then because of (4.5.7),

$$\mathrm{Re}\left(\frac{m_1 z_1}{|z_3 - z_1|^3} + \frac{m_2 z_2}{|z_3 - z_2|^3} \right) = 0.$$

But since also $\mathrm{Re}(m_1 z_1 + m_2 z_2) = -\mathrm{Re}\, m_3 z_3 = 0$, it must be true that $\mathrm{Re}\, z_1 = \mathrm{Re}\, z_2 = 0$, and all three particles are in a line. These are the same as the equilibrium configurations of the special case treated in §4.4.

It is now pertinent to ask whether particles invariably run off to infinity when $E \geq 0$ and the center of mass is fixed. This is in fact so, according to the virial theorem mentioned above:

The Virial Theorem (4.5.10)

Let $I \equiv \sum_j m_j |\mathbf{x}_j|^2$, $T \equiv \sum_j |\mathbf{p}_j|^2/2m$, and $V \equiv -\kappa \sum_{i \neq j} m_i m_j/|\mathbf{x}_i - \mathbf{x}_j|$. Then $\ddot{I} = 2(E + T)$.

Proof

See Problem 1. $\qquad\qquad\qquad\qquad\qquad\qquad\qquad\qquad\qquad\qquad\qquad\qquad\qquad$ □

Corollaries (4.5.11)

1. *Since* $T \geq 0$, $I(t) \geq I(0) + t\dot{I}(0) + t^2 E$. *Thus if* $E > 0$, *then* $\lim_{t \to \infty} I(t) = \infty$, *and at least one particle travels arbitrarily far away.*
2. *If* $I(t)$ *approaches zero, meaning that the system collapses, it must reach zero in a finite time, and can not do it asymptotically. This is because* $I \to 0$ *only if* $V \to -\infty$ (*Problem 2*), *and if* $\lim_{t \to \infty} V = -\infty$, *then, because* $\ddot{I} = 4E - 2V$, *there exists a time* t_0 *such that* $\ddot{I} > 1 \; \forall t > t_0$. *But then* $\lim_{t \to \infty} I(t)$ *can not be zero.*
3. *If the motion is almost periodic, then for any* $\varepsilon > 0$ *there exists a sequence* $\tau_i \to \infty$ *with* $|\dot{I}(0) - \dot{I}(\tau_i)| < \varepsilon \; \forall i$. *Then*

$$\lim_{i \to \infty} \frac{1}{\tau_i} \int_0^{\tau_i} dt(2T + V) = \lim_{i \to \infty} \frac{1}{\tau_i} (\dot{I}(\tau_i) - \dot{I}(0)) = 0.$$

According to this, the average of the potential energy is twice the total energy, which is obviously only possible if $E < 0$.

If $N \geq 3$, it is energetically possible that one of the particles gets catapulted off to infinity. The requisite energy can always be produced if the other particles come close enough together. It might be supposed that whenever the kinetic energy of a particle exceeds its potential energy, the particle flies off, never to be seen again. However, the energy of an individual particle is not conserved, and a closer analysis is needed to see if this is really true.

A Lower Bound for the Kinetic Energy (4.5.12)

For simplicity we look at the situation where $N = 3, \kappa = m_1 = m_2 = m_3 = 1$. Let $E < 0$, and

$$\bar{r} \equiv \min_{i,j} |\mathbf{x}_i - \mathbf{x}_j|. \qquad\qquad\qquad\qquad (4.5.13)$$

Then, because $|V| \geq |E|$, we have the bound

$$\bar{r} \leq r^* \equiv \frac{3}{|E|}. \qquad\qquad\qquad\qquad (4.5.14)$$

To separate off the center-of-mass motion, we introduce the coordinates

$$\mathbf{s} = \frac{1}{\sqrt{3}}(\mathbf{x}_1 + \mathbf{x}_2 + \mathbf{x}_3)$$

$$\mathbf{u} = \frac{1}{\sqrt{2}}(\mathbf{x}_1 - \mathbf{x}_2) \qquad\qquad (4.5.15)$$

$$\mathbf{x} = \frac{1}{\sqrt{6}}(\mathbf{x}_1 + \mathbf{x}_2 - 2\mathbf{x}_3)$$

(see Figure 42). Since this transformation is orthogonal on $\mathbb{R}^9$, the kinetic energy is simply

$$T = \tfrac{1}{2}(|\dot{\mathbf{s}}|^2 + |\dot{\mathbf{u}}|^2 + |\dot{\mathbf{x}}|^2), \qquad\qquad (4.5.16)$$

and the potential energy is

$$V = -\frac{1}{\sqrt{2}|\mathbf{u}|} - \frac{\sqrt{2}}{|\sqrt{3}\mathbf{x} + \mathbf{u}|} - \frac{\sqrt{2}}{|\sqrt{3}\mathbf{x} - \mathbf{u}|}. \qquad (4.5.17)$$

If particle #3 escapes, then the interesting coordinate is x, the motion of which is governed by

$$\ddot{\mathbf{x}} = -\sqrt{\frac{2}{3}}\left(\frac{\mathbf{x} + \mathbf{u}/\sqrt{3}}{|\mathbf{x} + \mathbf{u}/\sqrt{3}|^3} + \frac{\mathbf{x} - \mathbf{u}/\sqrt{3}}{|\mathbf{x} - \mathbf{u}/\sqrt{3}|^3}\right).$$

$$(4.5.18)$$

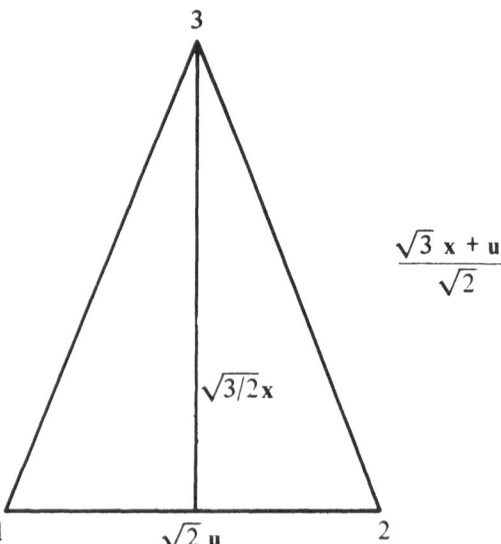

Figure 42   Center-of-mass and relative coordinates for three bodies.

Of course, $\mathbf{u}$ depends on the time in some unknown way, but because of (4.5.14), if $|\mathbf{u}|$ is initially $< r^*/\sqrt{2}$, and if

$$\sqrt{\frac{3}{2}}r - \frac{r^*}{2} > r^* \Leftrightarrow r > \sqrt{\frac{3}{2}}r^*, \tag{4.5.19}$$

where $r = |\mathbf{x}|$, then $|\mathbf{u}|$ must always be less than $r^*/\sqrt{2}$. Therefore, $\dot{r}$ is bounded below as

$$\frac{\dot{r}(t)^2}{2} \ge \frac{\dot{r}(0)^2}{2} - \sqrt{\frac{2}{3}}\left(\frac{1}{r(0) + r^*/\sqrt{6}} + \frac{1}{r(0) - r^*/\sqrt{6}}\right)$$

$$+ \sqrt{\frac{2}{3}}\left(\frac{1}{r(t) + r^*/\sqrt{6}} + \frac{1}{r(t) - r^*/\sqrt{6}}\right) \tag{4.5.20}$$

(see Problem 3).

This produces a crude

Criterion for Escaping to Infinity (4.5.21)

If at any time $|\mathbf{u}| < r^*/\sqrt{2}$ and

$$r > \sqrt{\frac{3}{2}}r^*, \dot{r} > 0, \frac{\dot{r}^2}{2} > \sqrt{\frac{2}{3}}\left(\frac{1}{r + r^*/\sqrt{6}} + \frac{1}{r - r^*/\sqrt{6}}\right),$$

then $\dot{r}$ is greater than some positive number at all later times, and particle #3 can not be prevented from escaping.

Remarks (4.5.22)

1. It is possible to relax the conditions so that $\dot{r}$ is not necessarily greater than 0; even an initially incoming particle escapes if its energy and momentum are great enough (Problem 4). The other particles stay in Kepler trajectories, so no collisions take place. This shows that there are open regions in phase space, of infinite measure, in which particles do not collide, and X_H is complete. Furthermore, if we use an H_0 equal to H with the potential replaced by

$$-\frac{m_1 m_2}{|\mathbf{x}_1 - \mathbf{x}_2|} - \frac{m_3(m_1 + m_2)}{|\mathbf{x}_3 - (m_1\mathbf{x}_1 + m_2\mathbf{x}_2)/(m_1 + m_2)|},$$

then we find that the Møller transformations (3.4.4) exist on these parts of phase space. The reason is that for large $|\mathbf{x}_3|$, the difference in the potentials is $\sim |\mathbf{x}_3|^{-3} \sim t^{-3}$, which decreases sufficiently fast to ensure the convergence of $\Phi_{-t} \circ \Phi_t^0$ and $\Phi_{-t}^0 \circ \Phi_t$.

2. To get a feeling for the numbers that come up in (4.5.21), let us rewrite the last condition with $\rho = \sqrt{\frac{3}{2}} r = $ the distance from particle #3 to the center of mass of #1 and 2:

$$\frac{2}{3}\frac{\dot{\rho}^2}{2} \ge \frac{1}{\rho + r^*/2} + \frac{1}{\rho - r^*/2}.$$

Thus the condition means that the potential energy of particle #3 is less than its kinetic energy with a reduced mass $\frac{2}{3}$. The reduced mass

$$\frac{m_3(m_1 + m_2)}{m_1 + m_2 + m_3},$$

in this case of particle #3 and the pair (1, 2), is already familiar from (4.2.3; 4). With this correction, our initial supposition about the energetics is correct.

3. When $N = 4$, there are unbound trajectories for which particles can reach infinity in a finite time[15]. These involve a linear configuration of the particles #1, 2, 3, and 4 (in that order), in which #3 and 4 draw steadily nearer together. The energy thereby released is transmitted to #2, which runs faster and faster between particle #1 and the pair (3, 4), forcing them apart. (The orbit of particle #2 through two reversals is regarded as the limit of a Kepler ellipse with infinite eccentricity. The particles are reflected by the $1/r$ potential, and do not pass through each other.)

4. Computer studies of the three-body problem indicate that sooner or later some particle gains enough energy that (4.5.21) holds, and the system breaks up. This instability, known in atomic physics as the Auger effect, may well be characteristic for all systems with $1/r$ potentials. It is even suspected that, in the equal-mass case, the trajectories for which the system breaks apart may be dense in large parts of phase space. Of course, the physically relevant question is how large the probability of a break-up in a realistic time is. Unfortunately, present analytic methods fail to give an answer, and we must have recourse to the calculating machines, according to which there is a large probability that the system breaks up within 100 natural periods, $\tau \sim R^{3/2}/\sqrt{\kappa M}$ [8].

The meagerness of these results makes it clear that the system of equations, (1.1.1) and (1.1.2) for large N, is too difficult for present-day mathematics to handle effectively. Though it is considered the correct expression of the laws of nature, its useful content is slight, because only a very few relevant propositions can be derived from it. Later, in the context of quantum mechanics, we shall return to the same equations and get useful information from them from another point of view. It will not be possible—or even desirable—to calculate the details of all the trajectories; yet one can predict quite a bit about the statistical behavior of the system.

Problems (4.5.23)

1. Derive (4.5.10).

2. Show that $I \to 0 \Rightarrow V \to -\infty$.

3. Prove (4.5.20). Hint: Use

$$\ddot{r} = \frac{\mathbf{x} \cdot \ddot{\mathbf{x}}}{r} + \frac{|\dot{\mathbf{x}}|^2}{r} - \frac{(\mathbf{x} \cdot \dot{\mathbf{x}})^2}{r^3} \geq \frac{(\mathbf{x} \cdot \ddot{\mathbf{x}})}{r}.$$

4. Use (4.5.18) to estimate how $L \equiv [\mathbf{x} \times \dot{\mathbf{x}}]$ varies in time and, with Problem 3, to show that particles with sufficiently large $r_0 \equiv r(0)$, $\mathbf{L}_0 \equiv \mathbf{L}(0)$, and $|\dot{r}_0|$ escape even if $\dot{r}_0 < 0$.

Solutions (4.5.24)

1. $\dot{I} = 2 \sum_j (\mathbf{x}_j \cdot \mathbf{p}_j)$ is twice the generator of a dilatation, and is known as the virial. We know that $\{\dot{I}, T\} = 4T$ and $\{\dot{I}, V\} = 2V$, from which (4.5.10) follows.

2. Since the function $1/x$ is convex for $x > 0$, Jensen's inequality implies that

$$|V| \left(\sum_{i \neq j} m_i m_j \right)^{-1} = \frac{\kappa}{2} \sum_{i \neq j} \frac{m_i m_j}{|\mathbf{x}_i - \mathbf{x}_j|} \left(\sum_{i \neq j} m_i m_j \right)^{-1} \geq \frac{\kappa}{2} \left(\sum_{i \neq j} m_i m_j |\mathbf{x}_i - \mathbf{x}_j| \right)^{-1} \sum_{i \neq j} m_i m_j.$$

Then with the triangle and Cauchy–Schwarz inequalities we get

$$\sum_{i \neq j} m_i m_j |\mathbf{x}_i - \mathbf{x}_j| \leq \sum_{i \neq j} m_i m_j (|\mathbf{x}_i| + |\mathbf{x}_j|)$$

$$\leq 2M \sum_i m_i |\mathbf{x}_i| \leq 2M \left(\sum_i m_i \right)^{1/2} \left(\sum_i m_i |\mathbf{x}_i|^2 \right)^{1/2}, \text{ where } M = \sum_i m_i.$$

So finally,

$$V \leq \frac{-\kappa}{4M^{3/2}} \left(\sum_{i \neq j} m_i m_j \right)^2 \left(\sum_i m_i |\mathbf{x}_i|^2 \right)^{-1/2}.$$

3. The inequality of the hint implies that

$$\ddot{r} \geq -\sup_{|\mathbf{u}| < r^*/\sqrt{2}} \frac{\mathbf{x}}{r} \cdot \sqrt{\frac{2}{3}} \left[\frac{\mathbf{x} + \mathbf{u}/\sqrt{3}}{|\mathbf{x} + \mathbf{u}/\sqrt{3}|^3} + \frac{\mathbf{x} - \mathbf{u}/\sqrt{3}}{|\mathbf{x} - \mathbf{u}/\sqrt{3}|^3} \right]$$

$$\geq -\sup_{|\mathbf{u}| < r^*/\sqrt{2}} \sqrt{\frac{2}{3}} \left[\frac{1}{|\mathbf{x} + \mathbf{u}/\sqrt{3}|^2} + \frac{1}{|\mathbf{x} - \mathbf{u}/\sqrt{3}|^2} \right]$$

$$= -\frac{\sqrt{2/3}}{(r + r^*/\sqrt{6})^2} - \frac{\sqrt{2/3}}{(r - r^*/\sqrt{6})^2} = \frac{\partial}{\partial r} \sqrt{\frac{2}{3}} \left(\frac{1}{r + r^*/\sqrt{6}} + \frac{1}{r - r^*/\sqrt{6}} \right),$$

because $(\mathbf{a} \cdot \mathbf{b}) \geq -|\mathbf{a}||\mathbf{b}|$, and it is clear that the greatest forces occur when some particle approaches as near as possible to particle #3. Multiplication by $\dot{r}$ and integration produce (4.5.20).

4. Let $L_m \in \mathbb{R}^+$ be such that $|L(t)| \geq L_m$ for all t. From (4.5.18),

$$\ddot{r} = \frac{\mathbf{x} \cdot \ddot{\mathbf{x}}}{r} + \frac{L^2}{r^3} \geq -\sqrt{\frac{8}{3}} \frac{1}{(r - r^*/\sqrt{6})^2} + \frac{L_m^2}{r^3} = -\frac{\partial}{\partial r} V_m(r),$$

where

$$V_m \equiv -\sqrt{\frac{8}{3}}\frac{1}{r - r^*/\sqrt{6}} + \frac{L_m^2}{2r^2}.$$

Choose the initial values r_0 and $\dot{r}_0$ so that if v and r_m are defined by

$$\tfrac{1}{2}v^2 = \tfrac{1}{2}\dot{r}_0^2 + V_m(r_0) = V_m(r_m),$$

then the larger of the two solutions for r_m is greater than $2r^*/\sqrt{6}$. Then (because the force from V_m is always less than the actual force) $r > r_m \; \forall t$; and we can calculate that $V_m(r) < v^2 r_m^2/2r^2$; that $|\dot{r}| > v\sqrt{1 - r_m^2/r^2}$; and so, finally, that

$$vt < \left(\int_{r_m}^{r_0} + \int_{r_m}^{r}\right) \frac{dr\, v}{\sqrt{v^2 - 2V_m(r)}} < \left(\int_{r_m}^{r_0} + \int_{r_m}^{r}\right) \frac{dr\, v}{\sqrt{v^2 - v^2 r_m^2/r^2}}$$

$$= \sqrt{r_0^2 - r_m^2} + \sqrt{r^2 - r_m^2} < r_0 + r.$$

We still need to show that L_0 can be chosen consistently with these calculations, in which case we conclude that r gets arbitrarily large. Since (4.5.18) tells us that

$$|\dot{\mathbf{L}}| = |[\mathbf{x} \times \ddot{\mathbf{x}}]| \le \tfrac{2}{3}\frac{rr^*}{(r - r^*/\sqrt{6})^3},$$

the inequality

$$L > L_0 - \tfrac{4}{3}\int_{r_m}^{\infty} \frac{dr\, rr^*}{\dot{r}(r - r^*/\sqrt{6})^3} > L_0 - \tfrac{2}{3}\int_{r_m}^{\infty} \frac{\sqrt{6}\, dr\, r^2 r_m/v}{(r - r_m/2)^3 \sqrt{r^2 - r_m^2}}$$

$$= L_0 - \frac{C}{v}, \quad \text{where } C \equiv \tfrac{2}{3}\int_{1}^{\infty} \frac{dx\sqrt{6}x^2}{(x - \tfrac{1}{2})^3 \sqrt{x^2 - 1}},$$

holds. Using $L_m \le L_0 - C/v$ and observing that $|\mathbf{L}| \le |\dot{r}r|$, we reduce the problem to satisfying the condition

$$r_0^2(v^2 - \dot{r}_0^2) + \sqrt{\frac{2}{3}}\frac{4r_0^2}{r_0 - r^*/\sqrt{6}} \le \left(L_0 - \frac{C}{v}\right)^2 \le \left(\dot{r}_0 r_0 - \frac{C}{v}\right)^2.$$

Since r^* depends on the total energy, and can be chosen independently of r_0, $\dot{r}_0$, and L_0, we discover that there are open regions of phase space, with $v^2 \lesssim \dot{r}_0^2 \gg 1/r_0 \ll 1/r^*$, where this condition is satisfied.

Relativistic Motion

5

5.1 The Hamiltonian Formulation of the Electrodynamic Equations of Motion

The theory of special relativity replaces the Galilean group with the Poincaré group. This makes the equations of motion of a particle in an external field only slightly more complicated. However, physics at high velocities looks quite different from its nonrelativistic limit.

Newton's equations, as we know, are only an approximation, and have to be generalized to (1.1.4) or (1.1.6) when the speed of a particle approaches the speed of light. In order to solve these equations in some physically interesting cases, we first put (1.1.4) into Hamiltonian form. This means that the motion takes place in extended configuration space, which is a particular subset of $\mathbb{R}^4$. We shall just concern ourselves with one-body problems, since even when there are only two bodies only special solutions are known if the interaction is relativistic—and therefore not instantaneous (cf. [12]).

Relativistic Notation (5.1.1)

(a) Let $x^0 \equiv t$ and $\mathbf{x} \equiv (x^1, x^2, x^3)$, choosing the units of time so that $c = 1$. The x^α, $\alpha = 0, 1, 2, 3$, are Cartesian coordinates on extended phase space M_e, and we define a pseudo-Riemannian tensor field (see (2.4.14)),

$$\eta = \eta_{\alpha\beta}\, dx^\alpha\, dx^\beta, \qquad \eta_{\alpha\beta} = \begin{array}{cccc} {\scriptstyle 0} & {\scriptstyle 1} & {\scriptstyle 2} & {\scriptstyle 3} \\ \left| \begin{array}{cccc} -1 & & & \\ & 1 & & \\ & & 1 & \\ & & & 1 \end{array} \right| \end{array} \tag{5.1.2}$$

185

on this chart. This corresponds to a scalar product on $T(M_e)$,

$$\langle \partial_\alpha | \partial_\beta \rangle = \eta_{\alpha\beta}, \qquad \partial_\alpha = \frac{\partial}{\partial x^\alpha},$$

and hence to a bijection $T(M_e) \to T^*(M_e)$. The inverse mapping is given by the contravariant tensor field $\eta^{\alpha\beta} \partial_\alpha \partial_\beta$, where $\eta_{\alpha\beta} \eta^{\beta\gamma} = \delta_\alpha^\gamma$. Since $\eta_{\alpha\beta} \eta_{\beta\gamma} = \delta_{\alpha\gamma}$, the matrix $\eta^{\alpha\beta}$ is numerically the same as $\eta_{\alpha\beta}$ in (5.1.2).

(b) The motion of a particle is described by writing its coordinates x^α as functions $x^\alpha(s)$ of a parameter s, the proper time. We shall denote differentiation by s with a dot: $\dot{x}^\alpha \equiv dx^\alpha/ds$, and normalize s so that $\dot{x}^\alpha \dot{x}^\beta \eta_{\alpha\beta} = -1$; if $|\dot{x}| \ll 1$, then $\dot{x}^0 \approx 1$, and s becomes equal to t (cf. Problem 2).

The Equations of Motion (5.1.3)

By (1.1.4), the equations

$$m\ddot{x}^\alpha \eta_{\alpha\beta} = e\dot{x}^\alpha F_{\alpha\beta}, \qquad F \equiv \begin{vmatrix} 0 & E_1 & E_2 & E_3 \\ -E_1 & 0 & -B_3 & B_2 \\ -E_2 & B_3 & 0 & -B_1 \\ -E_3 & -B_2 & B_1 & 0 \end{vmatrix},$$

hold for the motion of a particle in an electric field $\mathbf{E}$ and a magnetic field $\mathbf{B}$.

Remarks (5.1.4)

1. Since $F_{\alpha\beta} = -F_{\beta\alpha}$,

$$\ddot{x}^\alpha \dot{x}^\beta \eta_{\alpha\beta} = \frac{1}{2} \frac{d}{ds} \dot{x}^\alpha \dot{x}^\beta \eta_{\alpha\beta} = 0.$$

The normalization of (b) of (5.1.1) is consistent with (5.1.3).

2. If $\gamma = dt/ds$, then the three spatial coordinates of (5.1.3) can be wrtiten as

$$\frac{d}{dt}\left(m\gamma \frac{d\mathbf{x}}{dt}\right) = e\left(\mathbf{E} + \left[\frac{d\mathbf{x}}{dt} \times \mathbf{B}\right]\right).$$

In other words, the rate of change of the momentum, using the relativistic mass $m\gamma$, equals the Lorentz force, which, as well as causing an acceleration in the direction of $e\mathbf{E}$, causes a positive (i.e., counter-clockwise) torque, looking in the direction of $e\mathbf{B}$. Since a stationary electric current $\mathbf{j}$ at the origin induces a magnetic field

$$\mathbf{B} = \frac{[\mathbf{j} \times \mathbf{x}]}{r^3}$$

at the point $\mathbf{x}$, this means that parallel currents attract (see figure).

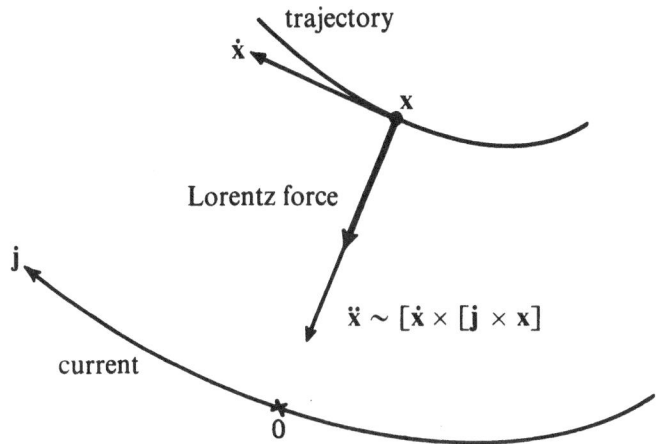

The direction of the Lorentz force.

3. The time component of (5.1.3) expresses conservation of energy:

$$\frac{d}{dt} m\gamma = e\left(\mathbf{E} \cdot \frac{d\mathbf{x}}{dt}\right).$$

4. The nonrelativistic limit, $\gamma \to 1$ and $\mathbf{B} \to \mathbf{0}$, reproduces the earlier equations (1.1.1).

The Lagrangian and Hamiltonian (5.1.5)

The electromagnetic field is a 2-form on M_e, though not an arbitrary one, as it satisfies the homogeneous form of Maxwell's equations,

$$dF = 0. \tag{5.1.6}$$

We shall concern ourselves only with manifolds on which

$$F = dA, \qquad A \in E_1(M_e), \tag{5.1.7}$$

follows from (5.1.6) (cf. (2.5.6; 3)). This makes equations (5.1.3) the Euler–Lagrange equations of the Lagrangian

$$L(x(s), \dot{x}(s)) = \frac{m}{2} \dot{x}^\alpha \dot{x}^\beta \eta_{\alpha\beta} - e\dot{x}^\alpha A_\alpha(x) \tag{5.1.8}$$

(cf. (2.3.22)). The canonically conjugate momenta $p_\alpha = \partial L/\partial \dot{x}^\alpha$ and the Hamiltonian are

$$p_\alpha = m\dot{x}^\beta \eta_{\alpha\beta} - eA_\alpha(x)$$

$$\mathscr{H} = \frac{1}{2m}(p_\alpha + eA_\alpha(x))(p_\beta + eA_\beta(x))\eta^{\alpha\beta}. \tag{5.1.9}$$

Remarks (5.1.10)

1. With Cartesian coordinates, equation (5.1.7) reads more explicitly

$$A = A_\alpha \, dx^\alpha, \qquad F = \tfrac{1}{2} F_{\alpha\beta} \, dx^\alpha \wedge dx^\beta, \qquad F_{\alpha\beta} = A_{\beta,\alpha} - A_{\alpha,\beta},$$

or, separating the space and time coordinates,

$$A_\alpha = (V, -\mathscr{A}): \mathbf{B} = \nabla \times \mathscr{A}, \qquad \mathbf{E} = -\dot{\mathscr{A}} - \nabla V.$$

2. The canonical form ω on $T^*(M_e)$ is

$$\omega = \sum_\alpha dx^\alpha \wedge dp_\alpha$$

or, written differently,

$$\{x^\alpha, p_\beta\} = \delta^\alpha_\beta.$$

3. Equation (5.1.7) determines A only up to a gauge transformation $A \to A + d\Lambda$, where $\Lambda \in C^\infty(M_e)$, which leaves the equations of motion, but not $\mathscr{H}$, invariant. If the gauge transformation is combined with the canonical transformation $x^\alpha \to x^\alpha$, $p_\beta \to p_\beta - e\Lambda_{,\beta}(x)$ (cf. Problem 4), then $\mathscr{H}$ is left unchanged. The canonical momenta p have no gauge-invariant meaning (and thus neither does the origin of $T^*(M_e)$), although $\dot{x}^\alpha$ is gauge-invariant.

4. Conversely, according to (2.5.6; 3), A's that produce the same F differ on starlike regions at most by a gauge transformation.

5. The Poisson brackets between x^α and $\dot{x}^\beta$ are still

$$\{x^\alpha, \dot{x}^\beta\} = \frac{\eta^{\alpha\beta}}{m},$$

though

$$\{\dot{x}^\alpha, \dot{x}^\beta\} = \frac{-e}{m^2} \eta^{\alpha\gamma} \eta^{\beta\delta} F_{\gamma\delta}(x)$$

is now not zero, but depends only on the gauge-invariant quantity F.

6. Since $\mathscr{H}$ does not depend explicitly on s, it is a constant of the motion. Expressed in terms of $\dot{x}$, it is

$$\mathscr{H} = \frac{m\dot{x}^\alpha \dot{x}^\beta}{2} \eta_{\alpha\beta},$$

so we shall always work with the submanifold $\mathscr{H} = -m/2$, in accordance with the normalization of s in (b) of (5.1.1).

7. The diffeomorphism φ of extended phase space: $\varphi(x) = x$, $\varphi(p) = p - eA$, casts $\mathscr{H}$ into the form $\varphi(\mathscr{H}) = p^\alpha p^\beta \eta_{\alpha\beta}/2m$. This diffeomorphism is not canonical, as $\varphi(\omega) = \omega + eF$, and $\varphi(\{A, B\}) = \{\varphi(A), \varphi(B)\}_{\varphi(\omega)}$, where $\{\ \ \}_{\varphi(\omega)}$ is the Poisson bracket calculated with $\omega + eF$. This means that if extended phase space is equipped with the symplectic form $\omega + eF$, then the free Hamiltonian generates a motion in the field F. In this formulation the gauge-dependent A does not appear, and F is not even required to be exact. For (3.2.1) we only needed to know that $d\omega = 0$; therefore,

as long as F is closed we have a canonical formulation of the equations of motion. But since A is generally simpler than F, we nevertheless prefer to use (5.1.5).

Example: Free Particles (5.1.11)

We would like to compare the case $e = 0$ with the results of §4.1. This time,

$$\mathscr{H} = \frac{1}{2m} p_\alpha p_\beta \eta^{\alpha\beta} = \frac{1}{2m}(|\mathbf{p}|^2 - p_0^2).$$

And now $E = mi$ is $-p_0$, so the additional term E in (3.2.12) is changed by the factor $-p_0/2m$. Consequently, the largest group of point transformations on $T^*(M_e)$ that leave $\mathscr{H}$ invariant is the **Poincaré group** ($=$ displacements $+$ Lorentz transformations):

$$x^\alpha \to \Lambda^\alpha_\beta x^\beta + \lambda^\alpha, \qquad \lambda \in \mathbb{R}^4, \qquad \Lambda' \eta \Lambda = \eta, \qquad (5.1.12)$$
$$p_\alpha \to \eta_{\alpha\gamma} \Lambda^\gamma_\delta \eta^{\delta\beta} p_\beta.$$

Let ε be the parameter of the one-dimensional subgroup with the infinitesimal elements

$$\lambda^\alpha = \varepsilon e^\alpha, \qquad \Lambda^\alpha_\beta = \delta^\alpha_\beta + \varepsilon L^\alpha_\beta, \qquad (\eta L)' = L'\eta = -\eta L; \quad (5.1.13)$$

then the generator of the subgroup is

$$p_\alpha e^\alpha + L^\alpha_\beta p_\alpha x^\beta \qquad (5.1.14)$$

(cf. (3.2.6)). This gives us $4 + 6 = 10$ constants: the 6 generators of the Lorentz transformation corresponding to L^α_β (cf. Problem 5) can also be written as

$$\mathscr{M}^{\alpha\beta} \equiv p^\alpha x^\beta - p^\beta x^\alpha, \quad \text{where } p^\alpha \equiv \eta^{\alpha\beta} p_\beta, \qquad (5.1.15)$$

on account of the antisymmetry required in (5.1.13); the 4 generators corresponding to the e^α unite the energy and the momentum as the energy-momentum, in which $-p_0 = p^0 = m\gamma$ is the relativistic energy. The three spatial components of $\mathscr{M}^{\alpha\beta}$ are the angular momentum

$$\mathbf{L} = [\mathbf{x} \times \mathbf{p}]; \qquad (5.1.16)$$

while the center-of-mass theorem, coming from $\mathscr{M}^{0\beta}$, $\beta = 1, 2, 3$, now reads:

$$\mathbf{K} \equiv \mathbf{p}t - p^0\mathbf{x} \quad \text{is constant.} \qquad (5.1.17)$$

On the 7-dimensional submanifold where $-2m\mathscr{H} = p_0^2 - |\mathbf{p}|^2 = m^2$, there are three relationships of interdependence,

$$[\mathbf{p} \times \mathbf{K}] = p^0 \mathbf{L}, \qquad (5.1.18)$$

as there must be. Thus, concerning the number of independent constants of the motion and the structure of the trajectories the same general facts hold as in the nonrelativistic case of §4.1 despite the change of the invariance group.

Problems (5.1.19)

1. Calculate the Poisson brackets of the generators of the Poincaré group, and compare with those of the Galilean group.

2. Show that $\dot{x}^\alpha \dot{x}^\beta \eta_{\alpha\beta} = -1$ implies $|dx/dt| < 1$. Thus (5.1.4; 1) shows that electromagnetic forces can never cause particles to move faster than light.

3. Suppose that the Lagrangian for relativistic motion in a scalar field $\Phi \in \mathcal{T}^0_0(M_e)$ is

$$L = \tfrac{1}{2}\dot{x}^\alpha \dot{x}^\beta \eta_{\alpha\beta} - \Phi(x).$$

   Is it possible for a particle to be accelerated to faster than the speed of light?

4. Show that (5.1.10; 3) is a canonical transformation. (Check the Poisson brackets.)

5. Show that condition (5.1.13) defines a 6-dimensional submanifold of the 4×4 matrices (cf. (2.1.10; 3)).

Solutions (5.1.20)

1. $\{p_i, p_j\} = \{p_i, p_0\} = \{L_i, p_0\} = 0,$
   $\{L_i, L_j\} = \varepsilon_{ijk} L_k, \qquad \{p_i, L_j\} = \varepsilon_{ijk} p_k, \qquad \{K_i, L_j\} = \varepsilon_{ijk} K_k,$
   $\{p^0, K_j\} = p_j, \qquad \{K_i, K_j\} = -\varepsilon_{ijm} L_m, \qquad \{p_i, K_j\} = p^0 \delta_{ij}.$

   This differs from the Galilean group in the last two relationships; the Galilean group has m instead of p_0 on the right side (see (4.1.10; 3)). In the nonrelativistic limit, p_0 approaches m, and so in this sense the Poincaré group goes over to the Galilean group; but note that although the elements of the Poincaré group are point transformations on $T^*(M_e)$, those of the Galilean group are not.

2.
$$dt^2 - |dx|^2 = ds^2 \Rightarrow \left|\frac{dx}{dt}\right|^2 = 1 - \left(\frac{ds}{dt}\right)^2 < 1.$$

3. Yes. $\ddot{x}^\alpha \eta_{\alpha\beta} = \Phi_{,\beta}$. If, say, $\Phi = -|x|^2/2$, then $\ddot{x}^0 = 0$, and one solution is $x^0 = s$, and $x(t) = x(0) \cosh t + \dot{x}(0) \sinh t$. The velocity $\dot{x}(t)$ gets arbitrarily large.

4. The only nontrivial Poisson bracket is

$$\{p_\alpha + e\Lambda_{,\alpha}, p_\beta + e\Lambda_{,\beta}\} = e(\Lambda_{,\beta\alpha} - \Lambda_{,\alpha\beta}) = 0.$$

5. Let $L_{\alpha\beta}$ be the components of ηL. For the 16 functions $N_{\alpha\beta} \equiv L_{\alpha\beta} + L_{\beta\alpha}$, equation (5.1.13) implies $N_{\alpha\beta} = 0$. Because $dN_{\alpha\beta} = dN_{\beta\alpha}$, only 6 of the differentials $dN_{\alpha\beta}$ are linearly independent.

5.2 The Constant Field

This is an integrable system, the relativistic generalization of the elementary example of motion in a constant field. It also contains the Larmor precession in a constant magnetic field as a special case.

In this section we discuss the motion in an electromagnetic field the Cartesian components of which are constant. That a field $F_{\alpha\beta}$ has constant strength means that the potentials are linear in x:

$$A_\beta = \tfrac{1}{2} x^\alpha F_{\alpha\beta}. \qquad (5.2.1)$$

Since $A_\beta \in C(\mathbb{R}^4)$, we may set $M_e = \mathbb{R}^4$. The invariance of (5.1.9) under the Poincaré group is broken by the addition of A. Specifically, the p_α are no longer constant, even though the equations of motion (5.1.3) are translation-invariant. A displacement $x^\alpha \to x^\alpha + \lambda^\alpha$ causes $A_\beta \to A_\beta + \tfrac{1}{2}\lambda^\alpha F_{\alpha\beta}$, which is a gauge transformation with $\Lambda = \tfrac{1}{2}\lambda^\alpha F_{\alpha\beta} x^\beta$ (cf. (5.1.10; 4)). Adding the gauge function $\tfrac{1}{2}F_{\alpha\beta}x^\beta$ to the generators of the displacements produces (cf. Problem 4)

The Constants of the Motion (5.2.2)

$$\frac{d}{ds}\left(p_\alpha + \frac{e}{2} F_{\alpha\beta} x^\beta\right) = 0.$$

Although the Lorentz transformations do not even leave the equations of motion invariant, they can still be used to put the problem into a convenient form.

The Transformation Relations of the Field Tensor (5.2.3)

Under a Lorentz transformation

$$\bar{x} = \Lambda x, \qquad \Lambda^t \eta \Lambda = \eta,$$

the field strength, being a 2-form, must transform as

$$\bar{F} = \eta \Lambda \eta F \eta \Lambda^t \eta. \qquad (5.2.4)$$

The 1-form A transforms as p:

$$\bar{A} = \tfrac{1}{2}\bar{x}\bar{F} = \tfrac{1}{2}x\Lambda^t\eta\Lambda\eta F\eta\Lambda^t\eta = \tfrac{1}{2}xF\eta\Lambda^t\eta = \eta\Lambda\eta A. \qquad (5.2.5)$$

Under spatial rotations, $\mathbf{E}$ and $\mathbf{B}$ change as vectors, but the transformation to a moving reference frame, generated by the $\mathbf{K}$ of (5.1.17), causes them to mix. For instance, K_i generates the one-parameter group

$$\Lambda = \begin{vmatrix} \dfrac{1}{\sqrt{1-v^2}} & \dfrac{-v}{\sqrt{1-v^2}} & 0 & 0 \\[2mm] \dfrac{-v}{\sqrt{1-v^2}} & \dfrac{1}{\sqrt{1-v^2}} & 0 & 0 \\[2mm] 0 & 0 & 1 & 0 \\[2mm] 0 & 0 & 0 & 1 \end{vmatrix} \qquad (5.2.6)$$

for $v \in (-1, 1)$. Transformation formula (5.2.4), when expressed in terms of **E** and **B**, reads:

$$\bar{E}_1 = E_1, \qquad\qquad \bar{B}_1 = B_1,$$

$$\bar{E}_2 = \frac{E_2 - B_3 v}{\sqrt{1 - v^2}}, \qquad \bar{B}_2 = \frac{B_2 + E_3 v}{\sqrt{1 - v^2}}, \qquad (5.2.7)$$

$$\bar{E}_3 = \frac{E_3 + B_2 v}{\sqrt{1 - v^2}}, \qquad \bar{B}_3 = \frac{B_3 - E_2 v}{\sqrt{1 - v^2}}.$$

The Normal Form of the Field Tensor (5.2.8)

The first point to recall in the discussion of the possible ways that **E** and **B** can be changed by Λ is that the Poincaré transformation (5.1.12) multiplies the 4-form $F \wedge F$ by $\text{Det}(\Lambda)$. Since $\Lambda^t \eta \Lambda = \eta$, we conclude that $\text{Det}(\Lambda) = \pm 1$, and hence that $F \wedge F$ is invariant throughout any connected component of the group. It is easy to show that $F \wedge F$ is proportional to $(\mathbf{E} \cdot \mathbf{B})$ (Problem 3), which is therefore also invariant. Moreover,

$$F_{\alpha\beta} \eta^{\alpha\gamma} \eta^{\beta\delta} F_{\gamma\delta} = 2(|\mathbf{B}|^2 - |\mathbf{E}|^2) \qquad (5.2.9)$$

is invariant under (5.2.4). Hence the statements $|\mathbf{B}| \gtrless |\mathbf{E}|$, $|\mathbf{B}| = |\mathbf{E}|$, and $\mathbf{E} \perp \mathbf{B}$ are Poincaré-invariant. If both of our invariant quantities are zero, then $|\mathbf{E}| = |\mathbf{B}|$ and $\mathbf{E} \perp \mathbf{B}$. Otherwise, it is always possible to make **E** and **B** parallel with a Lorentz transformation: First, the plane of **E** and **B** can be rotated to the 2-3-plane, and then v can be chosen in (5.2.6) so that $\bar{E}_2/\bar{B}_2 = \bar{E}_3/\bar{B}_3$ (Problem 1).

Because of this, we shall assume that $E_2 = E_3 = B_2 = B_3 = 0$ in what follows, and let $E \equiv E_1$ and $B \equiv B_1$. The special case that both invariants vanish will be treated in §5.4. Now F and A have the components

$$F = \begin{vmatrix} & E & \\ -E & & -B \\ & B & \end{vmatrix}, \qquad A = \tfrac{1}{2}(-Ex, Et, zB, -yB). \qquad (5.2.10)$$

Accordingly, $\mathcal{H}$ divides into two parts, one of which depends only on the 0 and 1 coordinates, and the other only on the 2 and 3 coordinates:

$$\mathcal{H} = \frac{1}{2m}\left[\left(p_y + \frac{ezB}{2}\right)^2 + \left(p_z - \frac{eyB}{2}\right)^2 + \left(p_x + \frac{etE}{2}\right)^2\right.$$

$$\left. - \left(p_0 - \frac{exE}{2}\right)^2\right] \equiv \mathcal{H}_B + \mathcal{H}_E. \qquad (5.2.11)$$

The Flow of $\mathcal{H}_B$ (5.2.12)

$\mathcal{H}_B$ acts on a 4-dimensional phase space with coordinates $(y, z; p_y, p_z)$. Hence two additional constants of motion suffice to determine the trajectories completely. If $B = 0$, we simply get the free flow, so we suppose $B \neq 0$. Then the two constants of (5.2.2), with $\alpha = 2$ and 3, are proportional to

$$\bar{y} \equiv \frac{y}{2} + \frac{p_z}{eB}, \quad \text{and} \quad \bar{z} \equiv \frac{z}{2} - \frac{p_y}{eB}. \tag{5.2.13}$$

The two constants of motion determine the fixed centers of the Larmor orbits in the $y - z$-plane:

$$(\bar{y} - y)^2 + (\bar{z} - z)^2 = \frac{2m}{e^2 B^2} \mathscr{H}_B. \tag{5.2.14}$$

The trajectories are thus circles of radius $\sqrt{2m\mathscr{H}_B}/eB$ and centers $(\bar{y}, \bar{z})$.

Remarks (5.2.15)

1. If $(\bar{y}, \bar{z}) \neq (0, 0)$, the angular momentum $(y\dot{z} - z\dot{y})m$ is not conserved. However, since $\mathscr{H}$ is invariant under rotations about the x-axis, the generator of those rotations, known as the canonical angular momentum,

$$\bar{L}_1 \equiv yp_z - zp_y,$$

must be a constant. Yet $\bar{L}_1$ is not gauge-invariant, and normally has no physical significance. In the gauge we have chosen for A, $\bar{L}_1$ may be expressed in terms of the gauge-invariant constants we have already found:

$$2eB\bar{L}_1 = -2m\mathscr{H}_B + e^2 B^2 (\bar{y}^2 + \bar{z}^2).$$

This is as expected, since there is room for only three independent constants in the 4-dimensional phase space associated with $\mathscr{H}_B$.

2. There has to be a nonvanishing Poisson bracket between two of the three constants. and in fact,

$$\{\bar{z}, \bar{y}\} = \frac{1}{eB}.$$

Since $\{\bar{y}, \bar{z}\}$ does not depend on the coordinates and has a vanishing Poisson bracket with any other observable, nothing prevents the invariance group generated by $\bar{y}$ and $\bar{z}$ (Problem 5) from being Abelian. Jacobi's identity guarantees that $\forall f \in C(T^*(M))$, $\{\bar{z}, \{\bar{y}, f\}\} = \{\bar{y}, \{\bar{z}, f\}\}$ (cf. (3.2.8; 2)).

3. Since the Poisson brackets of $\bar{y}$ and $\bar{z}$ with the individual contributions to $\mathscr{H}_B$ vanish,

$$\{\bar{y}, \dot{y}\} = \{\bar{y}, \dot{z}\} = \{\bar{z}, \dot{y}\} = \{\bar{z}, \dot{z}\} = 0,$$

and since from (5.1.10; 5)

$$\{\dot{y}, \dot{z}\} = \frac{eB}{m^2},$$

the transformation

$$(y, z; p_y, p_z) \to \left(eB\bar{z}, \frac{m^{3/2}}{eB} \dot{y}; \bar{y}, m^{1/2}\dot{z}\right)$$

is canonical. Calling the second pair of these canonical coordinates,

$$\left(\frac{m^{3/2}}{eB} \dot{y}, m^{1/2}\dot{z}\right),$$

q and p makes $\mathcal{H}_B$ simply the Hamiltonian of a harmonic oscillator with the **cyclotron frequency** $\omega \equiv eB/m$:

$$\mathcal{H}_B = \tfrac{1}{2}(p^2 + \omega^2 q^2).$$

This is a degenerate case of the situation descirbed in (3.3.8). One of the two frequencies is necessarily zero and the other eB/m. The physical significance of this is that the circular orbits have a frequency that depends neither on their center nor on their radius.

4. The formulas for the trajectories in terms of the constants, and explicitly using the parameters, are

$$y(s) = \bar{y} + \frac{\sqrt{2m\mathcal{H}_B}}{eB} \cos \omega(s - s_0),$$

$$z(s) = \bar{z} - \frac{\sqrt{2m\mathcal{H}_B}}{eB} \sin \omega(s - s_0).$$

The Flow of $\mathcal{H}_E$ (5.2.16)

Up to some differences of signs, for $E \neq 0$ this is handled in much the same way. The remaining two constants from (5.2.2) are proportional to

$$\bar{x} = \frac{x}{2} + \frac{p_0}{eE} \quad \text{and} \quad \bar{t} = \frac{t}{2} - \frac{p_x}{eE}; \tag{5.2.17}$$

so the trajectories are the hyperbolas

$$(\bar{x} - x)^2 - (\bar{t} - t)^2 = -\frac{2m}{e^2 E^2} \mathcal{H}_E. \tag{5.2.18}$$

Remarks (5.2.19)

1. Since E_1 is unchanged by the Lorentz transformation (5.2.6) (cf. (5.2.7)), the generator of the transformation,

$$\bar{K}_1 = xp_0 + tp_x = \frac{m}{eE} \mathcal{H}_E + \frac{eE}{2}(\bar{x}^2 - \bar{t}^2),$$

is a constant of the motion, though it is not independent of the other constants already found.

2. By calculating the Poisson brackets,

$$\{\bar{x}, \bar{t}\} = \frac{-1}{eE}, \qquad \{\dot{x}, \bar{t}\} = +\frac{eE}{m^2},$$

and

$$\{\bar{x}, \dot{x}\} = \{\bar{x}, \bar{t}\} = \{\bar{t}, \dot{x}\} = \{\bar{t}, \bar{t}\} = 0,$$

we see that

$$\left(eE\bar{t}, \frac{m^{3/2}}{eE} \dot{x}; \bar{x}, m^{1/2}\bar{t} \right)$$

are canonical coordinates, and that

$$\mathscr{H}_E = -\tfrac{1}{2}(p^2 - v^2 q^2), \qquad q = \frac{m^{3/2}}{eE} \dot{x},$$

$$p = m^{1/2}\bar{t}, \qquad\qquad v = \frac{eE}{m}.$$

3. With the right changes of signs in the oscillator potential, the coordinates become hyperbolic functions of the proper time,

$$x(s) = \bar{x} + \frac{\sqrt{2m|\mathscr{H}_E|}}{eE} \cosh v(s - s_0)$$

$$t(x) = \bar{t} + \frac{\sqrt{2m|\mathscr{H}_E|}}{eE} \sinh v(s - s_0).$$

4. As regards the total number of constants of the motion, so far we have found 6 independent constants, but they are not independent of $\mathscr{H} = \mathscr{H}_E + \mathscr{H}_B$. Therefore there is a constant still to be found. We can find it by noting that the inverse function to the sinh, Arcsinh, exists globally, and that $s - \bar{s}_0$ can be expressed in terms of $t - \bar{t}$. Thus

$$y - \frac{\sqrt{2m\mathscr{H}_B}}{eB} \cos \left[\omega \operatorname{Arcsinh} \frac{eE}{\sqrt{2m|\mathscr{H}_E|}} (t - \bar{t}) \right]$$

is an additional independent constant. Consequently, the trajectories are one-dimensional submanifolds, which are diffeomorphic to $\mathbb{R}$, since in extended phase space there are no closed trajectories.

Motion as Seen From Another Frame of Reference (5.2.20)

If **E** and **B** are parallel, we have seen that the motion parallel to the field is hyperbolic, and the motion perpendicular to it is circular. Unless $|\mathbf{E}|^2 - |\mathbf{B}|^2 = (\mathbf{E} \cdot \mathbf{B}) = 0$, the general case can be obtained from this by making a

Lorentz transformation perpendicular to the field. For example, $\mathbf{E} = 0$, $\mathbf{B} = B(1, 0, 0)$ can be turned into

$$\mathbf{E} = \left(0, 0, \frac{Bv}{\sqrt{1 - v^2}}\right), \qquad \mathbf{B} = \left(\frac{B}{\sqrt{1 - v^2}}, 0, 0\right)$$

by a Lorentz transformation in the 2-direction. The circular orbits around the x-axis then move in the 2-direction, perpendicular to $\mathbf{E}$. In the new co-ordinate system this is interpreted by saying that there is an electric field present that the particle tries to move parallel to. At the same time, $\mathbf{B}$ causes the trajectory to bend, more strongly at lower speeds, and therefore smaller z, producing a drift velocity in the 2-direction, as shown in Figure 43.

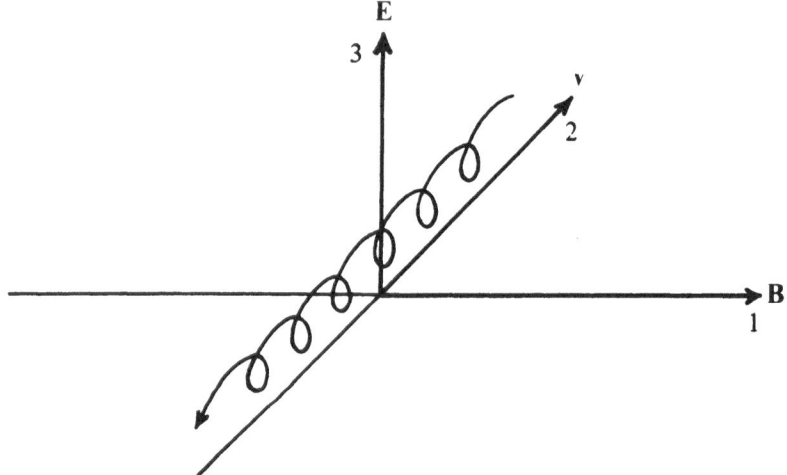

Figure 43 The influence of an electric field perpendicular to $\mathbf{B}$. The trajectory stays in the 2 − 3-plane, and slides along perpendicularly to $\mathbf{E}$.

Problems (5.2.21)

1. Find the v in the Lorentz transformation that makes $\mathbf{E}$ and $\mathbf{B}$ parallel, if it does not happen that $|\mathbf{E}|^2 - |\mathbf{B}|^2 = (\mathbf{E} \cdot \mathbf{B}) = 0$.

2. Discuss the equations of motion in a homogeneous magnetic field without using the constants of motion.

3. Express $F \wedge F$ in terms of $\mathbf{E}$ and $\mathbf{B}$.

4. Verify (5.2.2) by using the equations of motion.

5. What group of invariances of $\mathcal{H}_B$ (respectively $\mathcal{H}_E$) generates the $\bar{y}$ and $\bar{z}$ of (5.2.13) (resp. the $\bar{x}$ and $\bar{t}$ of (5.2.17))?

6. What are the shapes of the trajectories of the momenta and the velocities?

7. Write (5.2.9) as the $*$ of a 4-form (cf. (2.4.29; 4)).

Solutions (5.2.22)

1. By (5.2.7), the equation $\bar{E}_2/\bar{E}_3 = \bar{B}_2/\bar{B}_3$ implies

$$v^2 - v\left(\frac{E_2^2 + E_3^2 + B_2^2 + B_3^2}{E_2 B_3 - E_3 B_2}\right) + 1 = 0.$$

The equation $v^2 - 2\alpha v + 1$ has a solution $|v| < 1$ if $|\alpha| > 1$, which is the case, as $|\mathbf{E}|^2 + |\mathbf{B}|^2 > 2|\mathbf{E}| \cdot |\mathbf{B}| > 2|[\mathbf{E} \times \mathbf{B}]|$.

2. The equation $\ddot{x}(s) = M\dot{x}(s)$, $M_k^i = (e/m)\eta^{ij}F_{jk}$, has the solution $\dot{x}(s) = \exp(sM)\dot{x}(0)$. To integrate this once more, we look for the subspace where M is nonsingular, on which $x(s) = M^{-1}\exp(sM)\dot{x}(0) + \text{constant}$. On the subspace where M is singular, $x(s) \sim s$.

3. $$F \wedge F = -dx^0 \wedge dx^1 \wedge dx^2 \wedge dx^3 \cdot (\mathbf{E} \cdot \mathbf{B}).$$

4. $$\dot{p}_\alpha + \frac{e}{2}F_{\alpha\beta}\dot{x}^\beta = m\ddot{x}_\alpha + eF_{\alpha\beta}\dot{x}^\beta = 0.$$

5. $$(y, z; p_y, p_z) \rightarrow \left(y + \frac{a_3}{eB}, z - \frac{a_2}{eB}, p_y + \frac{a_2}{2}, p_z + \frac{a_3}{2}\right),$$

or respectively,

$$(x, t; p_x, p_0) \rightarrow \left(x + \frac{a_0}{eE}, t - \frac{a_1}{eE}; p_x + \frac{a_1}{2}, p_0 + \frac{a_0}{2}\right), \quad a_i \in \mathbb{R}.$$

6. Both are circles (or respectively hyperbolas), because

$$\dot{y}^2 + \dot{z}^2 = \frac{2}{m}\mathcal{H}_B = \frac{1}{m^2}\left[\left(p_z - \frac{eB}{2}\bar{y}\right)^2 + \left(p_y + \frac{eB}{2}\bar{z}\right)^2\right]$$

or respectively

$$\dot{x}^2 - \dot{t}^2 = \frac{2}{m}\mathcal{H}_E = \frac{1}{m^2}\left[\left(p_x - \frac{eE}{2}\bar{t}\right)^2 - \left(p_t + \frac{eE}{2}\bar{x}\right)^2\right].$$

7. $|\mathbf{E}|^2 - |\mathbf{B}|^2 = 2*(F \wedge *F)$.

5.3 The Coulomb Field

The 0(4) symmetry is broken by relativistic corrections, but the system remains integrable.

The motion described by (5.1.9) with $eA = (\alpha/r, 0, 0, 0)$, where $r = |\mathbf{x}|$, was extremely important in the infancy of atomic physics. Although one needs quantum mechanics to talk about atoms, the classical solution is still of interest—not merely to contrast with the result of quantum mechanics, but also to make the connection to the gravitational case, which we shall come to later.

The Hamiltonian (5.3.1)

$$\mathcal{H} = \frac{1}{2m}\left(|\mathbf{p}|^2 - \left(p_0 + \frac{\alpha}{r}\right)^2\right)$$

is continuous on $\mathbb{R} \times (\mathbb{R}^3 \backslash \{0\})$, and we can use polar coordinates to transform the extended configuration space into $\mathbb{R} \times \mathbb{R}^+ \times S^2$. Since the only spatial coordinate that shows up explicitly in $\mathcal{H}$ is r, we know the

Constants of the Motion (5.3.2)

$$\dot{\mathbf{L}} = \frac{d}{ds}[\mathbf{x} \times \mathbf{p}] = 0, \qquad \dot{p}_0 = 0, \quad \text{and} \quad \dot{\mathcal{H}} = 0.$$

Remarks (5.3.3)

1. From these five constants of the motion it is possible to construct four with vanishing Poisson brackets. Although the Poisson brackets of different components of $\mathbf{L}$ do not vanish, $|\mathbf{L}|^2$, as a scalar, is rotationally invariant, and so

$$\{L_i, |\mathbf{L}|^2\} = 0, \qquad i = 1, 2, 3.$$

Hence $\mathcal{H}$, p_0, $L \equiv |\mathbf{L}|$, and, e.g., L_3 are four independent constants with vanishing Poisson brackets, and the system is integrable.
2. The time-dependence can be determined from $\dot{t} = -(p_0 + \alpha/r)/m$, once $r(s)$ is known.
3. In this section we only study the motion in ordinary phase space $T^*(\mathbb{R}^3 \backslash \{0\})$, which is governed by

$$H = \frac{|\mathbf{p}|^2}{2m} + \frac{\alpha}{r}\frac{p^0}{m} - \frac{\alpha^2}{2mr^2}.$$

The connection between H and the energy $E = p^0$ is the equation

$$p^0 = m\sqrt{1 + \frac{2H}{m}} = \frac{\alpha}{r} + \sqrt{m^2 + |\mathbf{p}|^2},$$

since $\mathcal{H}$ has to equal $-m/2$. The energy p^0 is to be regarded as a constant in H, although its value is not independent of H.
4. Compared with (4.2.2), H has an extra negative contribution, because

$$\sqrt{m^2 + |\mathbf{p}|^2} - m \leq \frac{|\mathbf{p}|^2}{2m}.$$

If $H < 0$, the trajectories in $T^*(\mathbb{R}^3 \backslash \{0\})$ remain in compact sets, and Arnold's theorem applies. In order to compute the frequencies, we construct the

Action and Angle Variables (5.3.4)

We have to form combinations of H, L, and L_3 the conjugate variables of which are the three angle variables on T^3. As we have seen, L_3 generates a rotation about the z-axis. To ascertain what angle is conjugate to L, first note that

$$\{|\mathbf{L}|^2, x_i\} = 2L\{L, x_i\} = 2(\mathbf{L} \cdot \{\mathbf{L}, x_i\}) = 2[\mathbf{L} \times \mathbf{x}]_i \qquad (5.3.5)$$

(cf. (3.1.12; 3)). This implies that L generates a rotation about $\mathbf{L}$, i.e., in the plane of motion:

$$\{L, \mathbf{x}\} = \left[\frac{\mathbf{L}}{L} \times \mathbf{x}\right],$$

and   (5.3.6)

$$\{L, \mathbf{p}\} = \left[\frac{\mathbf{L}}{L} \times \mathbf{p}\right].$$

If we assume that $\mathbf{L}$ does not point in the z-direction, and define the angles φ and χ by

$$\cos \varphi = \frac{L_2}{\sqrt{L^2 - L_3^2}},$$

$$\cos \chi = \frac{[\mathbf{L} \times \mathbf{x}]_3}{\tau \sqrt{|\mathbf{L}|^2 - L_3^2}},$$

then (see Figure 44),

$$\{L_3, \varphi\} = \{L, \chi\} = -1, \quad \text{and} \quad \{L_3, \chi\} = \{L, \varphi\} = 0; \qquad (5.3.7)$$

because a rotation about the z-axis changes φ while leaving χ unchanged, whereas a rotation about $\mathbf{L}$ leaves φ alone but changes χ.

We may choose r and $p_r \equiv \mathbf{x} \cdot \mathbf{p}/r$ as the third pair of canonical coordinates. Although they are not action and angle variables, their Poisson brackets with the observables in (5.3.7) are all zero: Since both of them are rotationally invariant, their Poisson brackets with all the L's vanish. The only fact that remains to be checked for them to be canonical coordinates is that $\{p_r, \chi\} = 0$. This follows from the observation that $rp_r = (\mathbf{x} \cdot \mathbf{p})$ generates a dilatation (4.1.13; 3), under which $\mathbf{L}$ and $\mathbf{x}/r$ are invariant. With these variables H can be written

$$H = \frac{1}{2m}\left(p_r^2 + \frac{L^2}{r^2}\right) + \frac{\alpha p^0}{rm} - \frac{\alpha^2}{2mr^2} \qquad (5.3.8)$$

(cf. (4.2.12)). The action variables defined in (3.3.14) exist if $H < 0$ and $L > \alpha$, and are

$$I_\varphi = L_z, \qquad I_\chi = L,$$

and   (5.3.9)

$$I_r = \frac{1}{2\pi} \oint dr\, p_r = -\sqrt{L^2 - \alpha^2} + \frac{\alpha p^0}{2}\sqrt{\frac{2}{m|H|}}$$

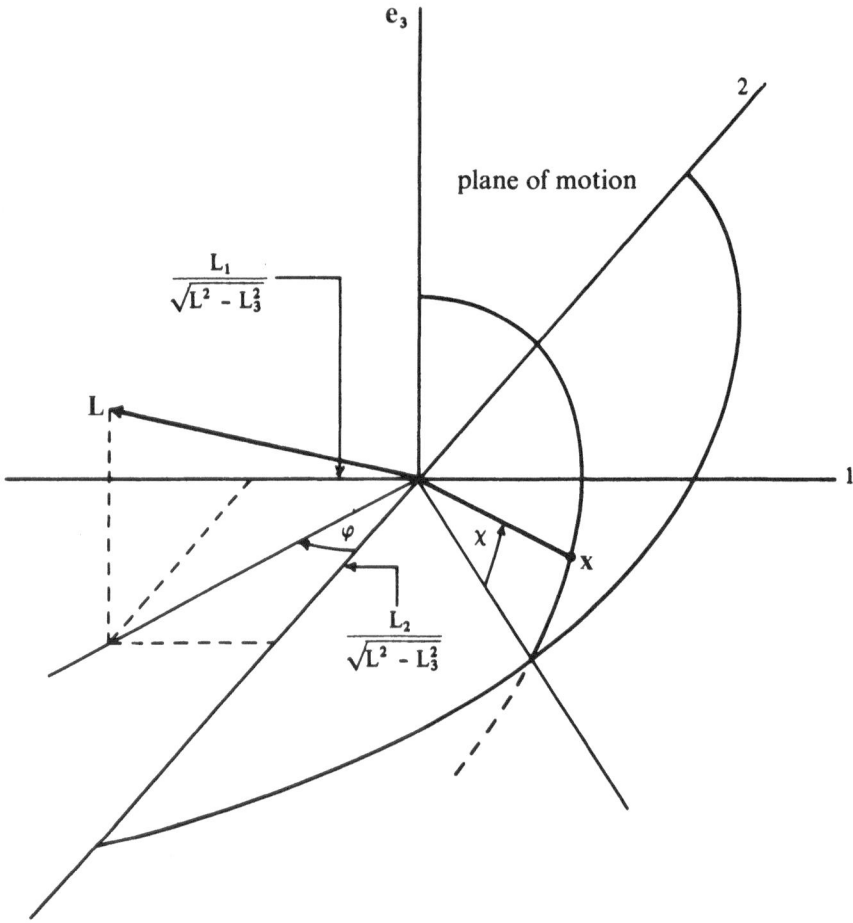

Figure 44   Action and angle variables.

(Problem 1). The Hamiltonian H can easily be expressed in terms of the action variables:

$$H = - \frac{\alpha^2 p^{02}}{2m(I_r + \sqrt{L^2 - \alpha^2})^2},\qquad (5.3.10)$$

and an explicit calculation of the frequencies (in s) (3.3.15; 4) gives

$$\omega_\varphi = 0, \qquad \omega_\chi = \frac{\alpha^2 p^{02}(L/\sqrt{L^2 - \alpha^2})}{m(I_r + \sqrt{L^2 - \alpha^2})^3},$$

$$(5.3.11)$$

$$\omega_r = \frac{\alpha^2 p^{02}}{m(I_r + \sqrt{L^2 - \alpha^2})^3}.$$

Remarks (5.3.12)

1. Since $\omega_\varphi = 0$ (due to conservation of angular momentum the plane of motion is fixed), there is a two-dimensional invariant torus. This torus is densely filled by the orbit, however, unless

$$\frac{\omega_\chi}{\omega_r} = \frac{L}{\sqrt{L^2 - \alpha^2}}$$

   is rational, in which case the orbits are closed.

2. In the nonrelativistic limit, $\alpha^2/r^2 \to 0$, we had $L/\sqrt{L^2 - \alpha^2}$ equal to 1, and there were invariant one-dimensional tori, the Kepler orbits. In the present case the projection of the orbit to configuration space is rosette-shaped (Figure 45), where the angle of each successive perihelion increases by $2\pi/\sqrt{1 - \alpha^2/L^2}$, and the orbit is in general dense in a ring-shaped region. Because of the relativistic mass increase near the center, r does not return to its initial value at the same time as χ, and the orbit precesses. The existence of a second frequency is also what gives rise to the fine structure of spectral lines in atomic physics, which was one of the early experimental confirmations of the theory of relativity.

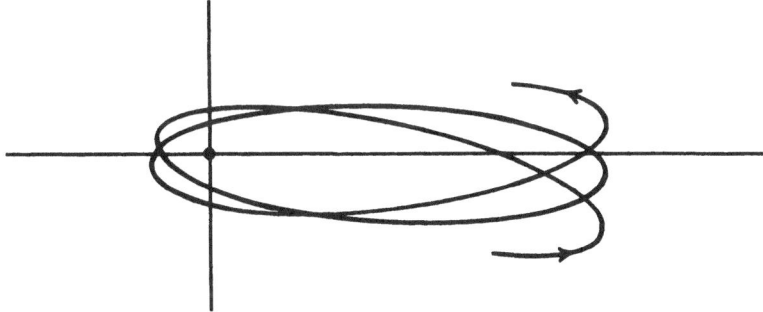

Figure 45   A rosette-shaped orbit in the relativistic Kepler problem.

3. From (5.3.9) we see that the motion is like one-dimensional motion with an effective potential

$$\frac{\alpha p^0}{rm} + \frac{L^2 - \alpha^2}{2mr^2}.$$

   If $\alpha < 0$ and $L \le |\alpha|$, then this potential is monotonic in r, and the particle spirals inevitably into the singularity, reaching it after a finite time. But if $\alpha > 0$ this can not happen, as follows from Remark (5.3.3; 3) (see Problem 4). If we are interested in the completeness of $X_{\mathscr{H}}$, we must reduce phase space in the attractive case to

$$T^*(\mathbb{R}^4) \backslash \{(x, p): L \le |\alpha|\};$$

   it is no longer sufficient to remove a submanifold of a lower dimension.

Unbound Trajectories (5.3.13)

In the part of phase space where $H > 0$ and $L > -\alpha$, the following three facts can be proved:

(a) $\lim_{t \to \pm \infty} r(t) = \infty$;

(b) $\mathbf{p}$ and $\mathbf{x}/r \in \mathcal{A}$ (see (3.4.1));

(c) $\Omega_{\pm} \equiv \lim_{t \to \pm \infty} \Omega_{-t} \circ \Omega_t^0$ exist, where Ω_t^0 is generated by $H + \alpha^2/2mr^2$, the Hamiltonian with a pure $1/r$ potential.

Remarks (5.3.14)

1. The scattering angle $\Theta = \measuredangle(\mathbf{p}_-, \mathbf{p}_+)$ can be written explicitly in terms of the constants of motion (Problem 3), although the function $\sigma(\Theta)$ of (3.4.12) can only be written implicitly in terms of them.

2. Fact (c) implies that this part of phase space has the maximal number of constants of motion, five. For instance, we could take $\mathbf{L}$ and $\mathbf{F} \circ \Omega_{\pm}^{-1} = \lim_{t \to \pm \infty} \mathbf{F}(t)$, where $\mathbf{F}$ was defined in (4.2.4), with $m \to p^0$. For $H > 0$, the invariance group is still $SO(3, 1)$, as in (4.2.7) (cf. (3.4.24; 6)).

Problems (5.3.15)

1. Calculate the integral (5.3.9) for I_r.

2. Write the angle variable φ_r conjugate to I_r in terms of r and the constants of motion.

3. Calculate the scattering angle Θ, and show that it approaches zero in the limit of large energies ($p^0 \to \infty$, $L \to \infty$, with $L/p^0 = b$ constant).

4. Calculate the maximum of the effective potential V_{eff} for $\alpha > 0$, and show that it is greater than $p_\infty^2/2m$, where p_∞ is the value of $|\mathbf{p}|$ at $r = \infty$.

Solutions (5.3.16)

1. The integral for I_r can be expressed as an indefinite intergral in terms of elementary functions (see Problem 2), but it is simpler to evaluate the definite integral with complex integration. In the complex r-plane the integral has a pole at the origin and a branch cut along an interval on the real axis, which is contained in the integration region:

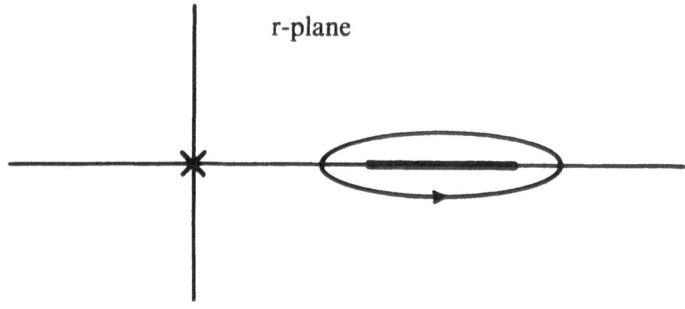

r-plane

Figure 46   Integration path in the complex r-plane.

By stretching the path of integration, we see that the integral just picks up the residues of the poles at $r = 0$ and $r = \infty$:

$$I_r = -\sqrt{L^2 - \alpha^2} + \frac{\alpha p^0}{2}\sqrt{2/m|H|}.$$

2. $\quad \varphi_r = \int dr \sqrt{2m\left(H - \frac{\alpha p^0}{mr} - \frac{L^2 - \alpha^2}{2mr^2}\right)} = \frac{1}{2}\sqrt{2mHr^2 - 2\alpha p^0 r - L^2 + \alpha^2}$

$$+ \frac{\alpha p^0}{\sqrt{-2mH}} \text{ arc sin } \frac{4mHr - 2\alpha p^0}{\sqrt{4\alpha^2 p^{02} + 4H(L^2 - \alpha^2)}}$$

$$+ \sqrt{L^2 - \alpha^2} \text{ arc sin } \frac{2(L^2 - \alpha^2)/r + 2\alpha p^0}{\sqrt{4\alpha^2 p^{02} + 4H(L^2 - \alpha^2)}},$$

because

$$\int dr \sqrt{\frac{a}{r^2} + \frac{b}{r} + c} = \frac{1}{2}\sqrt{a + br + cr^2} + \frac{b}{2}\frac{1}{\sqrt{-c}} \text{ arc sin } \frac{2cr + b}{\sqrt{b^2 - 4ac}}$$

$$- \sqrt{-a} \text{ arc sin } \frac{2a + br}{r\sqrt{b^2 - 4ac}}.$$

3. Because $dr/d\varphi = \dot{r}/\dot{\varphi} = r^2 p_r/L$,

$$\varphi = L \int \frac{dr}{r^2\sqrt{2m(H - (\alpha p^0/rm) - (L^2 - a^2)/2mr^2)}},$$

and if $u = 1/r$, then

$$\pi - \Theta = 2L \int_0^{u_0} \frac{du}{\sqrt{2mH - 2\alpha p^0 u - (L^2 - \alpha^2)u^2}}$$

$$= \frac{2L}{\sqrt{L^2 - \alpha^2}} \text{ arc cos}\left(1 + \frac{2mH(L^2 - \alpha^2)}{\alpha^2 p^{02}}\right)^{-1/2} \rightarrow 2 \text{ arc cos } 0 = \pi.$$

4. $V_{\text{eff}} = (\alpha/r)(p^0/m) - (\alpha^2/2mr^2)$ has its maximum $(p^0)^2/2m$ at $r = \alpha/p^0$, and according to (5.3.3; 3),

$$\frac{(p^0)^2}{2m} = \frac{m}{2} + \frac{p_\infty^2}{2m}.$$

5.4 The Betatron

Although this problem is not integrable, it is not only possible to solve for particular trajectories, but for suitable chosen magnetic fields it is even possible to determine the time-evolution for a larger class of initial conditions.

One of the most ingenious applications of the law of induction

$$\mathbf{V} \times \mathbf{E} = -\frac{\partial \mathbf{B}}{\partial t} \tag{5.4.1}$$

is the betatron. Its mechanism is based on the following ideas: If a current $\mathbf{j}$ starts to flow in a current loop, it induces a magnetic field $\mathbf{B}$ in the interior of the loop, and $\dot{\mathbf{B}}$ produces an electric field circulating about $\mathbf{B}$ in the same direction as a Larmor orbit, so as to oppose $\mathbf{j}$ (Problem 2). In order to see when the Lorentz force is able to counterbalance the centrifugal force, let us integrate (5.4.1) around a circle of radius a about the z-axis. If $B_z \equiv B$ depends only on the distance from the z-axis and t, then

$$E_\varphi = \frac{1}{2\pi a} \oint \mathbf{E} \cdot d\mathbf{s} = -\int \frac{\partial B}{\partial t} \frac{dS}{2\pi a} = -\frac{a}{2} \frac{d}{dt} \bar{B},$$

$$\bar{B} \equiv \frac{1}{\pi a^2} \int dS B. \tag{5.4.2}$$

On the other hand the (nonrelativistic) equation of motion for a particle in this field implies

$$\frac{d}{dt} mv = eE_\varphi = -\frac{ea}{2} \frac{d}{dt} \bar{B} \tag{5.4.3}$$

or, if the particle was at rest before the current was switched on,

$$mv = -\frac{ea}{2} \bar{B}. \tag{5.4.4}$$

Hence the centrifugal and centripetal forces will balance at a circular orbit whenever

$$\frac{mv^2}{a} = -evB(a) = -\frac{ev}{2} \bar{B}, \quad \text{which implies} \quad B(a) = \frac{\bar{B}}{2}; \tag{5.4.5}$$

that is, when the field at the orbit is half as strong as its average over the disc enclosed by the orbit. In this section we shall go through the details of the theory sketched above.

The Hamiltonian (5.4.6)

To exploit the symmetry of the problem we use cylindrical coordinates for space, or in other words a chart $\mathbb{R}^3 \backslash \{(0, 0, \mathbb{R})\} \to \mathbb{R}^+ \times S^1 \times \mathbb{R}$, with co-ordinates ρ, φ, and z. A field $B(\rho, t)$ in the z-direction comes from a vector potential $e \mathscr{A}$ with a covariant φ-component $A(\rho, t)$ such that

$$eB_z = \frac{1}{\rho} \frac{\partial}{\partial \rho} A, \quad eE_\varphi = -\frac{1}{\rho} \frac{\partial A}{\partial t} \tag{5.4.7}$$

(cf. Problem 3). This shows that a circulating electric field is created when **B** is switched on, according to the law of induction. The motion of a particle in extended phase space is then controlled by

$$\mathcal{H} = \frac{1}{2m}\left(p_z^2 + p_\rho^2 + \frac{1}{\rho^2}(p_\varphi - A)^2 - p_0^2\right). \tag{5.4.8}$$

Since this depends explicitly only on ρ and t (through A), we obtain the

Constants of the Motion (5.4.9)

$$\dot{p}_\varphi = \dot{p}_z = 0.$$

Remarks (5.4.10)

1. All together, including $\mathcal{H}$ we have three constants, one too few to integrate the problem completely. The complexity is comparable to that of the restricted three-body problem, since only one constant, $\mathcal{H}$, is available for the (ρ, t) motion.
2. We do not need to find the most general solution; the betatron is only operated when the particles are nearly at rest before it is switched on. For that reason we study the

Orbits with $p_\varphi = p_z = 0$ **(5.4.11)**

In this case, $A = 0 \Rightarrow \dot{\varphi} = 0$, as

$$\dot{\varphi} = \frac{p_\varphi - A}{m\rho^2}. \tag{5.4.12}$$

Then the ρ coordinate of the motion obeys the equations

$$\dot{\rho} = \frac{p_\rho}{m}, \qquad \ddot{\rho} = \frac{\dot{p}_\rho}{m} = -\frac{\partial}{\partial \rho}\frac{A^2}{2m^2\rho^2}. \tag{5.4.13}$$

As long as

$$\frac{\partial}{\partial \rho}\frac{A^2(\rho, t)}{2m\rho^2}\bigg|_a = 0 \tag{5.4.14}$$

for all t, $\rho = a = $ constant is a solution.

Remarks (5.4.15)

1. Equation (5.4.14) is equivalent to (5.4.5) (Problem 4), showing that the naïve arguments at the beginning of this section are correct, even relativistically. (Provided Stokes's theorem is applicable.)
2. A question of practical importance is whether the orbits are stable, since the initial condition chosen in (5.4.11) will never hold exactly. It is clear that the z-component of the motion is free, and therefore unstable. Some

z-dependence must be introduced into A to prevent this (see [10]), but that will not concern us here. Stability in the radial direction depends on the form of A.

Instead of making a thorough investigation of the stability question, we shall rely on the theorist's freedom to leave the construction of suitable fields to the experimenter, and look only at a soluble

Example (5.4.16)

$$A = \rho\sqrt{\mu^2(\rho - a)^2 + v^2t^2}.$$

This satisfies condition (5.4.14) and produces the fields

$$B_z = \frac{\mu^2(\rho - a)(2\rho - a) + v^2t^2}{\rho\sqrt{\mu^2(\rho - a)^2 + v^2t^2}},$$

$$E_\varphi = \frac{-v^2t}{\sqrt{\mu^2(\rho - a)^2 + v^2t^2}}. \tag{5.4.17}$$

At the orbit $\rho = a$, the electric field is constant, and the magnetic field grows linearly in t. The angular velocity (using s) then grows linearly with the ordinary time t according to (5.4.12) (cf. Problem 1):

$$\dot{\varphi}|_{\rho = a} = \frac{P_\varphi - avt}{ma^2}. \tag{5.4.18}$$

If $p_z = P_\varphi = 0$, then

$$\mathscr{H} = \frac{1}{2m}(p_\rho^2 - p_0^2 + \mu^2(\rho - a)^2 + v^2t^2). \tag{5.4.19}$$

Except for the sign of p_0^2, this brings us back to the two-dimensional harmonic oscillator. There are two constants of motion; the ρ and t contributions to $\mathscr{H}$ separate, and the equations can be integrated without difficulty (the coefficients being determined by $1 = \dot{t}^2 - \dot{\varphi}^2 - \dot{\rho}^2$):

$$\rho(s) = a + c \sin\frac{\mu}{m}(s - s_0), \qquad c = p_\rho(0)/\mu,$$

$$t(s) = \frac{\sqrt{c^2\mu^2 + m^2}}{v}\sinh\frac{v}{m}s. \tag{5.4.20}$$

Remarks (5.4.21)

1. As for stability, the motion is stable if we only look at ρ, because the set where $|\rho - a| < c$ is invariant for all $c \in \mathbb{R}^+$. But if we look at both ρ and t (or ρ and φ), then it is not stable, because an arbitrarily small c can bring about arbitrarily large changes in $t(s)$, for s large enough.

2. Since a discussion of stability for arbitrary A is difficult, it is quite common to simply invoke the adiabatic theorem [11], a popular version of which goes roughly as follows: if the field is turned on very slowly, so that it changes by only a tiny fraction during each period, then it is safe to do calculations as if the field were constant. But it is not really possible to formulate the theorem in precisely this way, even though in the special example (5.4.10) it did turn out that the ρ and t dependences separated, and the growth-rate v did not enter into $\rho(s)$—showing up only in the connection between s and t.

Problems (5.4.22)

1. Calculate the speed $v = |dx/dt|$ of a particle following the trajectory (5.4.20) with $\rho = a$, and verify that $|v| < 1$.

2. How do the directions of the forces in the betatron square with the fact mentioned in §5.1, that antiparallel currents repel?

3. Calculate the coordinates of $\mathbf{B} = \mathbf{V} \times \mathcal{A}$ and $\mathbf{E} = -\dot{\mathcal{A}}$ in cylindrical coordinates, using the covariant components A of $\mathcal{A}$ (cf. (2.4.16)).

4. Show that (5.4.5) and (5.4.14) are equivalent. (Recall that B is an orthogonal component, while A is a covariant component.)

Solutions (5.4.23)

1.
$$v = a\frac{d\varphi}{dt} = \frac{a\dot{\varphi}}{t} = -\tanh\left(\frac{v}{m}s\right).$$

2.

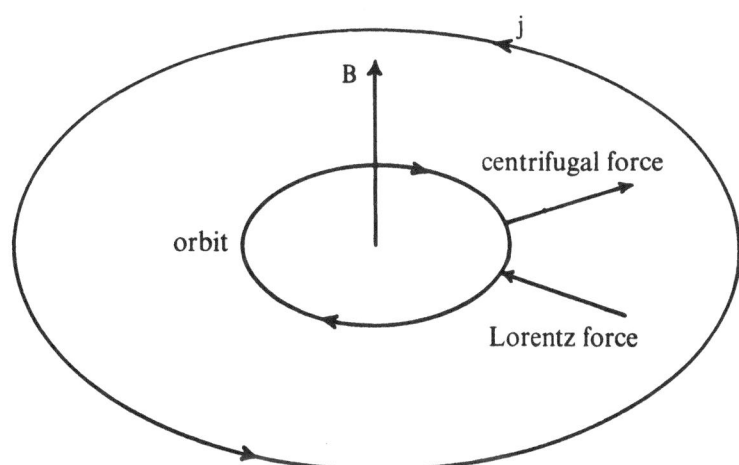

The directions of the forces in the betatron.

3. In cylindrical coordinates (z, ρ, φ), $g_{ii} = (1, 1, \rho^2)$, so $(\mathscr{A}_z, \mathscr{A}_\rho, \mathscr{A}_\varphi) = (A_z, A_\rho, A_\varphi/\rho)$. To calculate $\mathbf{V} \times \mathscr{A} = *(dA)$, we generalize the $*$ operation of (2.4.29; 4) to get $(dA)_i = \varepsilon_{ikj} g^{k\ell} g^{jm} (dA)_{\ell m} \sqrt{g}$, where $g \equiv \mathrm{Det}(g_{ik}) = \rho^2$. Thus

$$A = A_z \, dz + A_\rho \, d\rho + A_\varphi \, d\varphi,$$

$$dA = (A_{z,\rho} - A_{\rho,z}) d\rho \wedge dz + (A_{\rho,\varphi} - A_{\varphi,\rho}) d\varphi \wedge d\rho + (A_{\varphi,z} - A_{z,\varphi}) dz \wedge d\varphi,$$

$$*(dA) = \underbrace{\frac{1}{\rho}(A_{\varphi,\rho} - A_{\rho,\varphi}) dz}_{(\mathbf{V} \times \mathscr{A})_z} + \underbrace{\frac{1}{\rho}(A_{z,\varphi} - A_{\varphi,z}) d\rho}_{(\mathbf{V} \times \mathscr{A})_\rho} + \underbrace{\rho(A_{\rho,z} - A_{z,\rho}) d\varphi}_{(\mathbf{V} \times \mathscr{A})_\varphi},$$

which corresponds with (5.4.7).

4.
$$\bar{B} = \frac{1}{\pi a^2} \int_0^a 2\pi\rho \, d\rho \, \frac{A_{,\rho}}{\rho} = \frac{2}{a^2} A(a),$$

and

$$0 = \left.\frac{\partial}{\partial\rho}\frac{A}{\rho}\right|_a = \frac{A_{,\rho}}{\rho} - \left.\frac{A}{\rho^2}\right|_a = B(a) - \frac{\bar{B}}{2}.$$

Since there is still one arbitrary constant in A, we may set $A(0) = 0$.

5.5 The Traveling Plane Disturbance

The rich invariance group of this problem furnishes more constants of motion than are required for integrability. Nevertheless, the trajectory generally covers a two-dimensional submanifold of space.

Classical studies of the scattering of light by a charged particle deal with the motion of the particle in a plane electromagnetic wave. More recently, laser technology has made intense pulses of light available, and the interest in solving the equations of motion in the field of a disturbance moving in some direction at the speed of light has increased.

The Field and the Hamiltonian (5.5.1)

Let x be the direction of propagation, the fields depending only on the combination $t - x$. So that we can use the solutions of Maxwell's equations in a vacuum, $\mathbf{E}$ and $\mathbf{B}$ are taken perpendicular to each other and to the x-axis. Such a situation is described by the vector potential (5.1.7)

$$eA = (0, 0, f(t - x), g(t - x)).$$

This makes the fields

$$e\mathbf{E} = (0, f', g') \quad \text{and} \quad e\mathbf{B} = (0, -g', f');$$

and the motion of a particle in these fields is governed by

$$\mathscr{H} = \frac{1}{2m} [(p_y + f)^2 + (p_z + g)^2 + p_x^2 - p_0^2].$$

Remarks (5.5.2)

1. Both invariants $|\mathbf{E}|^2 - |\mathbf{B}|^2$ and $(\mathbf{E} \cdot \mathbf{B})$ vanish identically. If f' and g' are constant, we get the constant field with vanishing invariants.
2. We can assume that $f, g \in C^\infty(\mathbb{R})$, to be able to work on $M_e = \mathbb{R}^4$.

The Invariance Group (5.5.3)

The electromagnetic field tensor F is invariant under a 5-parameter sub-group of the Poincaré group. Since f and g depend only on $t - x$, it is clearly unchanged by displacements in the spatial directions y and z or in the "lightlike" direction $x + t$. It turns out (Problem 1) that the combined Lorentz transformations and rotations generated by $K_z + L_y$ and $K_y - L_z$ also do not affect F. Consequently, F is invariant under the group generated by $p_y, p_z, p_x + p_0, K_z + L_y$, and $K_y - L_z$ (Problem 2). Yet not all of these transformations leave A, and thus $\mathscr{H}$, invariant. Since M_e is starlike, Remark (5.1.10; 5) applies, and A can at most be regauged. In fact it can be calculated that the two kinds of Lorentz transformations change A only by the gauge transformations whose gauge functions are

$$\Lambda_g = \int_0^{t-x} du\, g(u) \quad \text{and} \quad \Lambda_f = \int_0^{t-x} du\, f(u).$$

We have thus accounted for all the

Constants of the Motion (5.5.4)

$$p_y, p_z, p_x + p_0,$$

$$K_z + L_y + \Lambda_g = p_z(t - x) + z(p_0 + p_x) + \int_0^{t-x} du\, g(u),$$

and

$$K_y - L_z + \Lambda_f = p_y(t - x) + y(p_0 + p_x) + \int_0^{t-x} du\, f(u)$$

are constant.

Remarks (5.5.5)

1. The group generated by the five constants of the motion is isomorphic to the invariance group of the field (Problem 2).
2. The Poisson brackets of p_y, p_z, and $p_x + p_0$ vanish; so, counting $\mathscr{H}$, we have all four constants necessary for integrability.

3. Counting $\mathcal{H}$ there are all together six constants, so one more would be needed to completely determine a trajectory in $T^*(M_e)$. As there is no other such constant, we have to resort to quadrature. Set the origin of s at the point where the trajectory crosses the plane $t = x$. Then from

$$\frac{d}{ds}(t - x) = -\frac{p_0 + p_x}{m} \equiv \alpha > 0,$$

it follows that

$$t(s) - x(s) = \alpha s.$$

Using $\mathcal{H} = -m/2$, we find that

$$\frac{d}{ds}(t + x) = \frac{p_x - p_0}{m} = \alpha^{-1}\left(1 + \frac{(p_y + f)^2 + (p_z + g)^2}{m^2}\right)$$

$$t(s) + x(s) = 2t(0) + \alpha^{-2}\int_0^{\alpha s} du\left[1 + \frac{(p_y + f(u))^2 + (p_z + g(u))^2}{m^2}\right].$$

If we collect all these results, we obtain the

Explicit Solution for the Coordinates as Functions of Proper Time (5.5.6)

$$t(s) = t(0) + \frac{s}{2}\left(\frac{1}{\alpha} + \alpha\right) + \frac{1}{2m^2\alpha^2}\int_0^{\alpha s} du[(p_y + f(u))^2 + (p_z + g(u))^2],$$

$$x(s) = x(0) + \frac{s}{2}\left(\frac{1}{\alpha} - \alpha\right) + \frac{1}{2m^2\alpha^2}\int_0^{\alpha s} du[(p_y + f(u))^2 + (p_z + g(u))^2],$$

$$y(s) = y(0) + s\frac{p_y}{m} + \frac{1}{m\alpha}\int_0^{\alpha s} du\, f(u),$$

$$z(s) = z(0) + s\frac{p_z}{m} + \frac{1}{m\alpha}\int_0^{\alpha s} du\, g(u).$$

Because of the absence of the seventh constant, the trajectory is generally a Lissajou figure.

Example (5.5.7)

The superposition of two plane waves:

$$f = A_1 \cos \omega_1 u, \qquad g = A_2 \cos(\omega_2 u + \delta),$$

$$e\mathbf{E} = (0, -A_1\omega_1 \sin \omega_1 u, -A_2\omega_2 \sin(\omega_2 + \delta)),$$

$$e\mathbf{B} = (0, A_2\omega_2 \sin(\omega_2 u + \delta), -A_1\omega_1 \sin \omega_1 u).$$

The solution (5.5.6) with $x(0) = t(0)$ is computed as

$$\begin{Bmatrix} t(s) \\ x(s) \end{Bmatrix} = t(0) + \frac{s}{2}\left(\frac{1}{\alpha} \pm \alpha + \frac{p_y^2 + p_z^2}{m^2}\right) + \frac{p_y A_1}{\omega_1 \alpha^2 m^2} \sin \omega_1 \alpha s$$

$$+ \frac{p_z A_2}{\omega_2 \alpha^2 m}(\sin(\omega_2 \alpha s + \delta) - \sin \delta) + \frac{A_1^2 + A_2^2}{4m^2 \alpha} s + \frac{A_1^2}{8\alpha^2 m^2 \omega_1}$$

$$\cdot \sin 2\omega_1 \alpha s + \frac{A_2^2}{8\alpha^2 m^2 \omega_2}(\sin 2(\omega_2 \alpha s + \delta) - \sin 2\delta),$$

$$y(s) = y(0) + s\frac{p_y}{m} + \frac{A_1}{\omega_1 \alpha m} \sin \omega_1 \alpha s,$$

$$(5.5.8)$$

$$z(s) = z(0) + s\frac{p_z}{m} + \frac{A_2}{\omega_2 \alpha m}(\sin(\omega_2 \alpha s + \delta) - \sin \delta).$$

For ω's with irrational ratios, the trajectory fills a two-dimensional region.

Special Cases (5.5.9)

1. Circularly polarized waves: $A_1 = A_2 = A$, $\omega_1 = \omega_2 = \omega$, and $\delta = \pi/2$. The invariance group of this field has an extra parameter, and there is an extra constant,

$$2L_x + \frac{p_1 - p_0}{\omega} \tag{5.5.10}$$

(Problem 3). The solution (for $x^\beta(0) = 0$) accordingly simplifies to

$$\begin{Bmatrix} t(s) \\ x(s) \end{Bmatrix} = \frac{s}{2\alpha}\left(1 \pm \alpha^2 + \frac{A^2 + p_y^2 + p_z^2}{m^2}\right)$$

$$+ \frac{A}{m^2 \alpha^2 \omega}(p_y \sin \omega \alpha s + p_z(\cos \omega \alpha s - 1)),$$

$$y(s) = \frac{A}{\omega \alpha m} \sin \omega \alpha s + s\frac{p_y}{m},$$

$$(5.5.11)$$

$$z(s) = \frac{A}{\omega \alpha m} \cos \omega \alpha s + s\frac{p_z}{m}.$$

If $p_y = p_z = 0$, then the particle describes a circular orbit in the plane perpendicular to the wave, with its velocity in the direction of **B** and perpendicular to **E**:

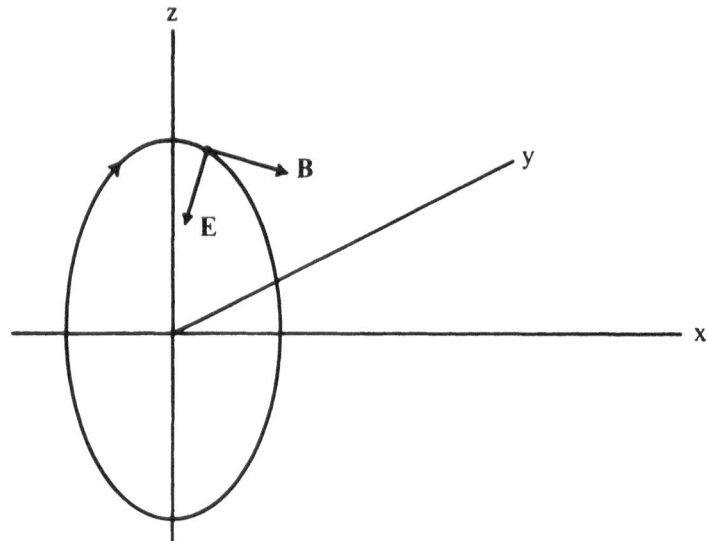

Figure 47   A circularly polarized wave.

2. Linearly polarized waves: $A_2 = 0$. If $p_y = p_z = y(0) = z(0) = t(0) = 0$, and $\alpha^2 = 1 + A^2/2m^2$, then the orbit is shaped like a bow tie:

$$\left\{ \begin{matrix} t(s) \\ x(s) \end{matrix} \right\} = \frac{s}{2\alpha} \left(1 \pm \alpha^2 + \frac{A^2}{2m^2} \right) + \frac{A^2}{8\alpha^2 m^2 \omega} \sin 2\,\omega\alpha s,$$

$$y(s) = \frac{A}{\omega\alpha m} \sin \omega\alpha s,$$

$$z(s) = 0.$$

(5.5.12)

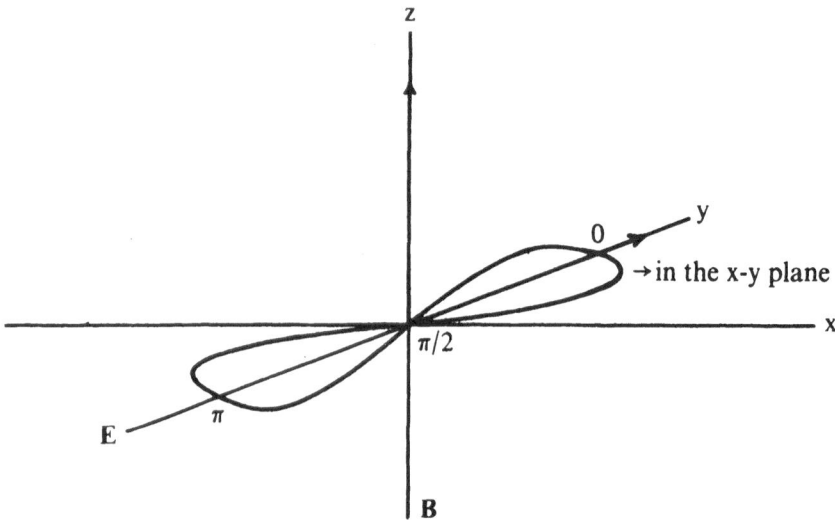

Figure 48   Motion in a linearly polarized wave.

Remark (5.5.13)

To understand this motion, recall the result of §5.2, where the trajectory was as shown in Figure 49 when **E** and **B** were related in the same way but were constant fields. For smaller y, this trajectory has a smaller velocity, and thus a smaller Larmor radius. In a plane wave, the fields start to change direction as soon as the particle goes through the origin. The radius of curvature decreases, and the particle returns to the origin, where it encounters fields of the opposite polarity and follows a mirror-image path.

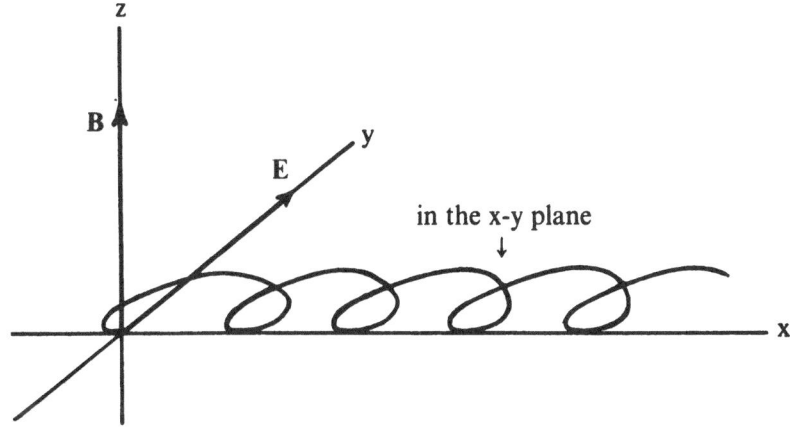

Figure 49   Motion in a constant field.

Problems (5.5.14)

1. Calculate the effect of the infinitesimal transformation generated by $K_z + L_y$, on A, **E**, and **B**. How do the equations in (5.5.1) change?

2. Calculate the Poisson brackets of the generators of the Poincaré group that leave F invariant, and of the invariance group of $\mathscr{H}$.

3. Show that (5.5.10) is a constant for $A = (0, 0, \cos(t - x), -\sin(t - x))$ and $\omega = 1$. Then convince yourself that the Poisson brackets of (5.5.10) with the other constants can all be written in terms of the other constants.

Solutions (5.5.15)

1. Let ε be the infinitesimal parameter. Then from (5.2.6) and (5.2.7) with the appropriate renormalization of the coordinates, we can read off that

$$A \to (A_0 + \varepsilon A_3, A_1 - \varepsilon A_3, A_2, A_3 + \varepsilon(A_0 + A_1)),$$
$$\mathbf{E} \to (E_1 - \varepsilon(E_3 + B_2), E_2 + \varepsilon B_1, E_3 + \varepsilon E_1),$$
$$\mathbf{B} \to (B_1 + \varepsilon(E_2 - B_3), B_2 - \varepsilon E_1, B_3 + \varepsilon B_1).$$

For (5.5.1) this means that $A \to A + \varepsilon \, d\Lambda$, were $\Lambda = \int_0^{t-x} du \, g(u)$, and **E** and **B** remain unchanged.

2. $\{K_z + L_y, p_z\} = p_0 + p_x = \{K_y - L_z, p_y\} = \{K_z + L_y + \Lambda_g, p_z\} = p_0 + p_x$
$\qquad = \{K_y - L_z + \Lambda_f, p_y\}.$

3. $\{(p_y + \cos(t - x))^2, 2L_x + p_1 - p_0\} = 4(p_y + \cos(t - x))(-p_z + \sin(t - x))$
$\qquad\qquad\qquad\qquad = -\{(p_z - \sin(t - x))^2, 2L_x + p_1 - p_0\}.$

5.6 Relativistic Motion in a Gravitational Field

In the nonrelativistic limit the equations are very similar to the electro-dynamic equations. On the other hand, in their exact form they have a simple geometrical interpretation.

In order to feel comfortable with the complicated-looking system of equations (1.1.6) and (1.1.7) and to see how it compares with its electromagnetic analogue, we start with

The Nonrelativistic Limit (5.6.1)

By this phrase we mean that $|d\mathbf{x}/dt| \ll 1$, but that terms of first order in $d\mathbf{x}/dt$ are to be kept. Moreover, the statement that the gravitational field is weak will mean that $g_{\alpha\beta}$ equals the $\eta_{\alpha\beta}$ of (5.1.2) plus a small quantity. With these approximations, $g_{\alpha\beta}^{-1}$ equals $\eta_{\alpha\beta}$ minus the same small quantity and s and t can be identified (see (5.6.6; 3)). From field theory we shall learn that a "small" mass M moving with velocity $\mathbf{j}$, $|\mathbf{j}| \ll 1$, at the origin produces a gravitational potential

$$g_{\alpha\beta} = \eta_{\alpha\beta} + \frac{4M\kappa}{r}(j_\alpha j_\beta + \tfrac{1}{2}\eta_{\alpha\beta}), \qquad j_\alpha = (-1, v_1, v_2, v_3) \qquad (5.6.2)$$

at the point $\mathbf{x}$, if $|\mathbf{x}| = r \gg M\kappa$. Substituting (5.1.2) into (1.1.6) and (1.1.7) gives the equations of motion

$$\frac{d^2\mathbf{x}}{dt^2} = -M\kappa\frac{\mathbf{x}}{r^3} - \frac{4M\kappa}{r^3}\left[\frac{d\mathbf{x}}{dt} \times \left[\mathbf{j} \times \mathbf{x}\right]\right]. \qquad (5.6.3)$$

Remarks (5.6.4)

1. The mass m of the particle moving in the gravitational field does not appear in equation (5.6.3); Galileo's discovery that particles of all masses respond identically to a gravitational field is a universal law of nature.
2. A velocity-dependent term of the same form as the Lorentz force is added to the Newtonian force. Both force terms have the opposite effect to that of their electrodynamic counterparts. Masses of the same sign attract, and mass-currents in the same direction repel. This has been put forth as a

confirmation of Mach's principle: if, say, a rotating cylinder encloses another body that rotates along with it, then the forces between the mass-currents act to oppose the centrifugal force in the interior of the cylinder (the H. Thirring effect).† If there were nothing else in the universe, then, according to Mach, there could be no centrifugal force when the angular velocities were equal, since the statement that the two bodies rotate would be meaningless.

3. We shall see in §5.7 how (5.6.3) is altered if $r < M\kappa$ and $|dx/dt| \sim 1$.

So that we can discuss (1.1.6) and (1.1.7) in the framework of our formalism, we next write down

The Lagrangian Form of the Equations of Motion (5.6.5)

The Lagrangian

$$L = \frac{m}{2} \dot{x}^\alpha \dot{x}^\beta g_{\alpha\beta}(x(s))$$

has equations (1.1.6) and (1.1.7) as its Euler–Lagrange equations ((2.3.24) with s in place of t, and $i = 0, 1, 2, 3$).

Proof

Problem 2.   ☐

Remarks (5.6.6)

1. The factor $m/2$ is, of course, unimportant, and is there only to reproduce (5.1.11) for $g_{\alpha\beta} = \eta_{\alpha\beta}$.
2. We are not able to choose the normalization of (5.1.1), (b), as we did in (5.1.3), since these quantities are not independent of s. Moreover, in the case at hand L itself is a constant; it is quadratic in $\dot{x}$ and in fact equals $\mathscr{H}$. Hence we shall normalize s by requiring $\dot{x}^\alpha \dot{x}^\beta g_{\alpha\beta} = -1$.
3. The argument made in (5.1.20; 2) is no longer valid, and it does not follow from the equations of motion that $|dx/dt| < 1$ (cf. (5.1.19; 3)). If $\mathbf{j} = \mathbf{0}$ in the gravitational potential (5.6.2), we get

$$\dot{t}^2\left(1 - \frac{2M\kappa}{r}\right) - |\dot{\mathbf{x}}|^2\left(1 + \frac{2M\kappa}{r}\right) = 1,$$

so

$$\left|\frac{d\mathbf{x}}{dt}\right|^2 < \left(1 - \frac{2M\kappa}{r}\right)\left(1 + \frac{2M\kappa}{r}\right)^{-1}.$$

---

† In the electrodynamic case, the centrifugal force on the inner current is counterbalanced when the outer current flows in the opposite direction, as in the betatron.

Thus there still exists a maximum velocity, which depends on x through the $g_{\alpha\beta}$, but in other situations might not be less than 1. This could cause some uneasiness, as it sounds as though a gravitational field could accelerate a particle to faster than the speed of light. But note that the maximum velocity is a universal bound for particles of all masses, and is likewise a maximum for photons. As will later be discussed, in this case x and t do not gauge the same lengths and times as one would measure with real yardsticks and clocks. As it would actually be measured, in units where the speed of light is 1, the maximum velocity is also 1.

Using (2.3.26), we can immediately pass from (5.6.5) to

The Hamiltonian Form of the Equations of Motion (5.6.7)

The Hamiltonian

$$\mathscr{H} = \frac{1}{2m} p_\alpha p_\beta g^{\alpha\beta}(x), \quad g^{\alpha\beta} g_{\beta\gamma} = \delta^\alpha_\gamma,$$

generates a locally canonical flow equivalent to (1.1.6) and (1.1.7).

Remarks (5.6.8)

1. The g's ought to be at least C^1: at the points where they are singular, either the chart must be changed, or else extended configuration space must be restricted. One may either have a global flow and be able to extend it over the whole extended phase space or not, depending on the global structure of the extended configuration space as a manifold.
2. The normalization of (5.6.6; 2) is equivalent to $\mathscr{H} = -m/2$.
3. The quantity p/m is only loosely connected with the real velocity:

$$\frac{p_\alpha}{m} = \dot{x}^\beta g_{\beta\alpha}, \qquad \dot{\mathbf{x}} = \frac{d\mathbf{x}}{dt} \cdot \frac{dt}{ds},$$

and $d\mathbf{x}/dt$ is again different from the velocity as measured with real yardsticks and clocks.

According to (2.4.14), $g_{\alpha\beta}$, a symmetric tensor of degree two, gives the extended configuration space a pseudo-Riemannian structure—where we assume that g invariably has one negative and three positive eigenvalues. The universality of gravitation gives a real, physical meaning to the spatial and temporal intervals defined formally with g; in chapter 6 we shall discuss in detail how gravitation influences actual yardsticks and clocks just so that the distances and times they measure are the same as the ones coming locally from g. Put more concretely, the distance between a point (x_0, x_1, x_2, x_3) and a point $(x_0, x_1 + dx_1, x_2, x_3)$ goes as $\sqrt{g_{11}}\, dx_1$ rather than as dx_1 as

$dx_1 \rightarrow 0$. In Equation (5.6.2) with $\mathbf{j} = 0$, $g_{00} = -1 + 2M\kappa/r$ and $g_{11} = 1 + 2M\kappa/r$, and so the times and distances measured with actual clocks at this point are $\sqrt{1 - 2M\kappa/r}\ \dot{dt}$ and $\sqrt{1 + 2M\kappa/r}\ dx_1$, making the limiting velocity (5.6.6; 3) again 1. However, as $r \rightarrow \infty$, dx_1 and dt approach the real length and time elements as measured out there, giving an external observer the impression that yardsticks must contract and clocks run slow if they are at small r in a gravitational potential. But there are no such things as ideal clocks and yardsticks that could directly measure dt and dx at small r, because gravity affects all objects equally. Hence it only makes sense to speak of the metric structure determined by g (not η). If $ds^2 = -dx^\alpha\, dx^\beta\, g_{\alpha\beta} > 0$, then the points x^α and $x^\alpha + dx^\alpha$ have a timelike separation and ds has the significance of a proper time—it is the interval measured by a clock that is itself moving from x^α to $x^\alpha + dx^\alpha$ in such a way that only dx^0 is nonzero in its rest frame. This is the operational meaning of the

Geodetic Form of the Equations of Motion (5.6.9)

Equations (1.1.6) and (1.1.7) are the Euler–Lagrange equations of the variational principle

$$W = \int ds \sqrt{-g_{\alpha\beta}\dot{x}^\alpha \dot{x}^\beta}, \qquad DW = 0.$$

Proof

Problem 3.   □

Remarks (5.6.10)

1. To be more precise, W is determined as follows: Let u and v be two points in the extended configuration space such that there is a trajectory $x(s_0) = u$, $x(s_1) = v$, with $g_{\alpha\beta}(x(s))\dot{x}^\alpha(s)\dot{x}^\beta(s) < 0$ $\forall s$ such that $s_0 \leq s \leq s_1$. For all trajectories satisfying these conditions, W is defined as the above integral, i.e., $s_1 - s_0$. The choice of s is immaterial: if $s \rightarrow \bar{s}(s)$, where $\bar{s}$ is monotonic and differentiable, then W is unchanged.
2. The previous comment shows that W is precisely the time interval that would be read off a clock that moved along $x(s)$.
3. The condition (5.6.9) that W is stationary actually requires it to be a maximum. To see this, consider the case $g_{\alpha\beta} = \eta_{\alpha\beta}$. Let $u = (0, 0, 0, 0)$, and choose the coordinate system so that $v = (t_1, 0, 0, 0)$. Then

$$W = \int_0^{t_1} dt \sqrt{1 - \left|\frac{d\mathbf{x}}{dt}\right|^2}.$$

Obviously, $0 < W \leq t_1$. The maximum is achieved by the trajectory $x(s) = (s, 0, 0, 0)$, which satisfies the Euler–Lagrange equation $\ddot{x} = 0$. The infimum 0 is not actually achieved, although it is approached arbitrarily nearly by particles moving almost as fast as light, whose proper time

$s = t\sqrt{1 - |d\mathbf{x}/dt|^2}$ runs very slowly. At any point it is possible to put g in the form η by choosing the right coordinates (see (5.6.11)), and so the trajectory that satisfies the Euler–Lagrange equations locally maximizes the proper time when the points u and v are sufficiently close together. This is not necessarily so if u and v are far apart (cf. (5.7.17; 1)).

4. In mathematical terminology, the trajectories are characterized as the timelike geodesics in extended configuration space, given a pseudo-Riemannian structure by g.

If the coordinate system is changed, $x \to \bar{x}$, then g transforms as a tensor of degree two,

$$g_{\alpha\beta} \to g_{\gamma\delta} \frac{\partial x^\gamma}{\partial \bar{x}^\alpha} \frac{\partial x^\delta}{\partial \bar{x}^\beta},$$

according to Definition (2.4.19). Hence if $g_{\alpha\beta} \neq \eta_{\alpha\beta}$, then it is possible that $g_{\alpha\beta}$ is merely $\eta_{\alpha\beta}$ on some different chart, and not a true gravitational field. Then the Γ's in the equations of motion (1.1.6) are merely fictitious forces, like the ones encountered in an accelerating reference frame. In electrodynamics we met with gauge potentials $A_i = \Lambda_{,i}$, which do not produce any fields. Now we see that there are g's that produce the equations of free motion, written in different coordinates; in the volume on field theory we shall learn of criteria for when this happens. Of course, when we talk about such a possibility we are considering g throughout the whole manifold, for at any individual point there is always the

Principle of Equivalence (5.6.11)

For all $x \in M_e$ there are coordinates, the **Riemann normal coordinates**, such that $g_{\alpha\beta}(x) = \eta_{\alpha\beta}$ and $\Gamma^\alpha_{\beta\gamma}(x) = 0$, where Γ is defined as in (1.1.7).

Proof

Problem 4.   □

Remarks (5.6.12)

1. Fictitious forces counterbalance gravity in this coordinate system. This is what happens in Einstein's famous free-falling elevator. A passenger does not detect any gravity, because it affects all bodies in the elevator equally, irrespective of their masses or any other such properties.
2. In (3.2.15; 1) we eliminated a constant gravitational field by transforming to an accelerating system. The term in (5.6.3) that resembles the Lorentz force can be counterbalanced by the Coriolis force in some rotating system (3.2.15; 2).
3. Γ does not generally also vanish at nearby points, the principle of equivalence holds only in the infinitely small limit. A gradient in the gravitational field, a tidal force, would be detectable.

Problems (5.6.13)

1. Use (1.1.6), (1.1.7), and (5.6.2) to derive (5.6.3).

2. Calculate the Euler–Lagrange equations of L in (5.6.5).

3. Show that (1.1.6) and (1.1.7) satisfy the Euler–Lagrange equations of (5.6.5).

4. Prove (5.6.11).

Solutions (5.6.14)

1. Since the derivatives of g are of first order in κ, we can set $(g^{-1})_{\alpha\beta} = \eta_{\alpha\beta}$ to that order, and since α is a spatial index in (5.6.3),

$$\Gamma^{\alpha}_{\beta\gamma} = \tfrac{1}{2}(g_{\alpha\beta,\gamma} + g_{\alpha\gamma,\beta} - g_{\beta\gamma,\alpha}) = -\frac{2M\kappa}{r^3}\{(j_{\alpha}j_{\beta} + \tfrac{1}{2}\eta_{\alpha\beta})x^{\gamma}$$

$$+ (j_{\alpha}j_{\gamma} + \tfrac{1}{2}\eta_{\alpha\gamma})x^{\beta} - (j_{\beta}j_{\gamma} + \tfrac{1}{2}\eta_{\beta\gamma})x^{\alpha}\}, \qquad (x^0 = 0).$$

In the nonrelativistic limit, we get

$$\frac{d^2x^{\alpha}}{dt^2} = -\Gamma^{\alpha}_{00} - 2\Gamma^{\alpha}_{0\beta}\frac{dx^{\beta}}{dt} = \frac{M\kappa}{r^3}\left\{-x^{\alpha} - 4\frac{dx^{\beta}}{dt}(j_{\alpha}x_{\beta} - j_{\beta}x_{\alpha})\right\}, \qquad \alpha = 1, 2, 3,$$

which is the same as (5.6.3).

2. $$\frac{d}{ds}\frac{\partial L}{\partial \dot{x}^{\alpha}} = \frac{d}{ds}(m\dot{x}^{\beta}g_{\alpha\beta}(x)) = m\{\ddot{x}^{\beta}g_{\alpha\beta} + \dot{x}^{\beta}\dot{x}^{\gamma}\tfrac{1}{2}(g_{\alpha\beta,\gamma} + g_{\alpha\gamma,\beta})\} = \frac{\partial L}{\partial x^{\alpha}} = \frac{m}{2}\dot{x}^{\beta}\dot{x}^{\gamma}g_{\beta\gamma,\alpha}.$$

3. In general, a function $f(L)$ gives the Euler–Lagrange equations.

$$\frac{d}{ds}\left(\frac{\partial f}{\partial L}\frac{\partial L}{\partial \dot{x}^{\alpha}}\right) = \frac{\partial f}{\partial L}\frac{\partial L}{\partial x^{\alpha}}.$$

But by (5.6.8; 2) L, and hence also $\partial f/\partial L$, are independent of s for the solutions of the Euler–Lagrange equations of (5.6.5).

4. Let us first choose the point in question, x, as the origin of the coordinate system. The symmetric matrix $g_{\alpha\beta}(0)$ is orthogonally diagonable, and its eigenvalues can be re-normalized to ± 1 by scaling. From the assumptions on the g's we allow we thus find $g_{\alpha\beta}(0) = \eta_{\alpha\beta}$. The claim will be proved if we can show that $g_{\alpha\beta}(x) = \eta_{\alpha\beta} + O(x^2)$. If we expand g,

$$g_{\alpha\beta}(x) = \eta_{\alpha\beta} + A_{\alpha\beta\gamma}x^{\gamma} + O(x^2);$$

then a change of charts

$$x^{\gamma} = \bar{x}^{\gamma} + \tfrac{1}{2}c^{\gamma}_{\alpha\beta}\bar{x}^{\alpha}\bar{x}^{\beta}$$

produces a transformed matrix

$$g_{\alpha\beta} = (\eta_{\gamma\delta} + A_{\gamma\delta\rho}x^{\rho})(\delta^{\gamma}_{\alpha} + c^{\gamma}_{\alpha\sigma}\bar{x}^{\sigma})(\delta^{\delta}_{\beta} + c^{\delta}_{\beta\sigma}\bar{x}^{\sigma}) + O(x^2)$$

$$= \eta_{\alpha\beta} + x^{\sigma}(A_{\alpha\beta\sigma} + c_{\beta\alpha\sigma} + c_{\alpha\beta\sigma}) + O(x^2).$$

If $c_{\alpha\beta\sigma} \equiv \eta_{\alpha\tau}c^{\tau}_{\beta\sigma} = -\tfrac{1}{2}A_{\alpha\beta\sigma}$, then the linear term goes away because $A_{\alpha\beta\sigma} = A_{\beta\alpha\sigma}$. Note that the accelerating system used in (3.2.15) to get rid of gravity nonrelativistically is a special case of the above quadratic transformation.

5.7 Motion in the Schwarzschild Field

Relativistic motion in the gravitational field of a point mass is the same as in the analogous electromagnetic field as regards the structure of the invariance group. However, what goes on at small r is physically quite remarkable.

Soon after Einstein published his field equations, an exact solution was discovered, describing the field of a point mass.

This simple situation exhibits the essential peculiarities of the relativistic theory of gravitation, because of which it is highly significant both physically and astronomically. It is generally known as

The Schwarzschild Solution (5.7.1)

The fields $g_{\alpha\beta}$ created by a mass M at the origin provide extended phase space with the pseudometric

$$g = dx^\alpha \, dx^\beta \, g_{\alpha\beta} = \left(1 - \frac{r_0}{r}\right)^{-1} dr^2 + r^2(d\vartheta^2 + \sin^2 \vartheta \, d\varphi^2)$$

$$- \left(1 - \frac{r_0}{r}\right) dt^2,$$

where

$$r_0 \equiv 2M\kappa,$$

in spatial polar coordinates.

Remarks (5.7.2)

1. At the radius r_0 the gravitational energy of a mass is of the same order as its rest energy: $Mm\kappa/r_0 \sim mc^2$ (units with $c = 1$ were used in (5.7.1)). According to (1.1.3), $2m_p\kappa/c^2 \sim 10^{-52}$ in cgs. units for a proton, and the earth contains about 10^{51} protons, making r_0 on the order of millimeters for the earth. The sun is a million times heavier, and so its $r_0 \sim$ km.
2. If $r \gg r_0$, then $g_{\alpha\beta} \to \eta_{\alpha\beta}$, and the coordinates are the intervals one would actually measure (cf. Problem 4).
3. If $r = r_0$, then $g_{00} = 0$. This does not necessarily mean that anything special happens at such a point. For example, if $r = 0$, then in polar coordinates $g_{\vartheta\vartheta} = g_{\varphi\varphi} = 0$ as well, but all that has happened is that the chart has become unsuitable at that point.
4. A singularity in g_{rr} at $r = r_0$ seems more serious than the one just cited, but it need not be so. For instance, if the coordinate x is used on the circle $x^2 + y^2 = 1$, then the line element $ds^2 = dx^2 + dy^2 = dx^2/(1 - x^2)$ is singular at $x = \pm 1$. Yet these points are as good as any other points on the circle, and it is only a question of the chart failing there.

5. If instead of (t, r), the coordinates

$$u = \sqrt{\frac{r}{r_0} - 1} \exp\left(\frac{r}{2r_0}\right) \cosh\left(\frac{t}{2r_0}\right),$$

$$v = \sqrt{\frac{r}{r_0} - 1} \exp\left(\frac{t}{2r_0}\right) \sinh\left(\frac{t}{2r_0}\right),$$

are introduced, then the metric becomes

$$g = \frac{4r_0^3}{r} e^{-r/r_0}(du^2 - dv^2) + r^2(d\vartheta^2 + \sin^2 \vartheta \, d\varphi^2)$$

(Problem 3), and the singularity at $r = r_0$ magically disappears. The region $\{r > r_0, -\infty < t < \infty\}$ in the old chart is mapped to $I = \{|v| < u\}$:

This gives us access to new territory, and the solution can be extended to $r = 0$. By inverting the transformation,

$$t = 2r_0 \text{ Arctan} \left(\frac{v}{u}\right), \qquad \left(\frac{r}{r_0} - 1\right) \exp\left(\frac{r}{r_0}\right) = u^2 - v^2,$$

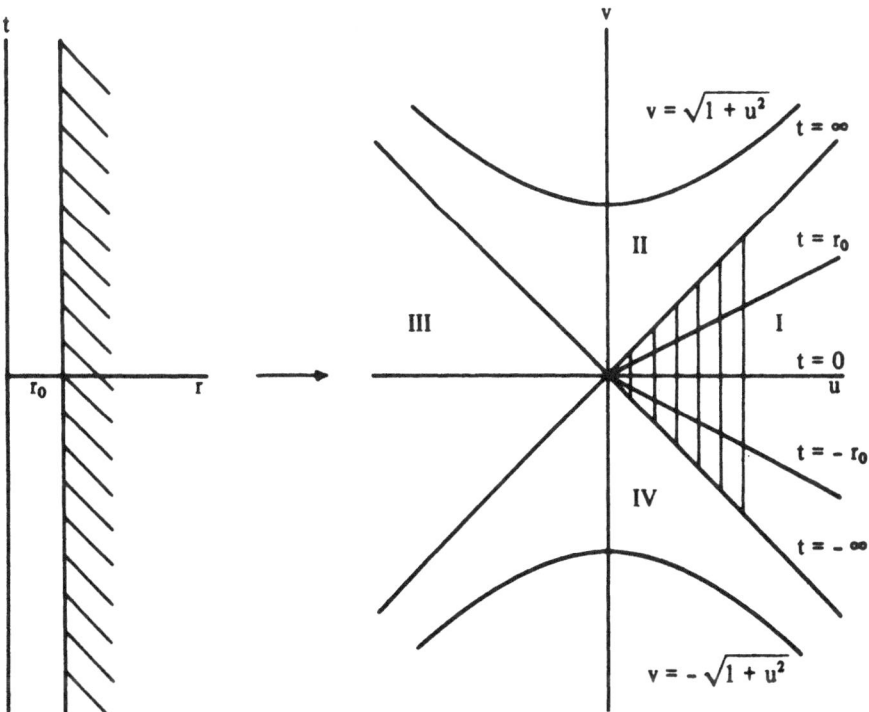

Figure 50   Extending the Schwarzschild solution.

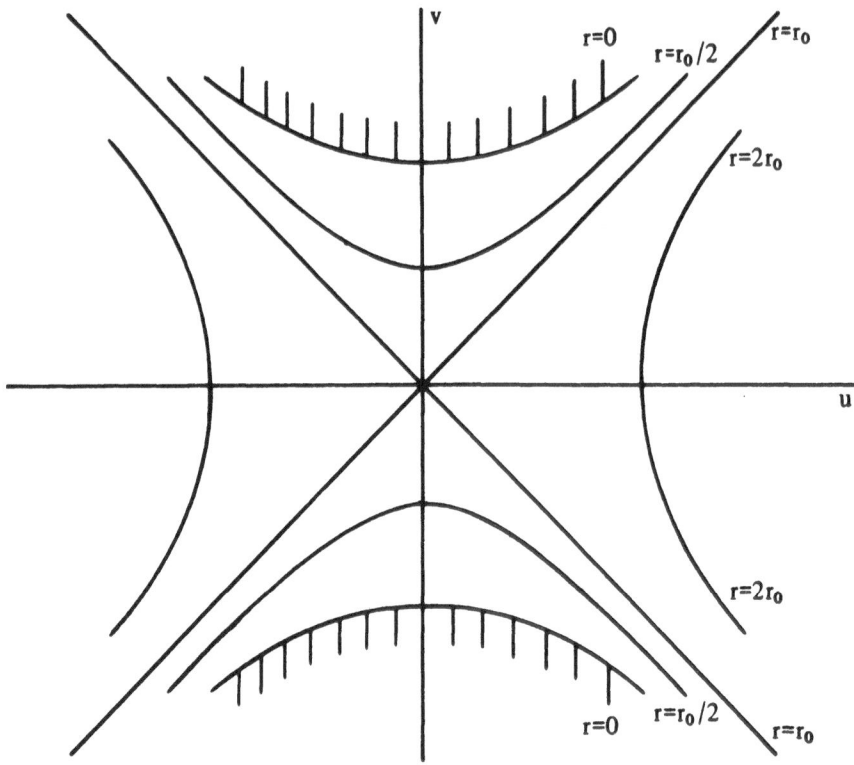

Regions where the complete Schwarzschild solution is valid.

we see that the territory gained is the region where $u^2 - v^2 > -1$. We shall later return to the physical significance of the new territory we have opened up.

In order to make the comparison with §§4.2 and 5.3 easier, we next solve the equations of motion in coordinates $(t, r, \vartheta, \varphi)$. The coordinate t can only be used in region I of Figure 50. But r can also be used in region II until $r = 0$ ($v = \sqrt{u^2 + 1}$). Accordingly, we turn our attention to the determination of $r(s)$, $\vartheta(s)$, and $\varphi(s)$. Substitution of (5.7.1) into (5.6.7) produces

The Hamiltonian (5.7.3)

$$\mathcal{H} = \frac{1}{2m}\left(|\mathbf{p}|^2 - \frac{r_0}{r} p_r^2 - \left(1 - \frac{r_0}{r}\right)^{-1} p_0^2\right).$$

Since the only coordinate used in $\mathcal{H}$ other than the momenta is r,

The Constants of the Motion (5.7.4)

(in s) are

$$\mathbf{L} = [\mathbf{x} \times \mathbf{p}], \qquad p_0, \quad \text{and} \quad \mathcal{H}.$$

These are just the same as in the electrical problem of §5.3. The construction of action and angle variables requires only a minor modification: In polar coordinates,

$$\mathcal{H} = \frac{1}{2m}\left(p_r^2\left(1 - \frac{r_0}{r}\right) + \frac{L^2}{r^2} - \frac{p_0^2}{1 - r_0/r}\right) = -\frac{m}{2}. \qquad (5.7.5)$$

Hence (cf. (5.6.8; 3))

$$p_r = \frac{m\dot{r}}{1 - r_0/r}.$$

Thus we can write

$$\frac{m}{2}\dot{r}^2 - \frac{mr_0}{2r} + \frac{L^2}{2mr^2} - \frac{L^2 r_0}{2mr^3} = \frac{m}{2}\left(\frac{p_0^2}{m^2} - 1\right) \equiv E = \text{constant.} \quad (5.7.6)$$

This is the equation of energy conservation in a one-dimensional system with an

Effective Potential (5.7.7)

$$V_{\text{eff}}(r) = -\frac{mM\kappa}{r} + \frac{L^2}{2mr^2} - \frac{L^2 r_0}{2mr^3}.$$

Remarks (5.7.8)

1. The first two terms are Newtonian and centrifugal potentials, as in the nonrelativistic theory (4.2.12).
2. The additional attractive term goes as r^{-3} (r^{-2} in (5.3.8)), so it eventually dominates the centrifugal term for sufficiently small r.
3. There are repulsive contributions to the effective potential for $dr/dt = \dot{r}(1 - r/r_0)$, and r_0 is only reached at $t = \infty$. Thus the vector field generated by $(1 - r_0/r)^{-1}$ on $r_0 < r$ is complete.
4. Using the dimensionless quantities $u \equiv r_0/r$ and $\ell \equiv L/mr_0$,

$$\frac{2}{m}V_{\text{eff}} = -u + \ell^2(u^2 - u^3).$$

If $\ell < \sqrt{3}$, then this is a monotonic function; if $\ell = \sqrt{3}$, a turning point appears; and if $\ell > \sqrt{3}$, then there are a maximum and minimum at

$$u_{\pm} = \frac{1}{3}\left[1 \pm \sqrt{1 - \frac{3}{\ell^2}}\right]$$

$$\frac{2}{m}V_{\text{eff}}(u_+) = \left(\frac{2\ell^2}{27} - \frac{1}{3}\right)\left(1 + \sqrt{1 - \frac{3}{\ell^2}}\right) + \frac{1}{9}\sqrt{1 - \frac{3}{\ell^2}}:$$

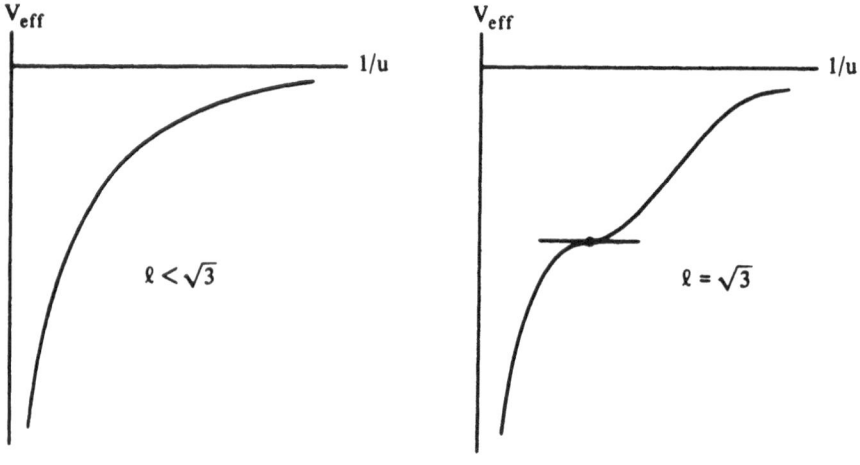

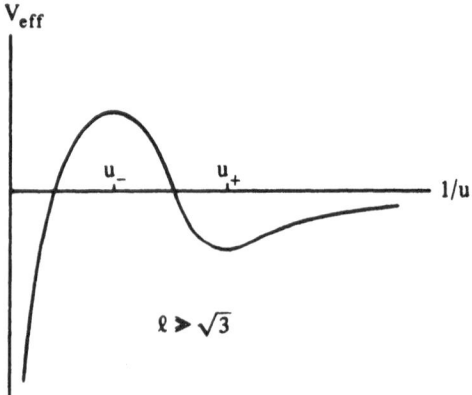

Figure 51   The effective potential for the Schwarzschild field.

5. For sufficiently large ℓ (i.e., impact parameter $\times$ speed$/c \gg r_0$), the centrifugal barrier is large enough to keep the particle from falling into the black hole. If $E < V_{\text{eff}}(u_+)$, it can no longer do so.

As in the electrical Kepler problem of §5.3, the elliptic orbits fill up a two-dimensional region in the plane of motion, which is perpendicular to L. To calculate the difference in the angular coordinate φ between successive maxima of r, it is most convenient to start with

$$\frac{dr}{d\varphi} = \frac{\dot{r}}{\dot{\varphi}} = \frac{mr^2}{L} \sqrt{\frac{p_0^2}{m^2} - 1 + \frac{r_0}{r} - \frac{L^2}{m^2 r^2}\left(1 - \frac{r_0}{r}\right)}, \qquad (5.7.9)$$

from which we obtain an elliptic integral for the

Precession Angle (5.7.10)

$$\Delta\varphi \equiv \oint \frac{dr\, L}{mr^2 \sqrt{\dfrac{p_0^2}{m^2} - 1 + \dfrac{r_0}{r} - \dfrac{L^2}{m^2 r^2}\left(1 - \dfrac{r_0}{r}\right)}} - 2\pi.$$

Expanding in the $1/r^3$ term to first order gives

$$\Delta\varphi \cong \frac{3\pi}{2}\left(\frac{r_0 m}{L}\right)^2.$$

(Problem 5).

Remarks (5.7.11)

1. The radius of a nearly circular orbit is $R = 2L^2/r_0 m^2$, and if $R \gg r_0$, then $\Delta\varphi = 3\pi r_0/R$.
2. The α of (5.3.12; 2) corresponds to $mr_0/2$, making $\Delta\varphi$ six times as large as in the electrical problem. Again, $\Delta\varphi$ is caused by an increase in the effective mass at small r. An explanation for the increase in $\Delta\varphi$ in this case can also be made in the context of Mach's principle, according to which inertia is due to nearby masses, which increase the effective mass of a particle.
3. Since $r_0/R \sim 1$ km./10^8 km. for the motion of Earth around the sun, the precession is a tiny effect of a few seconds of arc per century, and much smaller than other perturbations of the orbit. However, the effect seems to be confirmed for the inner planets, after making every imaginable correction, to within one percent accuracy. The predictions of Einstein's theory have also been confirmed using lunar radar echoes, with such good accuracy that its validity can no longer be doubted.
4. Recently a double star was discovered, with an orbital period of a few hours and $R \sim 10^5$ km. For this orbit, the precession should be on the order of degrees per year.

Unbound Trajectories (5.7.12)

If $0 < E < V_{\text{eff}}(u_+)$, then the trajectories do not fall into the origin, but escape to infinity. As for the scattering theory of such trajectories, it can be shown that the statements of (5.3.13) again hold. All that will be done here is to calculate the high-energy behavior of the scattering angle Θ, as the difference from the nonrelativistic theory of §4.2 and the electrical case of §5.3 is important. In the situations already discussed Θ approaches 0 as the energy increases while the impact parameter is held constant (5.3.15; 3). In this case, (5.7.9) means that the angle at $r = \infty$, measured from the minimum radius $r_{\min}$, taken as $\varphi = 0$, is

$$\varphi = \int_0^{1/r_{\min}} \frac{du}{\sqrt{\dfrac{p_0^2 - m^2}{L^2} + \dfrac{r_0 u m^2}{L^2} - u^2(1 - r_0 u)}}. \tag{5.7.13}$$

In the limit as $p_0 = \gamma m \to \infty$, and $L = \gamma m v b \to \infty$, with $p_0/L \to 1/b$, the Newtonian term r_0/r becomes negligible, but the correction to the centrifugal term remains significant, and

$$\varphi \to \int_0^{1/r_{\min}} \frac{b\,du}{\sqrt{1 - b^2 u^2 (1 - r_0 u)}}.$$

To calculate the scattering angle when $r_0/b \ll 1$, we introduce $\sigma \equiv bu\sqrt{1 - r_0 u}$, expand the integrand in r_0/b, set $bu \cong (1 + r_0\sigma/2b)$, and integrate to the point $\sigma = 1$, which corresponds to $u = 1/r_{\min}$:

$$\varphi = \int_0^1 \frac{d\sigma\left(1 + \dfrac{r_0}{b}\sigma\right)}{\sqrt{1 - \sigma^2}} = \frac{\pi}{2} + \frac{r_0}{b}.$$

As shown in Figure 52,

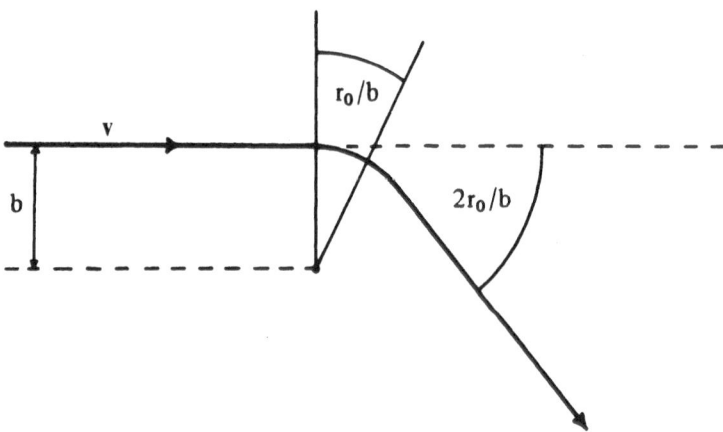

Figure 52   The impact parameter and the scattering angle.

The Scattering Angle for $p_0 \gg m$ and $b \gg r_0$ (5.7.14)

is

$$\Theta = -\frac{2r_0}{b}.$$

Remarks (5.7.15)

1. The negative sign means that gravity is attractive.
2. Light with a wavelength much less that r_0 behaves like a particle with $\gamma \to \infty$ in the Schwarzschild field. Formula (5.7.14) has been verified with good accuracy for the deflection of light when it passes near the sun.

As we have seen, a large enough angular momentum (and hence impact parameter) can keep a particle from falling into the black hole. But radial trajectories are not restricted to the region $r > r_0$, so we need to change to the variables u and v of Figure 50 to discuss them. If a radial line is timelike, then $|dv| > |du|$, and its slope in the diagram is necessarily steeper than $45°$. No creature that was once inside r_0 could ever contrive to get to the other side of a pulse of light emitted from one of these lines. Bearing these facts in mind, let us imagine an

Expedition to $r < r_0$ (5.7.16)

1. In region II, every trajectory, whether subject to gravity alone or in combination with an electromagnetic force, reaches $r = 0$. The electromagnetic forces cannot prevent the trajectories from being timelike lines. Thus ∂r is a timelike direction, and the fall into the center is as inevitable as aging is for us in region I. It is no more possible to stay at a fixed r inside r_0 than it is to make time stand still outside.
2. The line dividing regions I and II, $v = u > 0$, is the same as the curve $t = \infty$; this only means that signals sent by someone falling into the center appear to take an infinitely long time to an observer at $r > r_0$, which corresponds to the fact mentioned in §5.6, that clocks appear to run slow in gravitational fields. As measured by proper time, the fall is of short duration: $\sim r_0 \sim 10^{-5}$ seconds for stars. It is straightforward to figure out that light signals decrease in intensity as $\exp(-t/r_0)$, and so for practical purposes they die out immediately.
3. On the other side of r_0 there is another world symmetric to ours, region III. There is no way to know anything about it, as no trajectory can go from III to I. At best, if someone ventured into region II, he might learn about III just before it was all over.
4. Finally, there is region IV, symmetric in the time coordinate to II. The inhabitants of this region have a choice of emigrating from $r < r_0$ to either region I or III.

This example shows how puzzling other kinds of manifolds can be to someone used to thinking of $\mathbb{R}^n$. In chapter 6 we shall complete the re-education needed to understand these things.

Problems (5.7.17)

1. Show that the proper time required for a closed Kepler orbit to return to the initial position is not the maximum possible proper time. Compare with the elapsed proper time of an observer who remains fixed at the initial point.

2. For what $\bar{r}(r)$ does the speed of light in the radial direction equal 1 — and thus the metric can be written as $F(\bar{r})(d\bar{r}^2 - dt^2) + G(\bar{r}) \, d\Omega^2$?

3. With the coordinates

$$u = h(\bar{r} + t) + g(\bar{r} - t) \quad \text{and} \quad v = h(\bar{r} + t) - g(\bar{r} - t),$$

the metric of Problem 2 becomes

$$f^2(u, v)(du^2 - dv^2) + r^2(u, v)d\Omega^2, \quad \text{where} \quad f^2 = \frac{1 - r_0/r}{4h'g'}.$$

Find h and g so that the singularity at $r = r_0$ goes away, making sure that the answer is consistent with (5.7.2; 5) in the old variables.

4. Find coordinates $\bar{r}(r)$ such that the metric has the form

$$f(\bar{r})(d\bar{r}^2 + \bar{r}^2 \, d\Omega^2) - g(\bar{r})dt^2.$$

Verify that as $\bar{r} \to \infty$, $g(\bar{r}) \to 1 - r_0/\bar{r}$ and $f \to 1 + r_0/\bar{r}$, in agreement with (5.6.2).

5. Expand (5.7.10) to first order in $L^2 r_0/m^2 r^3$, and calculate the integral using complex integration as in (5.3.16; 1).

Solutions (5.7.18)

1. In the variables of (5.7.8; 3) the proper time s and the time t are related by

$$ds = dt \sqrt{\frac{1 - r_0/r}{1 + r^2\dot{\varphi}^2}} = dt \sqrt{\frac{1 - u}{1 + \ell^2 u^2}}$$

for circular orbits ($dr = 0$); whereas for the proper time s_0 of a stationary observer ($dr = d\varphi = 0$), $ds_0 = dt\sqrt{1 - u}$. If the fixed observer and an orbiting one meet again at the time t, then their proper times are in the ratio

$$\frac{s_0}{s} = \sqrt{1 + \ell^2 u^2} > 1.$$

2. $\bar{r} = \int dr/(1 - r_0/r) = r + r_0 \ln((r/r_0) - 1)$. The metric becomes

$$(1 - r_0/r)(d\bar{r}^2 - dt^2) + r^2 \, d\Omega^2.$$

3. Let $\{^h_g\} = \frac{1}{2} \exp((\bar{r} \pm t)/2r_0)$. Then

$$f^2 = (1 - r_0/r)4r_0^2 \exp(-(r + r_0 \ln((r/r_0) - 1))/r_0) = 4r_0^3 \exp(-r/r_0)/r.$$

4. The form of the metric requires that $r^2 = (dr/d\bar{r})^2(1 - r_0/r)$, which can be integrated to $r = \bar{r}(1 + (r_0/4\bar{r}))^2$. The metric then becomes

$$(1 + (r_0/4\bar{r}))^4(d\bar{r}^2 + \bar{r}^2 \, d\Omega^2) - (1 + (r_0/4\bar{r}))^{-2}(1 - (r_0/4\bar{r}))^2 \, dt^2.$$

5. It is only necessary to calculate the residue at $r = 0$:

$$\oint \frac{dr}{r\sqrt{-Ar^2 + Br - C + D/r}} = \oint \frac{dr}{r\sqrt{-Ar^2 + Br - C}}$$

$$-\frac{1}{2} \oint \frac{dr \, D}{r^2(-Ar^2 + Br - C)^{3/2}} + O(D^2) = \frac{2\pi}{\sqrt{C}} + 2\pi \frac{3}{4} \frac{DB}{C^{5/2}} + O(D^2).$$

If $A = 1 - p_0^2/m^2$, $B = r_0$, $C = L^2/m^2$, and $D = r_0 L^2/m^2$, this gives (5.7.10).

5.8 Motion in a Gravitational Plane Wave

The invariance group and methods of solution are the same as in the electromagnetic problem, but there are also some new aspects to consider.

Every reasonable gravitational field theory contains a counterpart to electromagnetic waves. An experiment detecting gravity waves would be one of the basic foundations of any theory of gravitation, but as yet they have not been convincingly detected. In order to observe the effects of these waves, we must first study how particles would behave in them. This chapter closes with a short discussion of the problem and a comparison with the electromagnetic counterpart of §5.5.

Shortly after Einstein wrote down the equations for $g_{\alpha\beta}$, approximate solutions were found exhibiting wave properties. But because of the non-linearity of the problem it was not obvious that these approximations closely resembled real solutions. Right up to the present day, no-one has succeeded in pushing through the details showing how a source produces gravity waves. Nonetheless, exact plane-wave solutions to the equations can be found fairly easily. They are as follows:

The Field and the Hamiltonian (5.8.1)

The gravitational potential

$$g_{\alpha\beta} = \begin{vmatrix} -1 & & & \\ & 1 & & \\ & & \dfrac{1}{f(t-x)} & \\ & & & \dfrac{1}{g(t-x)} \end{vmatrix}$$

satisfies Einstein's equations with no matter, if $f^{1/2}(f^{-1/2})'' + g^{1/2}(g^{-1/2})'' = 0$, and by (5.6.7) it makes the Hamiltonian

$$\mathscr{H} = \frac{1}{2m}[-p_0^2 + p_x^2 + p_y^2 f(t-x) + p_z^2 g(t-x)].$$

Remarks (5.8.2)

1. There exist fictitious fields of the form (5.8.1), which are only the fields of free motion in an accelerating coordinate system (5.1.11). In the second volume, "Classical Field Theory," we shall study Einstein's equations, and when we derive the above solutions we shall see that iff $(f^{-1/2})'' = (g^{-1/2})'' = 0$, then the field is necessarily fictitious (Problem 1).

2. Although f and g were not required to be continuous in §5.5, here we assume that $f, g \in C^2$ so that the above condition makes sense.

If we compare $\mathscr{H}$ with (5.5.1), we see that there is the same invariance under displacements as before, which ensures that (5.8.1) is integrable. If we slightly modify the two new quantities introduced in (5.5.3), we obtain the same number of

Constants of the Motion (5.8.3)

$$p_y, \qquad p_z, \qquad p_x + p_0,$$

$$p_z \int_0^{t-x} du \, g(u) + z(p_0 + p_x),$$

and

$$p_y \int_0^{t-x} du \, f(u) + y(p_0 + p_x)$$

are constant.

Proof

Follows immediately from

$$\dot{z} = p_z g/m,$$
$$\dot{y} = p_y f/m,$$

and

$$\dot{t} - \dot{x} = -(p_0 + p_x)/m. \qquad \square$$

Remarks (5.8.4)

1. The invariance group generated by the five constants is isomorphic to the electromagnetic invariance group (5.5.3) (Problem 4).
2. Just as before, one more constant would be needed to completely determine the trajectory. The remaining integration of the equations of motion is accomplished exactly as in (5.5.5; 3). We first deduce that

$$t(s) - x(s) = \alpha s,$$

and

$$t(s) + x(s) = t(0) + x(0) + \alpha^{-2} \int_0^{\alpha s} du \left[1 + \frac{f(u)p_y^2 + g(u)p_z^2}{m^2} \right],$$

where we have again taken $u(0) = 0$; and from this we arrive at

The Explicit Solution for the Coordinates as Functions of the Proper Time (5.8.5)

$$t(s) = t(0) + \frac{s}{2}\left(\frac{1}{\alpha} + \alpha\right) + \frac{1}{2m^2\alpha^2}\left[p_y^2 F(\alpha s) + p_z^2 G(\alpha s)\right],$$

$$x(s) = x(0) + \frac{s}{2}\left(\frac{1}{\alpha} - \alpha\right) + \frac{1}{2m^2\alpha^2}\left[p_y^2 F(\alpha s) + p_z^2 G(\alpha s)\right],$$

$$y(s) = y(0) + \frac{p_y}{m\alpha}F(s\alpha),$$

$$z(s) = z(0) + \frac{p_z}{m\alpha}G(s\alpha),$$

$$F(u) = \int_0^u du'\, f(u'), \qquad G(u) = \int_0^u du'\, g(u'),$$

$$\alpha = -\frac{p_0 + p_x}{m}.$$

Remark (5.8.6)

In particular, the solution with $\alpha = 1$ and $p_y = p_z = 0$ is $(t(s), x(s), y(s), z(s))$ $= (s + t(0), x(0), y(0), z(0))$. This might lead one to think that gravity waves, unlike light waves, have no effect on particles that are initially at rest, and thus are impossible to detect. But remember that the coordinates (t, x, y, z) are simply not the same as actually measured intervals; on the contrary, they are precisely the coordinates adjusted for the trajectories of initially stationary particles. The displacement of particles in the y (or z) direction is actually $dy\, f^{-1/2}$ (resp. $dz\, g^{-1/2}$), which varies with $t - x$ for fixed dy and dz. Thus a gravity wave causes accelerations perpendicular to its direction of propagation.

Example (5.8.7)

The condition of (5.8.1), that $(f^{-1/2})''f^{1/2} + (g^{-1/2})''g^{1/2} = 0$, is fulfilled, for example, by $f^{-1/2}(u) = \cos ku$ and $g^{-1/2}(u) = \cosh ku$. In order to construct a pulse with a length τ, we can piece it together from solutions of $(f^{-1/2})'' = (g^{-1/2})'' = 0$, such that f and g are twice differentiable:

	$f^{-1/2}(u)$	$g^{-1/2}(u)$
$u < 0$	1	1
$0 < u < \tau$	$\cos ku$	$\cosh(ku)$
$\tau < u$	$\cos k\tau + k(\tau - u)\sin k\tau$	$\cosh(k\tau) - k(\tau - u)\sinh(k\tau)$

The "δ pulse" would be a limiting case, as $\tau \to 0$ and $k = 1/\sqrt{\tau} \to \infty$, which means that:

	$f(u)$	$g(u)$	$F(u)$	$G(u)$
$u < 0$	1	1	u	u
$u > 0$	$\dfrac{1}{(1-u)^2}$	$\dfrac{1}{(1+u)^2}$	$\dfrac{u}{1-u}$	$\dfrac{u}{1+u}$

Remarks (5.8.8)

1. For $u \equiv t - x > 0$, the metric is

$$g = dx^2 + (1-u)^2\, dy^2 + (1+u)^2\, dz^2 - dt^2,$$

showing that the wave causes stationary bodies to be compressed in the y-direction and stretched in the z-direction; i.e., it is a quadrupole field.

2. Since g_{yy} is zero when $u = 1$, the chart (t, x, y, z) can only be used when $t - x < 1$. However, the singularity at $t - x = 1$ is only an apparent one, as by (5.8.2; 1) the g of Remark 1 is simply η written in another coordinate system. Specifically, if

$$\begin{aligned} T &= t - (1-u)y^2/2 + (1+u)z^2/2, \\ X &= x - (1-u)y^2/2 + (1+u)z^2/2, \\ Y &= (1-u)y, \\ Z &= (1+u)z, \end{aligned} \qquad u = t - x = T - X,$$

then

$$g = dX^2 + dY^2 + dZ^2 - dT^2$$

(Problem 1).

3. The curves $(x, y, z) = $ constant are the trajectories of particles initially at rest. These particles get focused in the y-direction when $u = 1$ so that the y-coordinate of their separation is zero. It sounds as if particles must be able to move faster than light, if they can be arbitrarily far apart at $u = 0$ and become focused like this, but if we look at what happens in the chart (T, X, Y, Z), which reproduces the actually measured distances, times, and velocities when $u > 0$, then in the new coordinates the trajectory $(t, x, y, z) = (s, 0, m, 0)$ becomes

$$X = -\frac{m^2/2}{1 + m^2/2}(1 - T), \qquad Y = \frac{m}{1 + m^2/2}(1 - T), \qquad Z = 0.$$

While it is true that a particle reaches the origin $\mathbf{x} = \mathbf{0}$ at time $T = 1$ for all m, it never moves faster than light:

$$\left(\frac{dX}{dT}\right)^2 + \left(\frac{dY}{dT}\right)^2 = \frac{m^2 + m^4/4}{(1 + m^2/2)^2} < 1.$$

Since the gravitational disturbance has reached a given trajectory at the time $T = -m^2/2$, particles that have started off farther away have had more time to approach the origin.

4. Remark 3 only appears to single out the x-axis to focus all particles onto. Actually, they are focused onto all trajectories $(t, x, y, z) = (s, 0, m, 0)$ for all m. Any point can of course be considered the origin, if the coordinates are displaced relative to (T, X, Y, Z).

A gravity wave excites quadrupole oscillations perpendicular to its direction of propagation in all objects it passes through. However, the intensities to be expected are so small that one should not be disturbed by the lack of evidence for them as yet.

Problems (5.8.9)

1. Verify (5.8.8; 2).

2. Check that $g_{\alpha\beta}\dot{x}^\alpha\dot{x}^\beta = -1$ for the solution (5.8.5).

3. Integrate $f^{1/2}(f^{-1/2})'' + g^{1/2}(g^{-1/2})'' = 0$ by making the ansatz that $f = L^{-2}\exp(2\beta)$ and $g = L^{-2}\exp(-2\beta)$.

4. Calculate the Poisson brackets of the constants (5.8.3) and compare with (5.5.14; 2).

Solutions (5.8.10)

1. With

$$U = u = T - X = t - x, \; V = X + T = x + t - (1 - u)y^2 + (1 + u)z^2, \; x + t = v,$$

we find

$$-dU\,dV + dY^2 + dZ^2 = -du[dv + 2y\,dy(1 - u) - 2z\,dz(1 + u) + du(y^2 + z^2)]$$
$$+ (dy(1 - u) - du\,y)^2 + (dz(1 + u) + du\,z)^2$$
$$= -du\,dv + dy^2(1 - u)^2 + dz^2(1 + u)^2.$$

2.
$$\dot{x}^2 - \dot{t}^2 + \frac{\dot{y}^2}{f} + \frac{\dot{z}^2}{g} = \frac{1}{4}\left(\frac{1}{\alpha} - \alpha + \frac{p_y^2 f + p_z^2 g}{\alpha m^2}\right)^2 - \frac{1}{4}\left(\frac{1}{\alpha} + \alpha + \frac{p_y^2 f + p_z^2 g}{\alpha m^2}\right)^2$$
$$+ \frac{p_y^2}{m^2}f + \frac{p_z^2}{m^2}g = -1.$$

3. The ansatz leads to $L'' + (\beta')^2 L = 0$, which is solved by

$$\beta(u) = \int_0^u du'\sqrt{-L''(u')/L(u')}.$$

Then we can take L with $L''/L < 0$ and calculate β.

4. $\{p_z G(u) + z(p_0 + p_x), p_z\} = p_0 + p_x = \{p_y F(u) + y(p_0 + p_x), p_y\}$, and the other Poisson brackets vanish. Therefore the invariance group is isomorphic to the one generated by the constants (5.5.4), and thus to the invariance group of the electromagnetic field tensor. ($G = \int g$ and $F = \int f$.)

Some Unsolved Problems (5.8.11)

1. The stability criterion (3.4.21) is practically useless, since it is difficult to prove that the limits $C_{\pm}$ exist. There also exist numerous other stability criteria, which, however, do not apply to Hamiltonian systems. Can they be put to any use?

2. The K–A–M theorem (3.6.19) guarantees that there exist invariant surfaces for systems of several degrees of freedom only when the perturbations are very small. Do bigger perturbations really destroy all invariant surfaces, or is the trouble only that the method of proof is poor?

3. If the perturbation is large, so that the K–A–M theorem does not hold, under what circumstances is a system in fact ergodic; in other words, when is a trajectory dense in the surface of constant energy?

4. Suppose that the whole energy surface is not filled densely. When is the trajectory dense in a sufficiently representative part of the energy surface so that the time-average of an observable equals its average over the energy surface?

5. With the same supposition as in Problem 4, how long is it until the time-average is, say, within $1\,\%_{\!oo}$ of its limit for infinitely long times?

6. In the N-body problem there were the following classes of trajectories:

   (a) trajectories with collisions;
   (b) trajectories where one particle escapes; and
   (c) trajectories for which all particles stay in a finite region.

   How large are the parts of phase space comprising each class, or comprising the closure of the trajectories in each class, when $N \geq 3$?

7. One thing that seems to happen in the gravitational N-body problem, $N \geq 3$, is that small clusters of tightly bound particles form. These have a tendency to become more tightly bound and to cause more loosely bound clusters to break up—which means that the specific heat is negative: the hot become hotter and the cold colder. Can such behavior be derived from the equations of motion?

8. In the relativistic case, even the two-body problem is still unsolved, because one has to worry about the infinite number of degrees of freedom of the field. The other choice, eliminating the infinite number of degrees of freedom, means that the force depends not on the position of a particle, but on its whole history. How can one cope with this mathematically? The question is not purely academic, since the relativistic two-body problem may be realistic for double stars with black holes or neutron stars.

The Structure of Space and Time 6

6.1 The Homogeneous Universe

In physics, space and time are defined by the way yardsticks and clocks behave, which in turn is determined by the equations of motion. It is this reasoning that gives a concrete significance to the mathematical structure of our formalism.

The first step towards a theory of relativity was the recognition that space and time are homogeneous. Homogeneity is expressed in the invariance of physical laws under spatial and temporal displacements, and implies that no point of the manifold is special. However, it is compatible with a structure in which certain directions are favored over others.

Mathematically, this is expressed by regarding

M_e as a Cartesian Product (6.1.1)

$$M_e = \mathbb{R} \times \mathbb{R} \times \mathbb{R} \times \mathbb{R}$$

$$= \text{time} \times (\text{up-down}) \times (\text{east-west}) \times (\text{north-south}).$$

In other words, canonical projections are specified onto the four coordinates. Such a situation results in a translation-invariant, not necessarily isotropic Hamiltonian,

$$H = \sum_{i=1}^{N} \varepsilon_i(\mathbf{p}_i) + \sum_{i<j} V_{ij}(\mathbf{x}_i - \mathbf{x}_j). \tag{6.1.2}$$

where ε_i is some function of $\mathbf{p}_i$.

Remarks (6.1.3)

1. M_e is still understood as the extended configuration space of a single particle, and the potentials are assumed regular enough that M_e can be taken as all of $\mathbb{R}^4$.
2. If V is of the form

$$V(\mathbf{x}) = \tfrac{1}{2}(\sqrt{x_1^2 \omega_1^2 + x_2^2 \omega_2^2 + x_3^2 \omega_3^2} - a)^2,$$

then molecules made up of such particles would oscillate at different rates in different orientations, so the directions of the coordinate axes could be determined by experiment.
3. In order to break the Galilean invariance, under $\mathbf{x} \to \mathbf{x} + \mathbf{v}t$, $t \to t$, the functions ε must be more complicated than $|\mathbf{p}|^2$, for example,

$$-\sum_{\alpha=1}^{3} k_\alpha^{-2} \cos(p_\alpha k_\alpha).$$

4. Hamiltonians like this are not merely to be found in science fiction, but are standard for the motion of electrons in anisotropic crystals.

If the H of (6.1.2) is not dilatation-invariant, then a variety of possible definitions of length are compatible with the fundamental laws. For example, if a molecule consisting of particles as in (6.1.3; 2) is brought to rest oriented in the i-direction, it has a natural length a/ω_i. It could be used as both a yardstick and a clock, and because of the homogeneity of M_e it could be used equally well at all places and times. It gives M_e an additional mathematical structure:

M_e as a Riemannian Space (6.1.4)

The Hamiltonian of (6.1.2) and Remarks (6.1.3; 2 and 3) provide M_e with the metric

$$g = dt^2 \left(\frac{\omega}{2\pi}\right)^2 + \frac{dx_1^2}{a_1^2} + \frac{dx_2^2}{a_2^2} + \frac{dx_3^2}{a_3^2},$$

$$\omega = \text{one of the } \omega_i, \qquad a_i = \frac{a}{\omega_i}.$$

Remarks (6.1.5)

1. This g uses the units defined by the molecule of (6.1.3; 2). After a time $\delta t = 2\pi/\omega$, the molecule has completed one cycle; and, once it is in position, it defines the unit lengths $\delta x_i = a_i$ in the various directions.
2. The metric (6.1.4) is an indisputable choice only if all yardsticks and clocks behave identically. And if that is the case, then we would not say that the molecule has different lengths in different directions, but would prefer to use scaled coordinates x_i/a_i.

3. A pseudo-Riemannian metric, $g = -dt^2(\omega/2\pi)^2 + \cdots$, could also be postulated. As long as the time axis is distinguished, this g contains the same information as the Riemannian one (6.1.4).
4. Even if H is not dilatation-invariant, it can happen that the equations of motion do not naturally define any yardsticks and clocks; invariances of the equations of motion do not necessarily correspond to invariances of H. E.g., the nonrelativistic equations of motion (1.1.1) and (1.1.2) are invariant under

$$\mathbf{x} \to \lambda^2 \mathbf{x} \quad \text{and} \quad t \to \lambda^3 t, \qquad \lambda \in \mathbb{R}^+,$$

which is not related to a one-parameter group of invariances of H. That matter actually does define distances and times is a quantum mechanical fact. The universal constants e, m, and $\hbar$ can be combined to form the Bohr radius $\hbar^2/me^2$ and the Rydberg frequency $me^4/2\hbar^3$. It is only in quantum theory that the solutions of the fundamental equations with realistic forces (1.1.2) naturally define the distances and times that make the physical M_e a Riemannian manifold.

6.2 The Isotropic Universe

On the surface of the earth the vertical direction is apparently more special than the compass directions. The isotropy that space would otherwise possess is destroyed when the direction up-down is singled out. Supposing that some scientists regarded this direction as really fundamental, they might well use the Hamiltonian

$$\mathcal{H} = \sum_i \varepsilon_i(|\mathbf{p}_i|) + \sum_{i>j} V_{ij}(|\mathbf{x}_i - \mathbf{x}_j|) + g \sum_i m_i z_i \qquad (6.2.1)$$

to describe the laws of nature.

Remarks (6.2.2)

1. Since the kinetic and potential energies depend only on the magnitudes of vectors, there is still a rotational invariance about the z-axis. The invariance group is thus extended from $\mathbb{R} \times \mathbb{R} \times \mathbb{R} \times \mathbb{R}$, as in §6.1, to $\mathbb{R} \times \mathbb{R} \times E_2$.
2. Accordingly, M_e would now be considered as the Cartesian product $\mathbb{R} \times \mathbb{R} \times \mathbb{R}^2 = $ time $\times$ (up-down) $\times$ the earth's surface. (We draw a distinction between $\mathbb{R} \times \mathbb{R}$ and $\mathbb{R}^2$; by writing the former we presuppose projections onto the two particular axes, while in the latter no direction is preferred.)

Our hypothetical scientists would eventually realize, by the time they invented space travel, anyway, that our position relative to the center of the earth is not really so important. They would then get rid of the g in their universal formula (6.2.1), and conclude that

$$M_e = \mathbb{R} \times \mathbb{R}^3 = \text{time} \times \text{space}. \qquad (6.2.3)$$

Remarks (6.2.4)

1. Mathematically, the product structure specifies canonical projections onto $\mathbb{R}$ and $\mathbb{R}^3$, as shown in Figure 53. Given any two points, the question whether they were at different places at the same time, or whether they were at the same place at different times, has a unique answer. All suitable coordinate systems differ from each other only by translations and spatial rotations, which do not affect these facts (Figure 54).

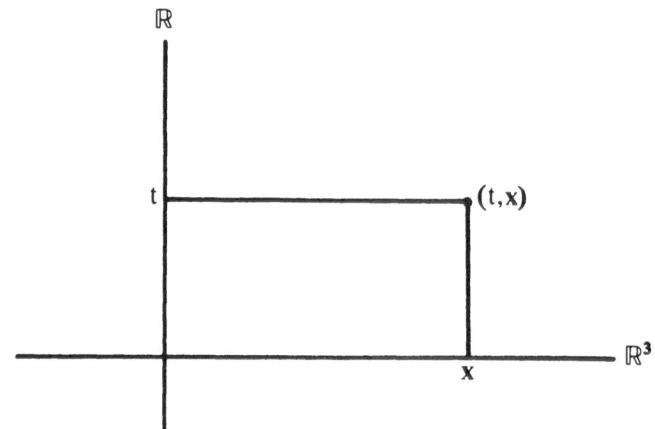

Figure 53   Canonical projections of M_e.

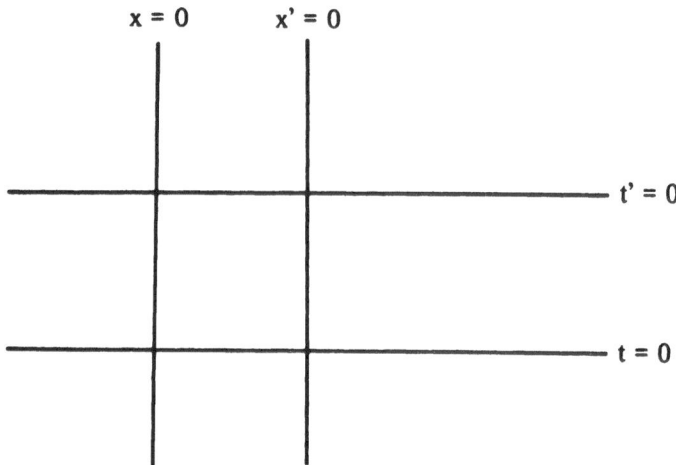

Figure 54   The translations.

2. The existence of the projection onto $\mathbb{R}^3$ distinguishes a system at rest from one that moves at a uniform velocity. One may then ask how bodies that are absolutely at rest can be distinguished from moving ones. Suppose that in (6.1.3; 3) all the k_α are equal, as in an isotropic crystal; then the kinetic energy (in some direction) of two bodies is $-k^{-2}$ ($\cos kp_1 + \cos kp_2$). When rewritten in center-of-mass and relative coordinates, as in (4.2.2), with $m_1 = m_2$, it becomes

$$-\frac{2}{k^2}\cos(pk)\cos\left(\frac{p_{cm}k}{2}\right).$$

Since V depends only on $x_1 - x_2$, $\mathbf{p}_{cm}$ is a constant, but

$$\dot{x}_{cm} = \mathbf{p}_{cm}(\tfrac{1}{2}|\mathbf{p}_{cm}|^2 + \sigma|\mathbf{p}|^2).$$

The momentum $\mathbf{p}$ is time-dependent, and x_{cm} is a constant only if $\mathbf{p}_{cm} = 0$. Hence the molecules at absolute rest are the only ones whose center-of-mass velocity is constant.

3. As mentioned in the previous section, a Hamiltonian of the form (6.2.1) can be used to define distances and times (possibly with the help of $\hbar$), and thereby give M_e a Riemannian structure. The sum of the scalar products in $\mathbb{R}$ and $\mathbb{R}^3$ produces a metric

$$g = v^2\,dt^2 + |d\mathbf{x}|^2 \qquad (6.2.5)$$

(and the difference produces a pseudometric), where v is a ratio between distances and times defined by (6.2.1).

Now that we have a picture of space and time as they would appear to a creature raised in a crystal, we turn our attention back to the world in which we ourselves live, at first as it appears without closer examination.

6.3 M_e According to Galileo

We now consider the standard laws (1.1.1) and (1.1.2), which are invariant under the Galilean group (4.1.9). With this invariance, the words "rest" and "uniform motion" have no absolute meaning, since the coordinate systems represented in Figure 55 are equally valid. There is no chart-independent way to say two events (elements of M_e) happen at the same place at different times. However, simultaneity still has an absolute meaning, unaffected by Galilean transformations.

Put mathematically (see (2.2.15)), we now consider

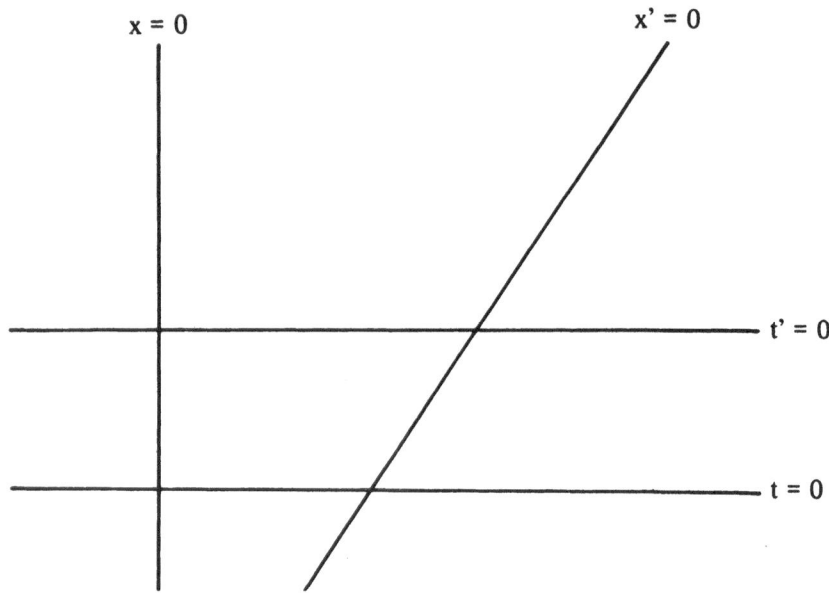

Figure 55   A Galilean transformation.

M_e as a Vector Bundle $\mathbb{R}^4$, with $\mathbb{R}$ (Time) as its Basis, and $\mathbb{R}^3$ (Space) as a Fiber (6.3.1)

Remarks (6.3.2)

1. There is no canonical projection $\mathbb{R}^4 \to \mathbb{R}^3$, onto space, and thus $M_e \neq \mathbb{R} \times \mathbb{R}^3$. While it is true that on a chart M_e has the form of a product, it is trivializable without being trivial. The physical meaning of this is that observers whose coordinate systems are in uniform relative motion can make their decomposition of M_e into space and time, but the submanifolds where $\mathbf{x} = 0$ determined in each system differ—though they are equally valid.
2. The fiber $\mathbb{R}^3$ is a vector space without a distinguished origin $\mathbf{x} = 0$. This is similar to the situation in phase space, $T^*(\mathbb{R}^3)$, where the origin $\mathbf{p} = 0$ of the fibers is undetermined, as it can be freely translated by Galilean transformations $\mathbf{p} \to \mathbf{p} + m\mathbf{v}$.
3. The projection onto the basis is a universal rule for synchronizing clocks. It is not difficult for all observers to agree on the rule in practice, because arbitrarily fast speeds can be attained: the submanifold $t = $ constant is uniquely characterized by the property that no trajectory can pass through two of its points, no matter how great the velocity is, a fact that would be perceived the same in all frames of reference.
4. Distances and times can be defined with the aid of $\hbar$ so as to give space and time a Riemannian structure. Since $M_e \neq \mathbb{R} \times \mathbb{R}^3$, there is no chart-independent way for M_e to inherit this structure. What M_e lacks is an

orthogonal coordinate system in which the metric could be written as a sum, as in (6.2.5). Although the time interval between two points can be defined by using the bundle projection, the spatial interval between two points at different times depends on the positions of the reference frames at those times, which are not independent of the chart.

6.4 M_e as Minkowski Space

We next investigate what structure the relativistic equations (1.1.3) give to space and time, united as **Minkowski space**. In addition, we suppose that the complete system of equations, (1.1.3) along with Maxwell's equations for F, is invariant under the Poincaré group (5.1.12). That is, everything looks the same in two reference frames after the coordinates have been transformed as in (5.1.12). The bundle structure of M_e is lost, because, compared with Figure 55, the special transformation (5.2.6) can be represented as follows:

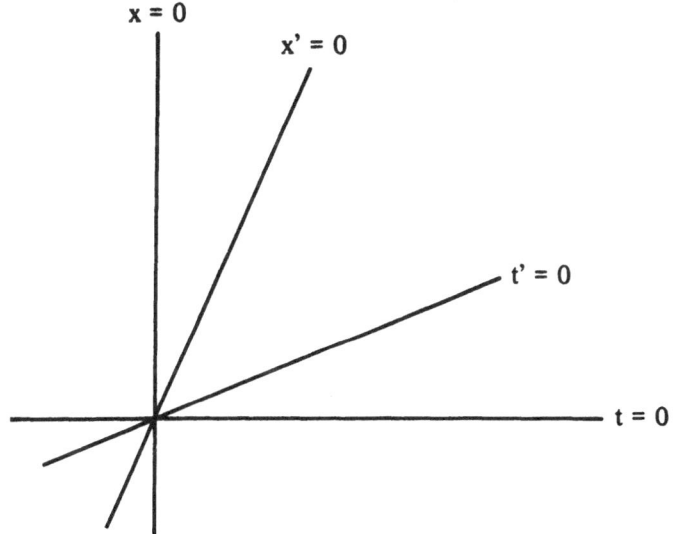

Figure 56   A Lorentz transformation.

Conclusion (6.4.1)

Simultaneity is defined in Minkowski space only with charts, and it is different with different charts.

Remarks (6.4.2)

1. Arbitrarily fast speeds are no longer available, as in (6.3.2; 3), to synchronize clocks with.

2. The synchronization defined in the charts (t, x) and (t', x') is the requirement that beams of light, ℓ_1 and ℓ_2, emitted at the points a and b simultaneously in one chart meet at the midpoint, and is known as Einstein's
synchronization:

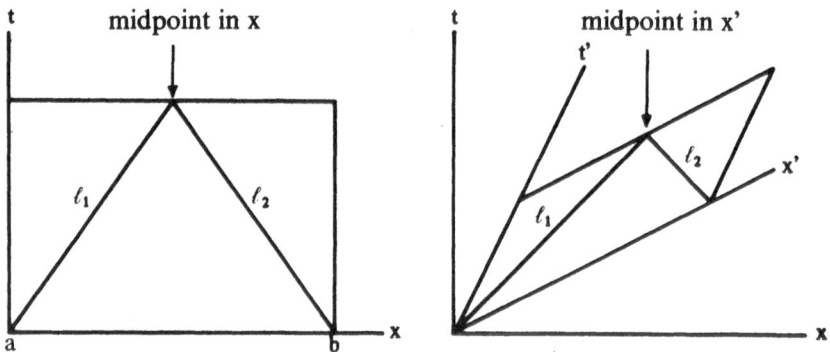

Einstein's synchronization with light beams.

3. It would of course be possible simply to decree that the projection onto
the t-axis defined by Einstein's synchronization in a certain system, the
"rest frame of the ether," shall be used in all systems. This is equivalent
to using the transformation,

$$x'' = \frac{x + vt}{\sqrt{1 - v^2}} = x', \qquad t'' = t\sqrt{1 - v^2} = t' - vx',$$

in which case $t = 0$ coincides with $t'' = 0$. The reasons not to do this are:

(a) Nature has no preference for any system.
(b) The users of (t'', x'') would feel put out for having to use such com-
    plicated equations.
(c) These transformations do not form a group.

But this chart cannot be excluded on purely logical grounds; even the
most peculiar kinds of coordinates are allowed by our definition of a
manifold. In this way it would be possible to save the notion of absolute
simultaneity, but no one is willing to make the necessary sacrifices any
more; philosophical principles are not as persuasive as mathematical
elegance.

We have remarked in (6.3.2; 4) that if there is no canonical projection to
$\mathbb{R}^3$, then spatial intervals are necessarily chart-dependent, unless the points
involved are simultaneous. Since now even simultaneity is chart-dependent,
so are all spatial and temporal intervals—just as the length of a body in the
1-direction lost its absolute meaning in the transition from §6.1 to §6.2. But
what has happened now is more unusual, and requires a more detailed discussion.

The first point to make clear is that the coordinates (t', x') reflect times and distances as actually measured just as much as (t, x) do: A yardstick of unit length, with one end at $(t, x) = (t, 0)$ and the other at $(t, x) = (t, 1)$, corresponds to a certain solution ℓ of the system of equations. The Lorentz-invariance of the equations ensures that there exists a Lorentz-transformed solution ℓ' with ends at $(t', x') = (t', 0)$ and $(t', 1)$. In other words, if the unit yardstick is moving, it is still a unit yardstick in the primed system. Of course, the acceleration to the primed system must not be too violent; the molecules must not be excited to new states, etc., or the original solution will not remain valid. Interestingly, both the length of ℓ' in the unprimed system and of ℓ in the primed one are equal to $\sqrt{1 - v^2} < 1$. The apparent contradiction is resolved if one notes that the length in each system is the distance between the ends at the same time, but "at the same time" means something different in the two systems:

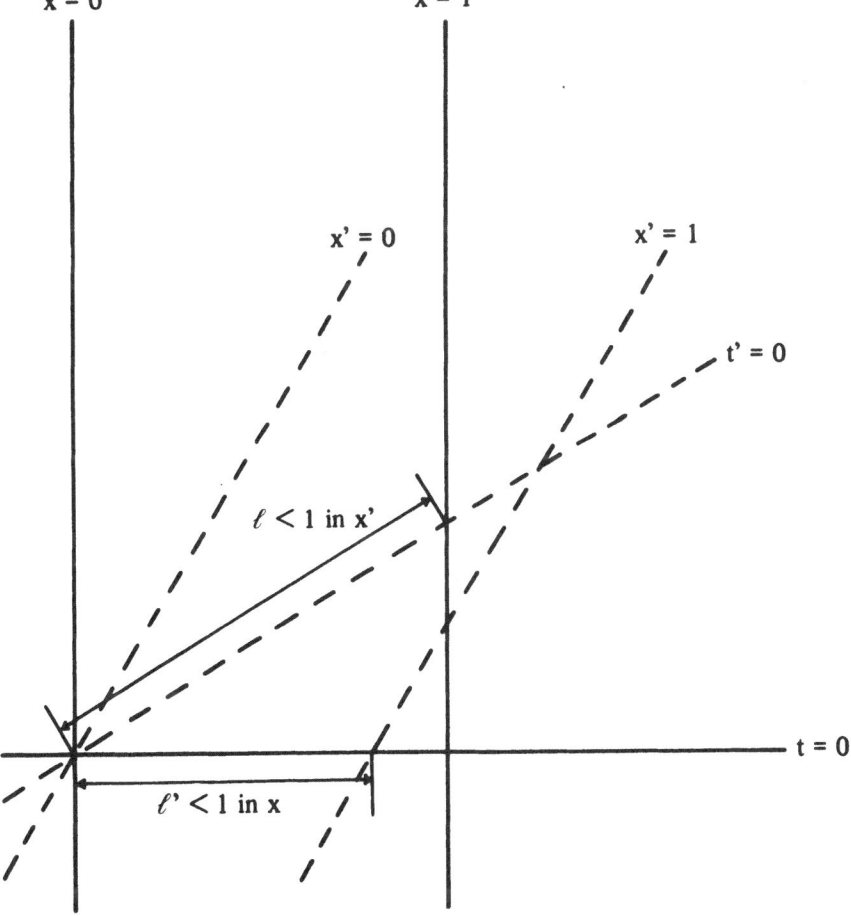

Figure 57   The Lorentz contraction.

Conclusion: The Lorentz Contraction (6.4.3)

In Minkowski space, moving bodies are contracted by the factor $\sqrt{1-v^2}$ in the direction of motion.

Remarks (6.4.4)

1. Einstein's synchronization is essential in this statement. In the coordinates (6.4.2; 3), $t'' = 0$ is a horizontal surface. Then, too, a moving yardstick is shortened, but a yardstick at rest is lengthened, as seen from a moving system. Notice that in the primed coordinate system the ends of the unit yardstick also have the world-lines $x'(t') = 0$ and $x'(t') = 1$; only the time-convention is different from the unprimed one, though, of course, a clock with world-line $x'(t') = 0$ goes around exactly once between $t' = 0$ and $t' = 1$.
2. If a picture is taken of a moving body, the time-delay of the light, $\sim v/c$, is a larger effect than the Lorentz contraction, $\sim (v/c)^2$. A picture would not show the instantaneous position of an object; instead, particles that are a distance L farther away are photographed at where they were at a time L/c earlier, i.e., at a position displaced by Lv/c. It can be shown that the net effect is that an object does not appear contracted, but rotated.
3. If one accepts that all systems in uniform relative motion are equally valid, then the Lorentz contraction can be demonstrated without reference to the synchronization of clocks. Imagine that two identical yardsticks ℓ_1 and ℓ_2 are sent by each other with equal but opposite velocities, and that the positions of their ends are marked at the instant that they coincide. From symmetry, one concludes that both ends get marked at the same time.

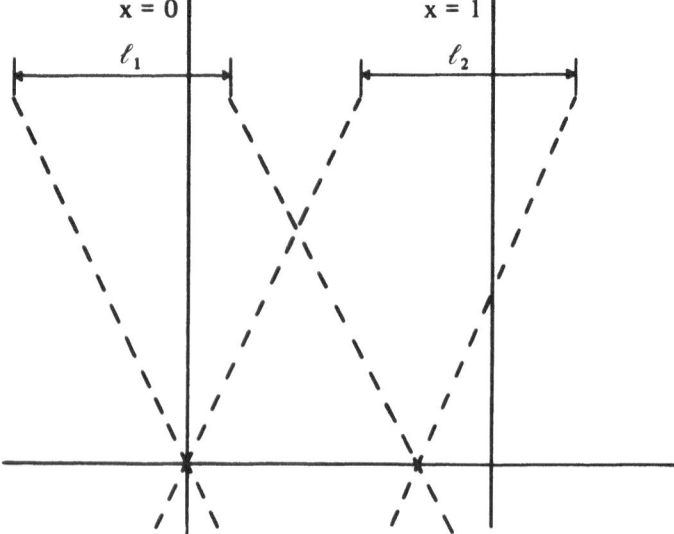

Demonstration of the Lorentz contraction of moving unit yardsticks.

The situation concerning the length of time between two events is similar. A periodic solution of the equations of motion can be used as a clock, and the Lorentz-transformed solution would describe a moving clock. If the two clocks are objects with the trajectories $x = 0$ and respectively $x' = 0$, and periods τ and τ', then what happens is depicted in Figure 58.

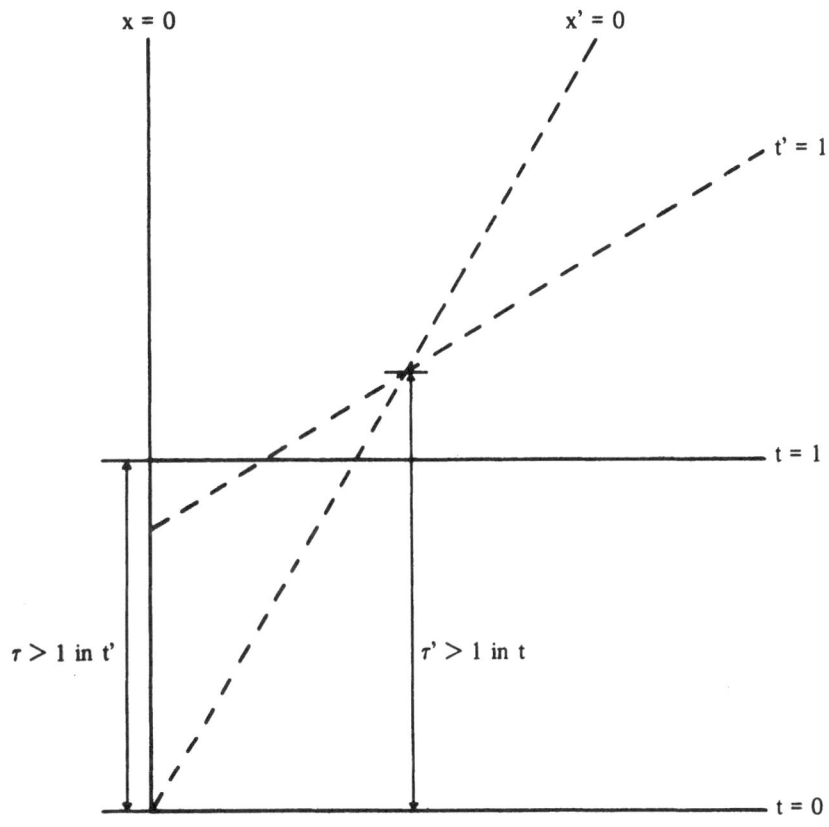

Figure 58   The time-dilatation.

Conclusion: Time-Dilatation (6.4.5)

In Minkowski space, moving clocks run slow by the factor $\sqrt{1 - v^2}$.

Remarks (6.4.6)

1. In the coordinate system (6.4.2; 3), moving clocks also run slow, but stationary clocks as seen from a moving frame of reference run fast. The apparent contradiction vanishes in the coordinates (t'', x''), too.
2. The slowing of time for moving bodies can be demonstrated without reference to synchronization, for instance for rotational motion. Fast-moving muons in a storage ring live much longer than stationary ones.

In fact, time-dilatations by a factor of 100 or more have been observed in storage rings; but it is also possible to measure the miniscule amount of retardation during an airplane's flight around the earth.

3. One might wonder whether a given clock, accelerated to some velocity, would actually run slow in consequence of the equations of motion. Let us take the Larmor motion as a model of a clock, and accelerate the particle with an electrical field parallel to **B**. We have already calculated in §4.2 that the frequency in s is always eB/m, so the frequency in t works out to be decreased by the factor $\sqrt{1 - v^2}$. The reason for this is the relativistic mass-increase, which makes a fast-moving particle go slower than it would otherwise. Care must be taken, of course, not to change the velocity much during one period, so that the motion will remain more or less periodic in t, and the whole system can reasonably be called a clock — just as in ordinary experience a clock cannot be accelerated too fast or it will break down. However, quantum mechanics makes atomic clocks so sturdy that hardly any care at all need be taken when they are accelerated.

In the models of space and time presented earlier, it was possible to use the inverse images of the natural divisions of the time axis under a distinguished projection to decompose all of M_e into past, present, and future with respect to any point. As we would like to preserve this decomposition so far as possible, we make

Definition (6.4.7)

For a and $b \in M_e$, $a \neq b$, we shall write $a > b$ if there exists a solution to the equations of motion that passes first through b and then later through a. If it never happens that $a > b$ and $b > a$, then we say that $>$ includes a **causal structure** on M_e, or, for short, that M_e is a **causal space**.

Remarks (6.4.8)

1. On the bundle of §6.3, we had $(t, x) > (t', x') \Leftrightarrow t > t'$.
2. We assume that arbitrarily strong fields are possible, so that the velocity of any solution can be instantaneously turned into the velocity of any other. This means that $a > b$ and $b > c \Rightarrow a > c$.
3. The existence of a closed trajectory in M_e would preclude this kind of order relation.
4. The past with respect to a is $\{b \in M_n : a > b\}$, and the future is $\{b \in M_e : b > a\}$, and the rest of M_e could be referred to as the present. What goes on at a can influence only its future, and be influenced only by its past.

The causal relationships on a more general manifold can be rather strange; there might, for example, be some point in the present for which no trajectory can pass through it and ever reach the future (cf. (6.5.5; 2)). The state of affairs is fortunately more clear-cut in

Minkowski Space as a Causal Space (6.4.9)

In Minkowski space,

$$(t, x, y, z) > (t', x', y', z') \Leftrightarrow$$
$$t > t' \quad \text{and} \quad (t - t')^2 > (x - x')^2 + (y - y')^2 + (z - z')^2.$$

Remarks (6.4.10)

1. In all the models through §6.3, the present was $\{(t', x', y', z'): t' = t\}$, a submanifold of lower dimensionality. In Minkowski space its interior is a 4-dimensional submanifold.
2. Unlike before, it is now possible to have two trajectories such that no point of either trajectory lies in the future (or the past) of any point of the other trajectory. An example would be the two hyperbolic trajectories

$$I = \{(\sinh s, \cosh s, 0, 0): s \in \mathbb{R}\},$$

and

$$II = \{(\sinh s, -\cosh s, 0, 0): s \in \mathbb{R}\},$$

of particles responding to the electric field $\mathbf{E} \cong (x_1/|x_1|, 0, 0)$ coming from a surface charge:

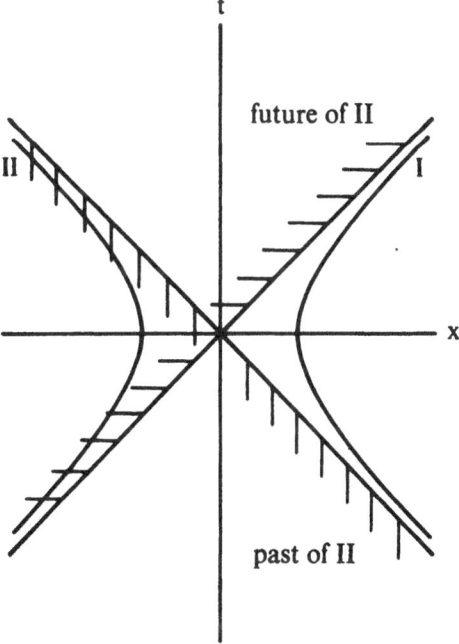

The future and past for hyperbolic trajectories.

Observers on these trajectories can never see each other directly, though people in between could know about both of them.

3. The causal structure defines a topology, in which the open sets are unions of $U_{ac} \equiv \{b \in M_e : a < b < c\}$, where a and c are any points of M_e. This topology is identical to the ordinary one.

4. Zeeman has proved a surprising theorem: every bijection $f : \mathbb{R}^4 \to \mathbb{R}^4$ that preserves the causal structure of (6.4.9) (i.e., $x > y \Leftrightarrow f(x) > f(y)$) is the product of a Poincaré transformation and a dilatation.

6.5 M_e as a Pseudo-Riemannian Space

Our final task is to take the influence of gravitation on space and time into account. The principle of equivalence (5.6.11) states that suitable co-ordinates can be used at any point to make $g_{\alpha\beta} = \eta_{\alpha\beta}$ and all the derivatives of $g_{\alpha\beta}$ zero. In other words, gravity is not detectable at a single point, and from the discussion of §6.4 we conclude that in these coordinates

$$g = dx_1^2 + dx_2^2 + dx_3^2 - dt^2 \tag{6.5.1}$$

reproduces distances and times as actually measured. Other coordinates could obviously be used, in which the actual metric of (6.5.1) is transformed from $\eta_{\alpha\beta}$ to $g_{\alpha\beta}$ at the point in question. At any rate, we can consider

M_e as a Space with a Pseudo-Riemannian Metric (6.5.2)

In a gravitational field, $g_{\alpha\beta} \, dx^\alpha \, dx^\beta$ gives the metric as actually observed.

Remarks (6.5.3)

1. This interpretation depends essentially on the universality of gravitation. If different particles acted differently in a gravitational field, then no single transformation could make the field vanish at a given point for all the particles. Universality also requires that Maxwell's equations in the coordinate system (6.5.1) have the same form as in the absence of gravity. If that is the case, then the forces which are at work on the yard-sticks and clocks are also unaffected by gravity, and everything works as if there were no gravity.

2. The attempt to interpret other fields geometrically has always failed from the lack of similar universal characteristics. If measuring instruments that have been built differently are affected differently by a field, then a geometrical interpretation is not convincing.

3. From time to time someone comes up with a "theory of gravitation in flat space." The established equations (1.1.6) and (1.1.7) are typically used, but in some coordinate system $\eta_{\alpha\beta} \, dx^\alpha \, dx^\beta$, rather than $g_{\alpha\beta} \, dx^\alpha \, dx^\beta$, is interpreted as the metric. Then it is explained that yardsticks and clocks

fail to measure this metric because they are influenced by the gravitational potential. But Nature does not select a special coordinate system, and real yardsticks and clocks in fact measure $g_{\alpha\beta}\, dx^\alpha\, dx^\beta$. Though logically possible, the attempt to save "absolute Minkowski space" is as artificial as the attempt in (6.4.2; 3) to choose an "absolute rest frame" arbitrarily.

4. Although the $g_{\alpha\beta}$ appear in the equations of motion, they are no more observable than the electrical potentials. Their influence can be regarded as that of a universal scaling transformation, which cannot be perceived locally.

One occasionally encounters the notion that one would need gravitational theory to be able to use an accelerating coordinate system. Actually, it goes the other way around. An accelerating reference system can always be used to reduce the situation in a gravitational field locally to that of Minkowski space. Differences from §6.4 occur only for nonlocal phenomena.

Alterations of the Geometry by Gravitation (6.5.4)

1. Although a universal dilatation could not be perceived locally, if the $g_{\alpha\beta}$ depend on x, then yardsticks at different places will have different lengths, and if they could be laid next to each other, the difference could be measured. Similarly, any differences between atomic oscillations used as clocks can be detected by a comparison of the frequencies of light emitted at different points. If we consider the Schwarzschild metric (5.7.1), which is time-independent, then electromagnetic waves are solutions of Maxwell's equations proportional to $\exp(i\omega t)$. The frequency in t is a constant throughout space, and consequently the frequency in s is different at different points (this is the origin of the gravitational red-shift). Modern experimental techniques can measure the red-shift from the change in the earth's gravitational potential due to a difference in altitude of only a few meters.

2. Variable $g_{\alpha\beta}$ of course destroy the large-scale Euclidean geometry of space. In particular, Pythagoras' theorem would no longer hold, as the sides of a triangle would be measured by yardsticks contracted by different amounts. These effects, however, are mainly of theoretical interest, since space on the earth is flat to within the precision of our measuring instruments; the only properties we can measure across greater distances are the direction, frequency, and intensity of light rays.

3. Previously, we always considered M_e as $\mathbb{R}^4$ with some additional structure, and thus $T(M_e)$ was always a Cartesian product—which we described by saying that M_e was parallelizable. That is, given two vectors at different points, it was possible to say whether they were parallel. In Minkowski space this meant that four-velocities at different points could be compared, in practice by using light beams. Two observers have the same velocity if and only if they do not appear Doppler-shifted to each other. But if

the manifold determined by g is not parallelizable, then there is no way to say whether two bodies are at rest with respect to each other unless they are at the same point. Light signals are no longer useful for determining whether vectors are parallel, because they are affected by gravity.

Alteration of the Causal Structure (6.5.5)

1. There exist manifolds that are not causal. For example, let

    $$M_e = \{\mathbf{x} \in \mathbb{R}^5 : x_1^2 + x_2^2 + x_3^2 - x_0^2 - x_4^2 = -r^2\},$$

    and induce a metric on it from $\mathbb{R}^5$ with $g = dx_1^2 + dx_2^2 + dx_3^2 - dx_0^2 - dx_4^2$. Then $(x_0, x_1, x_2, x_3, x_4) = r(\cos s, 0, 0, 0, \sin s)$ is a closed timelike geodesic, which, according to (6.4.8; 3), precludes a causal structure.
2. If M_e is Minkowski space with some parts removed, then there may exist two particles with disjoint futures. For example, in the Schwarzschild field there can be obesrvers with trajectories

    $$I = \{(u, v, \vartheta, \varphi) = (1 + \cosh s, \sinh s, 0, 0), s \in \mathbb{R}\}$$

    and

    $$II = \{(u, v, \vartheta, \varphi) = (-1 - \cosh s, \sinh s, 0, 0), s \in \mathbb{R}\}.$$

Unlike what we saw in (6.4.10; 2), no observer in between can have seen both of them.

Our discussion has shown how different laws of nature imprint their structure on the space-time manifold. It is clear that it would be presumptuous to try to state what the true essence of space and time is. The most we can discover is that facet of the essence which is reflected in our present knowledge of the laws of nature.

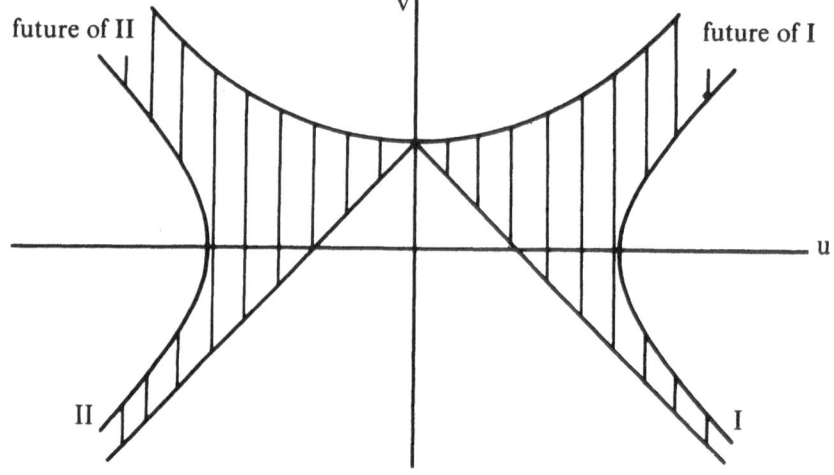

The futures of two uniformly accelerated observers.

Bibliography

Works Cited in the Text

[1]. J. Dieudonné. Foundations of Modern Analysis, in four volumes. New York: Academic Press, 1969–1974.

[2]. E. Hlawka. Differentiable Manifolds, pp. 265–307 in *Acta Phys. Austriaca* Suppl., vol. 7. Vienna: Springer-Verlag, 1970.

[3]. R. Abraham. Foundations of Mechanics. New York: Benjamin, 1967.

[4]. H. Flanders. Differential Forms. New York: Academic Press, 1963.

[5]. S. Sternberg. Lectures on Differential Geometry. Englewood Cliffs, New Jersey: Prentice-Hall, 1964.

[6]. C. L. Siegel and J. Moser. Lectures on Celestial Mechanics. New York and Berlin: Springer-Verlag, 1971.

[7]. V. Szebehely. Theory of Orbits, the Restricted Problem of Three Bodies. New York: Academic Press, 1967.

[8]. V. Szebehely. Families of Isoenergetic Escapes and Ejections in the Problem of Three Bodies. *Astronomy and Astrophysics* **22**, 171–177, 1973.

[9]. V. I. Arnold and A. Avez. Ergodic Problems of Classical Mechanics. New York: Benjamin, 1968.

[10]. N. Kerst and R. Serber. Electronic Orbits in the Induction Accelerator. *Phys. Rev.* **60**, 53–58, 1941.

[11]. V. Arnold. *Dok. Ak. Nauk.* **142**, 758–761, 1962. O Povedeniye Adiabaticheskogo Invarianta pri Medlennom Periodicheskom Izmeneniye Funkchiye Gamil'tona.

[12]. A. Schild. Electromagnetic Two-Body Problem. *Phys. Rev.* **131**, 2762–2766, 1963.

[13]. Y. Sinai. *Acta Phys. Austriaca* Suppl., vol. 10. Vienna: Springer-Verlag, 1973.

[14]. J. Moser. Stable and Random Motions in Dynamical Systems. Princeton: Princeton University Press, 1973.

[15]. R. McGehee and J. N. Mather. Solutions of the Collinear Four Body Problem which become Unbounded in Finite Time. In: Lecture Notes in Physics **38**, J. Moser, ed. New York: Springer-Verlag, 1975. (Entitled: Battelle Rencontres, Seattle 1974. Dynamical Systems: Theory and Applications.)

[16]. G. Contopoulos. The "Third" Integral in the Restricted Three-Body Problem. *Astrophys. J.* **142**, 802–804, 1965. G. Bozis. On the Existence of a New Integral in the Restricted Three-Body Problem: *Astronomical J.* **71**, 404–414, 1966.

[17]. V. Arnold. Small Denominators and Problems of Stability of Motion in Classical and Celestial Mechanics. *Russian Math. Surv.* **18**, 85–191, 1963.

[18]. R. C. Robinson. Generic Properties of Conservative Systems. *Amer. J. Math.* **92**, 562–603 and 897–906, 1970.

[19]. M. Breitenecker and W. Thirring. *Suppl. Nuovo Cim.*, 1978.

Further Reading

Chapter 2

W. M. Boothby. An Introduction to Differentiable Manifolds and Riemannian Geometry. New York: Academic Press, 1975.

Th. Bröcker and K. Jänich. Einführung in die Differentialtopologie. Heidelberger Taschenbücher 143. Heidelberg: Springer-Verlag, 1973.

Y. Choquet-Bruhat, C. DeWitt-Morette, and M. Dillard-Bleick. Analysis, Manifolds, and Physics. Amsterdam: North Holland, 1977.

V. Guillemin and A. Pollack. Differential Topology. Englewood Cliffs, New Jersey: Prentice-Hall, 1974.

R. Hermann. Vector Bundles in Mathematical Physics, vol. 1. New York: Benjamin, 1970.

H. Holman and H. Rummler. Alternierende Differentialformen. Bibliographisches Institut, 1972.

S. Kobayashi and K. Nomizu. Foundations of Differential Geometry, vol. 1. Interscience Tracts in Pure and Applied Mathematics No. 15, vol. 1. New York: Interscience, 1963.

L. H. Loomis and S. Sternberg. Advanced Calculus. Reading, Massachusetts: Addison-Wesley, 1968.

E. Nelson. Tensor Analysis. Princeton: Princeton University Press, 1967.

M. Spivak. Calculus on Manifolds; A Modern Approach to Classical Theorems of Advanced Calculus. New York: Benjamin, 1965.

Chapter 3

R. Barrar. Convergence of the von Zeipel Procedure. *Celestial Mechanics* **2**, 494–504, 1970.

N. Bogoliubov and N. Krylov. Introduction to Non-linear Mechanics. Princeton: Princeton University Press, 1959.

J. Ford. The Statistical Mechanics of Classical Analytic Dynamics. In: Fundamental Problems in Statistical Mechanics, vol. III, E. Cohen, ed. Amsterdam: North Holland, 1975.

G. Giacaglia. Perturbation Methods in Non-linear Systems. New York: Springer-Verlag, 1972.

M. Golubitsky and V. Guillemin. Stable Mappings and their Singularities. New York: Springer-Verlag, 1973.

V. Guillemin and S. Sternberg. Geometric Asymptotics. Providence: American Mathematical Society, 1977.

M. Hirsch and S. Smale. Differential Equations, Dynamical Systems, and Linear Algebra. New York: Academic Press, 1974.

W. Hunziker. Scattering in Classical Mechanics. In: Scattering Theory in Mathematical Physics, J. A. Lavita and J. Marchand, eds. Boston: D. Reidel, 1974.

R. Jost. Poisson Brackets (An Unpedagogical Lecture). *Rev. Mod. Phys.* **36**, 572–579, 1964.

G. Mackey. The Mathematical Foundations of Quantum Mechanics. New York: Benjamin, 1963.

J. Moser, ed. Dynamical Systems: Theory and Applications. New York: Springer-Verlag, 1975.

J.-M. Souriau. Structure des Systèmes Dynamiques: Maîtrises de Mathématiques. Paris: Dunod, 1970.

Chapters 4 and 5

A. Hayli, ed. Dynamics of Stellar Systems. Boston: D. Reidel, 1975.

L. Landau and E. Lifschitz. The Classical Theory of Fields. London and New York: Pergamon Press, 1975.

H. Pollard. Mathematical Introduction to Celestial Mechanics. Englewood Cliffs, New Jersey: Prentice-Hall, 1966.

S. Sternberg. Celestial Mechanics. New York: Benjamin, 1969.

K. Stumpff. Himmelsmechanik. Berlin: Deutscher Verlag der Wissenschaften, 1959.

Chapter 6

J. Ehlers. The Nature and Structure of Spacetime. In: The Physicist's Conception of Nature, J. Mehra, ed. Boston: D. Reidel, 1973.

E. H. Kronheimer and R. Penrose. On the Structure of Causal Spaces. *Proc. Camb. Phil. Soc.* **63**, 481–501, 1967.

C. Misner, K. Thorne, and J. Wheeler. Gravitation. San Francisco: W. H. Freeman, 1973.

S. Nanda. A Geometrical Proof that Causality Implies the Lorentz Group. *Math. Proc. Camb. Phil. Soc.* **79**, 533–536, 1976.

R. Sexl and H. Urbantke. Relativität, Gruppen, Teilchen. Vienna: Springer-Verlag, 1976.

A. Trautman. Theory of Gravitation. In: The Physicist's Conception of Nature, J. Mehra, ed. Boston: D. Reidel, 1973.

S. Weinberg. Gravitation and Cosmology. New York: Wiley, 1972.

E. C. Zeeman. Causality Implies the Lorentz Group. *J. Math. Phys.* **5**, 490–493, 1964.

Index

Texts and Monographs in Physics

Edited by **W. Beiglböck, M. Goldhaber, E. Lieb,** and **W. Thirring**

Texts and Monographs in Physics includes books from any field of physics that might be used as basic texts for advanced training and higher education in physics, especially for lectures and seminars at the graduate level.

Polarized Electrons
By **J. Kessler**
1976. ix, 223p. 104 illus. cloth

The Theory of Photons and Electrons
The Relativistic Quantum Field Theory of Charged Particles with Spin One-Half
Second Expanded Edition
By **J. Jauch** and **F. Rohrlich**
1976. xix, 553p. 55 illus. cloth

Essential Relativity
Special, General, and Cosmological
Second Edition
By **W. Rindler**
1977. xv, 284p. 44 illus. cloth

Inverse Problems in Quantum Scattering Theory
By **K. Chadan** and **P. Sabatier**
1977. xxii, 344p. 23 illus. cloth

Quantum Mechanics
By **A. Böhm**
1978. approx. 576p. approx. 85 illus. cloth

The Concepts and Logic of Classical Thermodynamics as a Theory of Heat Engines
Rigourously Constructed upon the Foundation Laid by S. Carnot and F. Reech
By **C. Truesdell** and **S. Bharatha**
1977. xxii, 154p. 15 illus. cloth

Principles of Advanced Mathematical Physics
Volume I
By **R.D. Richtmyer**
1978. approx. 448p. approx. 45 illus. cloth

Foundations of Theoretical Mechanics
Part I: The Inverse Problem in Newtonian Mechanics
By **R.M. Santilli**
1978. aprox. 304p. cloth
Part II: Generalizations of the Inverse Problem in Newtonian Mechanics
In preparation

Springer-Verlag   New York   Heidelberg   Berlin